AF581746

Sustainable Pavement Engineering and Road Materials

Sustainable Pavement Engineering and Road Materials

Editor

Edoardo Bocci

MDPI • Basel • Beijing • Wuhan • Barcelona • Belgrade • Manchester • Tokyo • Cluj • Tianjin

Editor
Edoardo Bocci
Università degli Studi eCampus
Italy

Editorial Office
MDPI
St. Alban-Anlage 66
4052 Basel, Switzerland

This is a reprint of articles from the Special Issue published online in the open access journal *Sustainability* (ISSN 2071-1050) (available at: https://www.mdpi.com/journal/sustainability/special_issues/road_materials).

For citation purposes, cite each article independently as indicated on the article page online and as indicated below:

LastName, A.A.; LastName, B.B.; LastName, C.C. Article Title. *Journal Name* **Year**, *Volume Number*, Page Range.

ISBN 978-3-0365-4745-9 (Hbk)
ISBN 978-3-0365-4746-6 (PDF)

Contents

About the Editor

Edoardo Bocci

Edoardo Bocci is an Associate Professor in the Faculty of Engineering at Università Telematica eCampus, Italy. In 2014, he received his PhD in Engineering Science from the Università Politecnica delle Marche of Ancona (Italy). He is a member of SIIV (Italian Society of Road Infrastructures) and RILEM (Réunion Internationale des Laboratoires et Experts des Matériaux, systèmes de construction et ouvrages). His research activity is focused on road pavements. In particular, he is an expert on special pavements for orthotropic steel deck bridges; pavements made from the hot and cold recycling of reclaimed asphalt; clear binders for road pavements; the recycling of steel slag; material from the construction and demolition of buildings; using fibers from end-of-life tires in hot bituminous mixtures; asphalt rubber; and damage modelling in asphalt concrete. He is the author of more than 40 papers published in international peer-reviewed journals or the proceedings of international conferences. In addition, he has supervised more than 100 bachelor's and master's theses.

Editorial

Sustainable Pavement Engineering and Road Materials

Edoardo Bocci

Faculty of Engineering, Università degli Studi eCampus, 22060 Novedrate, Italy; edoardo.bocci@uniecampus.it

Contents of the Special Issue

One of the most topical research areas currently concerns the identification and improvement of technologies against climate change, environmental pollution, exploitation of natural resources, and the economic crisis. Of course, the field of road pavement engineering has also been involved in this battle. In recent years, themes of sustainability and circular economy have become the main keywords for scientists, administrations, designers, and constructors worldwide. This is also testified by the 13 contributions published in this Special Issue, authored by European, Asian, African, and South American research teams.

In the field of pavement engineering, the word "sustainability" can be related to several aspects, including road materials and their performance, the management of temperatures and emissions during road construction and service life, and analyses of the socio-economic and environmental impacts generated by the road infrastructures.

Concerning the theme of road materials, five papers of this Special Issue (contributions 1, 7, 10, 11, and 12) deal with the use of wastes or by-products in different types of mixtures. In particular, contributions 1 and 6 regard porous cement concrete (or pervious concrete), a kind of sustainable urban drainage system that allows us to face the problems of storm water management, urban heat island, and air pollution. The paper by Elizondo-Martinez, Tataranni, Rodriguez-Hernandez, and Castro-Fresno (contribution 1, feature paper) investigates the possibility of substituting proportions of cement with metakaolin or geopolymer pastes, whose production is less harmful to the environment. The results showed that the replacement of 5% cement with metakaolin can increase both permeability and indirect tensile strength. In contrast, the geopolymer mixtures proved to require an accurate proportion of the mix components in order to balance the permeability and the mechanical properties. However, the findings are promising and demonstrated that these materials can be a suitable alternative to traditional porous cement concrete. The paper by Hung, Seo, Kim, and Lee (contribution 7) focuses on the influence of blended aggregate and blocking materials on the mechanical and permeability characteristics of porous cement concrete. The study provided the optimum proportions between coarse, intermediate, and fine aggregates in terms of strength and permeability. In particular, the results showed that porosity and permeability decrease when increasing the content of fines. Moreover, the research proved the efficacy of vacuum-cleaning and high-pressure spraying in partly restoring the porous concrete permeability when the voids become clogged. Vaitkus, Gražulytė, Kravcovas, and Mickevič (contribution 10) evaluated the role of foamed bitumen-treated base courses to the bearing capacity of the pavement structure using a falling weight deflectometer. The study showed that the foamed bitumen-treated layer has a mechanical behavior that evolves with time due to material curing. For this reason, the mechanistic–empirical pavement design approach has to be updated in order to account for these kinds of mixtures and their peculiar characteristics. Moreover, the effect of the water content in the layer should also be incorporated in the correction factors. Contributions 10 and 11 are review articles. The paper by Bocci and Prosperi (contribution 11) aims at describing the chemical phenomena that happen during bitumen aging, specifically, loss of volatiles, oxidation, and physical and steric hardening. To restore the aged bitumen properties, rejuvenators should rebalance the colloidal components, disrupt the asphaltene

Citation: Bocci, E. Sustainable Pavement Engineering and Road Materials. *Sustainability* **2022**, *14*, 2166. https://doi.org/10.3390/su14042166

Received: 1 February 2022
Accepted: 10 February 2022
Published: 14 February 2022

clusters, and re-establish the proper molecular mobility. The literature review allowed us to emphasize the importance of using different laboratory approaches for a precise and clear overview of how bitumen ages and rejuvenates in order to encourage more and more recycling of reclaimed asphalt pavement (RAP). The review by Mohammed, Koting, Katman, Babalghaith, Abdul Patah, Ibrahim, and Karim (contribution 12) describes the possible solutions to reuse coal bottom ash (CBA), a kind of hydraulic material from coal production. In particular, one of the most promising fields where CBA can be recycled is pavement construction, as it can be an aggregate replacement, cement replacement, additive in bitumen, and filler in hot-mix asphalt (HMA). All these solutions showed excellent outcomes, but in-depth preliminary analyses are required to characterize CBA, which is highly heterogeneous by nature, and CBA-containing mixtures.

Sustainability in road engineering can also be achieved by increasing the performance of the materials against traffic loads and environmental factors, thus extending the durability of the whole pavement system. From this perspective, the paper by Ragni, Canestrari, Allou, Petit, and Millien (contribution 2) investigated the shear resistance at the interface between old and new HMA layers in presence of geogrid reinforcement. The use of geogrids represents an extremely effective and sustainable solution for pavement maintenance, as it allows us to avoid the reconstruction of the entire thickness of the HMA layers and to increase the durability of the overlays, typically subjected to early reflective cracking. The research aimed at analyzing the performance of geogrid-reinforced interfaces against static and cyclic shear stresses. To this scope, three types of double-layer cores, i.e., unreinforced, reinforced with a carbon fiber geogrid, and reinforced with a glass fiber geogrid, were taken from a full-scale trial section and tested in the laboratory. The study showed different performances of the interfaces according to the reinforcement type and specimen temperature. However, the resistance against debonding of the reinforced interfaces was comparable to that of the unreinforced specimens, encouraging the adoption of this maintenance solution. The review by Fusco, Moretti, Fiore, and D'Andrea (contribution 4) presents the most commonly used nano-additives for HMA mixtures: nanoclays (NC), nanosilicates, carbon nanotubes (CNTs), graphene nanoplatelets (GNPs), nano-calcium oxide (CaO), and nano-titanium dioxide (TiO_2). The literature review highlighted that the performance of these mixtures strongly depends on type, concentration, and dispersal of the used nano-additive. Such variables influence the bonding properties, the viscosity, the resistance to aging, and the self-healing, consequently improving the rutting and fatigue behavior. Mokoena, Mturi, Maritz, Mateyisi, and Klein (contribution 13) studied the critical factors for the development of the temperature maps of Africa for the selection of the most suitable performance-graded bitumen in pavement design and construction. This choice is fundamental in providing an adequate duration of the HMA service life. In particular, the paper encourages the selection of a proper pavement temperature model for each climatic zone of a country, the evaluation of the urban heat island phenomenon in the bitumen choice, and the use of downscaled global climate models and pavement temperature maps.

For the sustainability of the road pavement, temperature is a key factor not only during the service life of the infrastructure but also when the HMA is produced. In fact, an incorrect management of the virgin aggregate, RAP, and bitumen temperatures at the mixing plant can determine excessive pollutant emissions and improper performance of the HMA. Calabi-Floody, Valdés-Vidal, Sanchez-Alonso, and Mardones-Parra (contribution 3) evaluated the gas emissions, the energy consumption, and the production costs of HMA manufactured at different temperatures, including different RAP contents (up to 30%). In particular, natural zeolite was used in the warm mix asphalt (WMA) and allowed us to reduce the mixing temperature from 155 °C to 125 °C. The laboratory results and analyses showed that the CO_2 emissions reduced by 23% and 37% for the WMA, respectively, without and with RAP, while the lower benefits were obtained in terms of CO emissions. The energy consumption decreased by 5–13%, but this did not allow total compensation of the cost of the zeolite. Bocci, Prosperi, Mair, and Bocci (contribution 5) investigated how keeping the HMA at high temperature for a long time influences the mix properties. The research demonstrated that HMA handling in the laboratory, during quality assurance and quality controls, is extremely important, since keeping the material in the oven for a prolonged time or re-heating it may significantly affect the test results. In contrast, in full-scale road construction, cooling and temperature segregation represent a higher risk for HMA than aging, but these can be limited using insulated trucks.

The sustainability of road infrastructure is also associated with pavement–tire interfaces, in particular, to the rolling resistance of vehicle tires, which is strictly related to fuel consumption. Guo, Li, Ran Zhou, and Yan (contribution 6) modelled the tire–pavement contact surface, with specific reference to trucks and buses, to predict stresses and rolling resistance under different rolling conditions. The proposed exponential equation describes a method that can forecast the rolling resistance related to the working conditions of truck–bus tires, allowing the estimation of fuel consumption and greenhouse gas emissions.

When administrations and designers have to plan construction or maintenance work, an accurate prediction of the environmental, economic, and social impact is fundamental in the optic of sustainability. Two papers of this Special Issue deal with this topic. Halámek, Matuszková, and Radimský (contribution 8) carried out an ex-post cost and benefit (CB) analysis on a set of 144 regional road modernization projects in Czech Republic. In particular, they compared the results of the ex-post CB analysis with those of ex-ante CB analysis in order to identify the best indicators for supporting administrations in defining where and how much to invest. The study showed that the money spent for regional road modernization was 11% lower than expected. The impact of the projects on traffic accidents was not as positive as foreseen in the ex-ante CB analysis, negatively influencing the net present value and the weighted profitability index. Therefore, the authors recommend eliminating the impact on traffic accidents in the ex-ante analysis. Balaguera, Alberti, Carvajal, and Fullana-i-Palmer (contribution 9) performed a life cycle analysis (LCA) to identify the recycled soil stabilizer with the lowest environmental impact for two rural roads in Colombia. In particular, they considered brick dust, fly ash, sulphonated oil, and recycled polymeric emulsion. The sulphonated oil was not suitable, as it required significant additional resources to become a stabilizer. Moreover, the manufacture of the polymer and its application as a stabilizer were also found to have great impacts. Brick dust and fly ash showed the highest impact in the stabilizer manufacture step but resulted the best solutions for the two investigated rural roads in terms of global warming potential.

All the valuable and interesting papers included in this Special Issue significantly contribute to the knowledge of road material behavior and the new frontiers of pavement engineering. Many solutions can be found, investigated, and validated in order to make the road infrastructure industry sustainable. The hope is that these innovations can soon become the normality.

Funding: This research received no external funding.

Institutional Review Board Statement: Not applicable.

Informed Consent Statement: Not applicable.

Data Availability Statement: Not applicable.

Conflicts of Interest: The author declares no conflict of interest.

List of Contributions:

1. Elizondo-Martinez, E.-J.; Tataranni, P.; Rodriguez-Hernandez, J.; Castro-Fresno, D. Physical and Mechanical Characterization of Sustainable and Innovative Porous Concrete for Urban Pavements Containing Metakaolin. *Sustainability* **2020**, *12*, 4243. [CrossRef]
2. Ragni, D.; Canestrari, F.; Allou, F.; Petit, C.; Millien, A. Shear-Torque Fatigue Performance of Geogrid-Reinforced Asphalt Interlayers. *Sustainability* **2020**, *12*, 4381. [CrossRef]
3. Calabi-Floody, A.T., Valdés Vidal, G.A.; Sanchez-Alonso, E.; Mardones-Parra, L.A. Evaluation of Gas Emissions, Energy Consumption and Production Costs of Warm Mix Asphalt (WMA) Involving Natural Zeolite and Reclaimed Asphalt Pavement (RAP). *Sustainability* **2020**, *12*, 6410. [CrossRef]
4. Fusco, R.; Moretti, L.; Fiore, N.; D'Andrea, A. Behavior Evaluation of Bituminous Mixtures Reinforced with Nano-Sized Additives: A Review. *Sustainability* **2020**, *12*, 8044. [CrossRef]

5. Bocci, E.; Prosperi, E.; Mair, V.; Bocci, M. Ageing and Cooling of Hot-Mix-Asphalt during Hauling and Paving—A Laboratory and Site Study. *Sustainability* **2020**, *12*, 8612. [CrossRef]
6. Guo, M.; Li, X.; Ran, M.; Zhou, X.; Yan, Y. Analysis of Contact Stresses and Rolling Resistance of Truck-Bus Tyres under Different Working Conditions. *Sustainability* **2020**, *12*, 10603. [CrossRef]
7. Hung, V.V.; Seo, S.-Y.; Kim, H.-W.; Lee, G.-C. Permeability and Strength of Pervious Concrete According to Aggregate Size and Blocking Material. *Sustainability* **2021**, *13*, 426. [CrossRef]
8. Halámek, P.; Matuszková, R.; Radimský, M. Modernisation of Regional Roads Evaluated Using Ex-Post CBA. *Sustainability* **2021**, *13*, 1849. [CrossRef]
9. Balaguera, A.; Alberti, J.; Carvajal, G.I.; Fullana-i-Palmer, P. Stabilising Rural Roads with Waste Streams in Colombia as an Environmental Strategy Based on a Life Cycle Assessment Methodology. *Sustainability* **2021**, *13*, 2458. [CrossRef]
10. Vaitkus, A.; Gražulytė, J.; Kravcovas, I.; Mickevič, R. Comparison of the Bearing Capacity of Pavement Structures with Unbound and Cold Central-Plant Recycled Base Courses Based on FWD Data. *Sustainability* **2021**, *13*, 6310. [CrossRef]
11. Prosperi, E.; Bocci, E. A Review on Bitumen Aging and Rejuvenation Chemistry: Processes, Materials and Analyses. *Sustainability* **2021**, *13*, 6523. [CrossRef]
12. Mohammed, S.A.; Koting, S.; Katman, H.Y.B.; Babalghaith, A.M.; Abdul Patah, M.F.; Ibrahim, M.R.; Karim, M.R. A Review of the Utilization of Coal Bottom Ash (CBA) in the Construction Industry. *Sustainability* **2021**, *13*, 8031. [CrossRef]
13. Mokoena, R.; Mturi, G.; Maritz, J.; Mateyisi, M.; Klein, P. African Case Studies: Developing Pavement Temperature Maps for Performance-Graded Asphalt Bitumen Selection. *Sustainability* **2022**, *14*, 1048. [CrossRef]

Article

Physical and Mechanical Characterization of Sustainable and Innovative Porous Concrete for Urban Pavements Containing Metakaolin

Eduardo-Javier Elizondo-Martinez [1,*], Piergiorgio Tataranni [2,*], Jorge Rodriguez-Hernandez [1] and Daniel Castro-Fresno [1]

[1] GITECO Research Group, Universidad de Cantabria, Avda de los Castros s/n, 39005 Santander, Spain; rodrighj@unican.es (J.R.-H.); castrod@unican.es (D.C.-F.)
[2] DICAM Department, University of Bologna, Via Terracini 28, 40131 Bologna, Italy
* Correspondence: eduardo-javier.elizondo@alumnos.unican.es (E.-J.E.-M.); piergiorg.tataranni2@unibo.it (P.T.)

Received: 6 May 2020; Accepted: 20 May 2020; Published: 21 May 2020

Abstract: Alternative materials to replace cement in pavements have recently been widely studied with the purpose of decreasing the environmental impacts that the construction industry generates. In this context, the implementation of sustainable urban drainage systems has grown, especially with porous pavements, with the intention to reduce water and environmental impacts. In the present investigation, the addition of alternative materials to minimize the use of cement in porous concrete pavements is evaluated. Starting from a partial substitution of Portland cement with metakaolin, experimental geopolymer concretes were produced with metakaolin and waste basalt powder according to several dosages. Two sets of mixtures were analyzed to evaluate the Porous Concrete Design (PCD) methodology for porous concrete mixtures with alternative materials. A deep analysis was proposed for the evaluation of the mechanical and volumetric properties of the mixtures. Results demonstrated that replacing 5% of cement with metakaolin can increase both permeability and indirect tensile strength. Geopolymer mixtures can achieve permeability significantly higher than the traditional porous concrete, but this decreases their indirect tensile strength. However, considering the promising results, an adjustment in the mix design of the geopolymer mixtures could increase their mechanical properties without negatively affecting the porosity, making these materials a suitable alternative to traditional porous cement concrete, and a solution to be used in urban pavements.

Keywords: porous concrete; metakaolin; geopolymers; permeable pavements; urban drainage systems

---

1. Introduction

Sustainable solutions are the main issue for researchers in the construction industry, this sector being responsible for 36% of global energy use and 40% of CO_2 emissions [1]. Pavements have gained considerable attention due to the impact they are causing, and the environmental benefits they can provide. By the early 2000s, around 3% of the total surface of the planet had been covered with pavements [2]. This brings many problems, such as the obstruction of the hydrological cycle, causing runoff and water pollution [3], and the increase of temperatures in urban areas due to the solar absorption of the pavements (Urban Heat Island, UHI) [2]. This high amount of pavement is also related to the use of motor vehicles, generating gas emissions into the air [4].

In this context, the concept of Sustainable Urban Drainage Systems (SUDS) has come up to deal with storm water management. Porous pavements are the most widely used type of SUDS that can, with a proper design, deal with the UHI effect and the air pollution, as well [4]. The most common materials employed in porous pavements are asphalt and cement concretes. The latter is recognized as a good solution to reduce both water and temperature environmental impacts [3] and has several advantages in the construction operations if compared to porous asphalt. The problem with the minor use of porous pavements remains in their structure, designed to maintain a high porosity, around 15–30% [5–7], which leads to a low load-bearing capacity that limits its ability to resist under traffic loads [8,9]. In addition, the energy and greenhouse gas emissions required for processing Portland cement are very high in comparison with other materials [10], although asphalt pavements have a greater impact on the environment because of the greenhouse gas emissions generated during the manufacturing of raw materials and the need of disposing the pavement in hazardous waste management facilities [11].

Thus, alternative materials have been recently studied to partially or totally replace the cement from porous concrete and obtain a more eco-friendly pavement. This is the case of geopolymer development [12–14], in which the use of specific powders (e.g., metakaolin, fly ash, etc.) called precursors, under strong alkaline conditions given by the activators, can generate a chemical reaction able to create a cementitious material [1]. Geopolymers have been widely studied in the last years primarily because of their early high strength [14,15], where metakaolin has attracted considerable attention because of its early resistance and good chemical resistance, among such other advantages as good fire-resistant behavior [15,16]. In addition, several studies highlighted that the greenhouse gas emissions generated during the geopolymer production can be around 40% lower than the ones related to Portland cement, making this material more environmentally friendly than traditional cement concrete [10]. The application of geopolymer as an alternative to cement concrete for construction and buildings is supported by well-established literature and experimental application. As for road pavements, the use of geopolymer mixtures is still under study. Metakaolin is a dehydrated form of the clay mineral kaolin, which is obtained by the calcination of this clay at temperatures of 500–800 °C. As a pozzolanic material, it is considered a good substitute for ordinary Portland cement [15]. However, in economic terms, the energy needed to produce the geopolymers is still an issue. For example, in the production of sodium silicate, one of the most common activators used for the chemical reaction, the energy demand is over 30% higher if compared to that needed to obtain the feedstock for Portland cement. Furthermore, for a metakaolin-based geopolymer mixture, the total cost of production is around 80% higher than common cement concrete mixtures, mainly because of the rare supplier mines of metakaolin, which make the transportation cost high [10].

In the light of the above, the present research introduces a comparison between porous mixtures made with cement and metakaolin, to understand the effect of these materials on the design parameters, as well as on the final functional and mechanical properties in terms of Indirect Tensile Strength (ITS) and permeability. In addition, some innovative and experimental mixtures were produced and tested with the same grading distribution but using the alkali-activation process with metakaolin and waste basalt powder for the production of alternative and eco-friendly mixtures. With this, two sets of mixtures were evaluated to observe the feasibility of designing geopolymers with the methodology explained in the following sections for porous concrete mixtures.

2. Materials and Methods

2.1. Methodology of Design

For the preparation of porous samples, the PCD (Porous Concrete Design) methodology was used, which process can be consulted in [17]. This method is based on the ACI 522R-10 standard, in which the relation between the coarse aggregates and sand is eliminated, introducing the latter into the cement paste design and obtaining a mortar. Therefore, the sand amount modifies the cement and the water quantities of the mixture. The coarse aggregate quantity depends on its particle density and porosity. Meanwhile, PCD starts with a proposed porosity (AV) amount of design, sand–cement (s/c) and water–cement (w/c) ratio. Then the coarse aggregate weight is determined. With this, the mortar volume and weight can be calculated to finally obtain the dosages of water, cement, and sand.

The EN 1097-3 standard was used to calculate the VMA (Voids in Mineral Aggregate) of the aggregate. The test was performed twice (compacted and loose aggregates) in order to establish a parameter for the VMA content in the mixture design. According to the PCD methodology, the VMA helps to establish the material amounts, where higher VMAs provide lower amounts of coarse aggregates and add higher amounts of mortar in the design of the mixture. The opposite performance is found when lower VMA amounts are considered [17].

2.2. Materials and Mixtures

Limestone aggregates were employed to elaborate all the mixtures in a gradation of 5–10 mm. For the Control mixture, Portland cement type I 42.5R was used, in a water to cement ratio (w/c) of 0.30. For the experimental mixtures, metakaolin was used to partially replace the cement in amounts of 5% and 10% of the cement weight, maintaining a constant w/c of 0.30. The first mixture was labelled 95C-5MK and the second 90C-10MK.

In addition, three other experimental mixtures were fabricated using the alkali-activation process to produce the binder for the final concrete. No water was used and the grading distribution was kept constant using the same aforementioned design parameters. The first mixture, labelled 100 MK, was produced with 100% metakaolin as precursor, employing a blend of sodium silicate and sodium hydroxide in a ratio of 3 to 1 as activator. The chemical composition of these materials can generate the process of alkali-activation as verified in previous studies [18]. An activator–precursor (A/P) ratio of 0.87 was used because of the workability presented by the material (at a lower A/P ratio, lower workability). Furthermore, basalt powder was applied to replace 50% and 75% of metakaolin in two additional experimental mixtures named 50MK-50Bas and 25MK-75Bas, respectively. A total of six different mixtures were produced and tested. The limestone aggregate and the material properties are reported in Tables 1 and 2, respectively.

Table 1. Limestone aggregate characteristics.

Property	Standard	Value
Bulk density (g/cm^3)	EN 1097-3	2.70
Specific gravity	EN 1097-6	2.80
Water absorption (%)	EN 1097-6	0.90
Voids in aggregate (%)	EN 1097-3	49.75
Compacted voids in aggregate (VMA) (%)	EN 1097-3	41.24
Gradation (mm)	EN 933-1	5–10

Table 2. Material characteristics.

Property	Standard	Cement	Metakaolin	Basalt Powder	Sodium Silicate	Sodium Hydroxide
Bulk density (g/cm^3)	EN 1097-3	3.14	2.40	1.36	2.40	2.13
Gradation (mm)	EN 933-1	<0.063	<0.063	<0.050	-	-

Samples were designed for a diameter of 10 cm and 6.5 cm height, compacted in the Marshall device at 35 blows on the surface, in compliance with the EN 12697-30 standard. Cement mixtures were submerged under water for 28 days for curing. The geopolymer mixtures were instead placed in the oven for 12 h at 70 °C and then cured for 21 days at room temperature. The curing in the oven is a common procedure for geopolymer mixtures and the setting time and temperature was validated in previous researches on the same synthetic mixture [1,18].

2.3. Geopolymer Mortar Characterization

In order to characterize the geopolymer paste, three 4 × 4 × 4 cm cubes were cast for each mixture and tested at different curing times (7 and 21 days). It is worth noting that there are no specific tests or standards for geopolymers. Thus, the mechanical properties are generally analyzed by evaluating the compressive strength of cubic samples in compliance with the EN 1015-11 standard, which refers to hardened mortars. It can be seen in Figure 1 that from curing day 7 to day 21 values seemed to be constant for the mixture with 100% metakaolin and the mixture with 50% metakaolin and 50% basalt powder, varying within 10 MPa. The mixture with 25% metakaolin and 75% basalt powder had a poor result at 21 days of curing because of the low amount of metakaolin, as basalt powder does not provide a good adherence for the aggregate particles. Overall, mixtures seemed to be cured just after 7 days. This was a further confirmation of the early strength of alkali-activate materials. Moreover, the 100MK showed a compressive strength even higher than the typical values registered for traditional Portland cement concretes after 7 days of curing. Finally, due to the difference in the performances the mixtures had because of the diverse materials and dosages employed, it was decided to test only one sample per mixture, as the objective was to observe the behaviors presented by the mixtures under the same methodology of design.

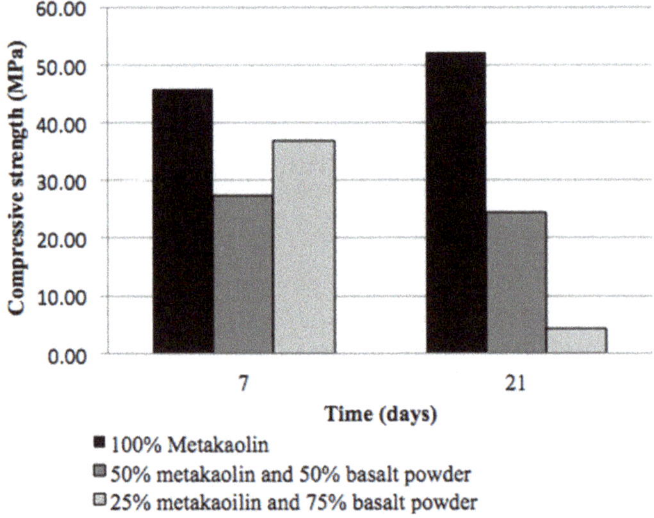

Figure 1. Compressive strength of geopolymer mixtures after different curing periods.

2.4. Determination of the Voids in Mineral Aggregate and the Porosity of Mixtures

Porous concrete pavements are normally designed to have a certain amount of porosity, between 15% and 30% [5–8]. Therefore, it was decided to establish a porosity target of 20% for the cement concrete mixtures designed in this investigation, as well as a VMA of 47% (in the range 41.24%–49.75% in Table 1), to increase the mortar amount in the mixtures, enhancing the mechanical strength. Considering that geopolymer mixtures do not have the same behavior, four different tentative mixtures were made, varying the VMA and the AV in order to decide the parameters for the final porous mixture design. VMAs of 41.52% and 47% and AVs of 20% and 30% were combined (mixtures A, B, C, and D in Figure 2) to define the best combination for the geopolymer mixtures, selected as values between the parameters obtained in the VMAs from Figure 1.

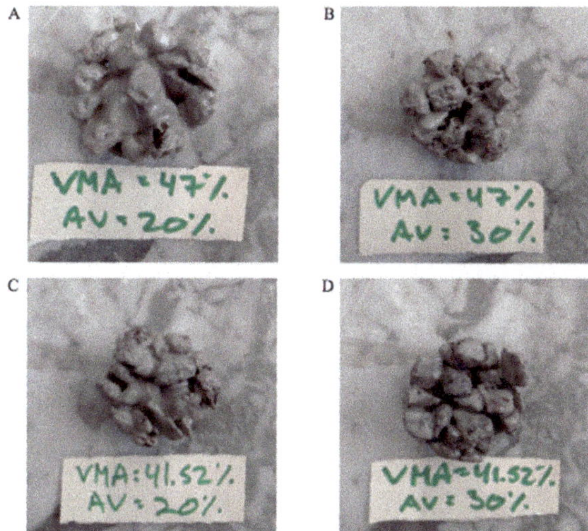

Figure 2. Four different dosages proposed for geopolymer mixtures, varying the voids in mineral aggregate (VMA) and porosity (AV), where (**A**) corresponds to a VMA of 47% and AV of 20% (highest amount of mortar); (**B**) to a VMA of 47% and AV of 30%; (**C**) to a VMA of 41.52% and AV of 20%; (**D**) to a VMA of 41.52% and AV of 30% (lowest amount of mortar).

As visible in Figure 2, the geopolymer paste used for mixture A, with the same design parameters adopted for the cement concrete, tended to clog the mixture completely. Therefore, for these mixtures, we considered employing opposite parameters (higher AV and lower VMA) than cement concrete, as showed by mixture D, which had the highest AV among the four mixtures. Mixture type D was selected for the final mix design of geopolymer mixtures to guarantee the presence of sufficient interconnected voids for infiltrating water. Mixtures B and C obtained a higher percentage of mortar in their designs than mixture D (lower than mixture A). This meant that both mixtures obtained fewer interconnected voids for permeability. In fact, they were almost clogged completely, not considered to fulfill a good permeability for porous mixtures.

The behavior of the mortars for each type of mixture is presented in Figure 3: the cement tended to create mortar bridges that bound together the aggregate particles, while the geopolymer paste seemed to coat completely the aggregates.

Figure 3. Mortar behavior between a cement porous concrete and a geopolymer porous concrete.

The final design parameters for all the mixtures are summarized in Table 3.

Table 3. Mixture compositions.

Mixture	Cementitious Material (g)			Agg[1] (g)	Water (g)	w/c	Activator (g)		A/P[1]	Porosity (%)
	C[1]	MK[1]	Bas[1]				SS[1]	SH[1]		
Control	237.50	-	-	771.80	80.60	0.30	-	-	-	20.00
95C-5MK	224.12	11.80	-	771.80	80.00	0.30	-	-	-	20.00
90C-10MK	210.77	23.42	-	771.80	79.50	0.30	-	-	-	20.00
100MK	-	77.60	-	855.40	-	-	57.20	19.10	0.87	30.00
50MK-50Bas	-	40.05	40.05	855.40	-	-	58.80	19.60	0.87	30.00
25MK-75Bas	-	20.33	60.99	855.40	-	-	59.50	19.80	0.87	30.00

[1] C: cement; MK: metakaolin; Bas: basalt powder; Agg: coarse aggregate; SS: sodium silicate; SH: sodium hydroxide; A/P: Activator–Precursor ratio.

2.5. Experimental Plan

The experimental plan was based on the evaluation of the total porosity (volumetric property), permeability (hydraulic property), and the Indirect Tensile Strength (mechanical property) of the mixtures. In compliance with the ASTM C1688 standard, the porosity amount (*AV*) can be calculated with the Equation (1):

$$AV = \left(\frac{\rho_t - \rho}{\rho_t}\right) \times 100, \tag{1}$$

where ρ_t corresponds to the bulk density, calculated by the sum of the total mass of the material proportions employed to elaborate the mixture in accordance to standard EN 1097-3, divided by the volume of the mold, and ρ is the real density obtained from the net mass of the sample divided by the volume of the container.

The permeability of the mixtures was obtained with a falling-head permeameter adapted for laboratory uses from the Spanish NLT 327/00 standard. It consists of a 10-cm diameter methacrylate tube, placed on top of the sample, which allows the water to flow into the sample to perform the test. The tube was calibrated for a 20-cm water column. Then, employing Darcy's law, the permeability capacity can be calculated, according to Equation (2):

$$k = \left[\frac{(A_{sample}) \times (h_{sample})}{(A_{tube}) \times (t)}\right]\left[ln\left(\frac{h_1}{h_2}\right)\right], \tag{2}$$

where k is the permeability capacity (cm/s), A_{sample} is the area of contact of the sample, h_{sample} is the height of the sample, A_{tube} is the area of the tube's gap, and t is the time that the water takes to flow from the highest point h_1 to the lowest point h_2.

The Indirect Tensile Strength (ITS) test was performed in order to analyze the mechanical properties of the mixtures according to the EN 12390-6 standard. This test enables an indirect assessment of the pavement behavior under traffic loads by the evaluation of the level of cohesion between aggregates, generated by the cement binder. The ITS test description, equipment, and equations used were in compliance with the EN 12390-6 and EN-12390-1 standards. Finally, the behavior of geopolymer mixtures after being exposed to water (permeability test) was measured through the ITS, as well. Once the permeability test was made, mixtures were dried at ambient temperature for 14 days to perform the ITS, evaluating one sample per mixture. A flow chart is presented in Figure 4 to summarize the experimental plan of the following research.

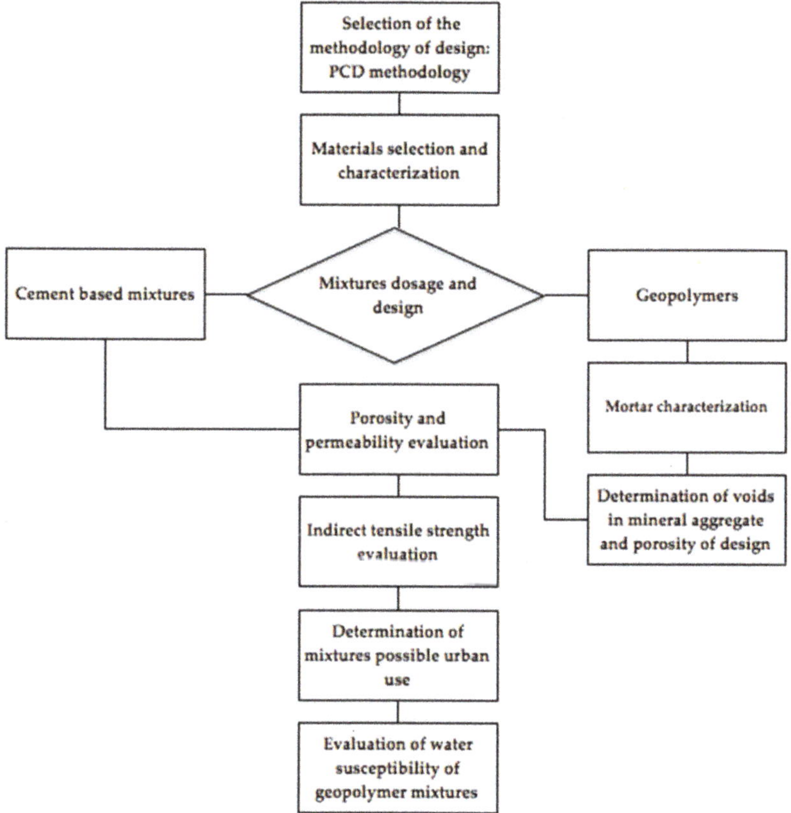

Figure 4. Experimental plan flow chart.

3. Results and Discussion

3.1. Porosity and Permeability

The total porosity (AV) and permeability (k) results are shown in Figure 5.

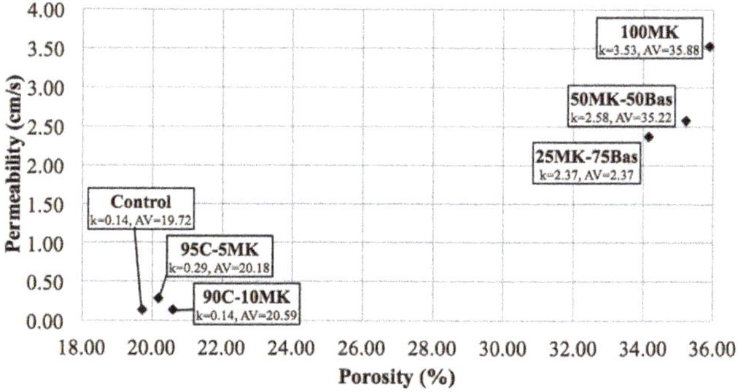

Figure 5. Porosity and permeability general results.

It can be observed that the geopolymer mixtures achieved permeability results over 90% higher than the Control mixture because of the porosity parameter used in order to control the VMAs. If the mixtures with cement (Control, 95C-5MK, and 90C-10MK) had been designed with a higher porosity, the mortar amount would have been lower, making the adhesion between the aggregate particles poor and, therefore, weak.

In addition, the use of basalt powder in the geopolymer mixtures tended to clog the air voids in the sample, decreasing the permeability capacity. The addition of 50% of basalt powder decreased the permeability by 27%, and with 75% of basalt powder the decrement was equal to 33%. Nevertheless, both AV and permeability results were considerably high for the geopolymer mixtures with basalt powder.

In the case of replacing part of the cement with metakaolin (95C-5MK and 90C-10MK), results demonstrated that the porosity increased and permeability resulted. In this scenario, replacing 5% of cement with metakaolin doubled the permeability. However, the increase in the metakaolin amount of 10% can be considered excess, where the mortar tended to cover the aggregates more and permeability started to decrease, showing the same results as the Control mixture. It can be stated that the replacement of cement with metakaolin over 5% seemed to negatively affect the permeability of the mixture. Nevertheless, according to the National Center for Asphalt Technology, a minimum permeability of 100 m/day (0.012 cm/s) is suggested for open-graded friction courses [19,20]. Therefore, all the mixtures overcame that parameter, even the Control mixture with the lowest permeability (0.14 cm/s). In addition, the behavior of the mixtures demonstrated that at higher porosity, permeability tended to increase, a performance in compliance with some authors' results [21,22].

3.2. Density and Indirect Tensile Strength

The density (ρ) and indirect tensile strength (IT) results are presented in Figure 6.

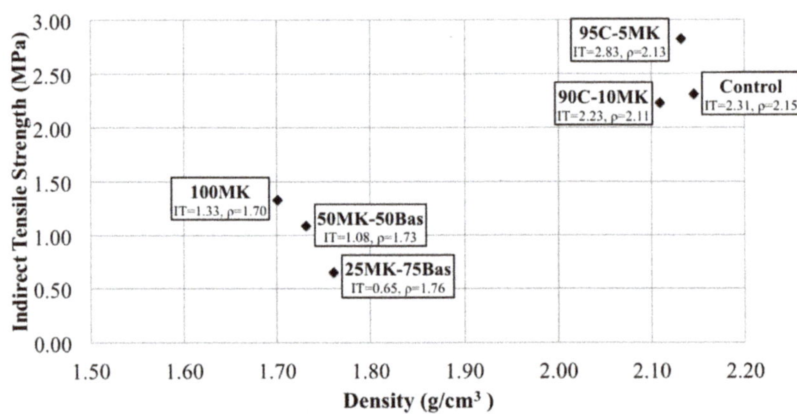

Figure 6. Indirect tensile and density general results.

Here, the geopolymer mixtures obtained lower results if compared to the mixtures with cement, because of the higher porosity. Between the geopolymer mixtures, 100MK showed the highest ITS values, and this was in line with the mechanical properties highlighted in the geopolymer paste characterization. However, in wider terms, the 100MK reached ITS values 42% lower than the Control mixture. Nevertheless, it can be observed that mixtures 100MK and 50MK-50Bas achieved acceptable values of ITS, over 1 MPa, despite the high porosity the sample presented.

In addition, the use of basalt powder demonstrated a weak bond between the geopolymer paste and the aggregate, as the paste became more fluent and went to the bottom of the mold. This clogged the mixture and decreased ITS: 50% of basalt powder in the mixture decreased the ITS by almost 19% if compared to mixture 100MK. Adding basalt powder up to 75% of the cementitious material weight reduced the strength around 51%. It was observed that the lowest cohesion between aggregate particles led to lower ITS results.

In the case of the cement mixtures, ITS results increased by 18% when replacing 5% of cement with metakaolin (95C-5MK) without a relevant variation in the density, compared to the Control mix. The ITS was reduced to almost the same values as the Control mixture when the metakaolin amount increased to 10% (90C-10MK).

Furthermore, to compare the obtained results with other experimental porous concretes, a detailed literature review was carried out. Bringing together the most recent studies on ITS in porous concrete pavements, Table 4 demonstrates the lower and higher results achieved by each author, as well as the w/c, aggregate size, and additional materials employed. ITS values range from 0.02 MPa to 3.09 MPa. Taking this into account, all the experimental mixtures included in this research paper showed acceptable results.

Table 4. Indirect tensile results and materials employed by some authors.

Author	Ref.	IT [1] (MPa)	w/c	Agg [1] (mm)	Note
Torres, A. et al., 2015	[23]	1.09–3.09	0.33	6.35–9.54	3 compaction levels and 2 aggregate sizes (limestone) evaluated.
Bonicelli, A. et al., 2015	[24]	0.02–0.21	0.27-0.35	3–10	Addition of sand.
Rangelov, M. et al., 2016	[5]	1.40–2.90	0.24	9.5	Addition of carbon fibers.
Adewumi, A. et al., 2016	[25]	0.21–1.32	0.35-0.40	4.50–22	Different mixtures of coarse aggregate, cement and w/c were used.
Brake, N. et al., 2016	[6]	0.98–3.04	0.27-0.30	10	Type I polycarboxylate superplasticizer and type S viscosity modifying agent were employed.
Bonicelli, A. et al., 2016	[26]	1.40–2.20	0.27-0.35	-	Monofilament polypropylene and polyethylene fibers used.
Hsin-Lun, H. et al., 2018	[27]	0.5–2.1	0.35	19–25	Portland cement, and co-fired fly ash and blast-furnace slag as replacement of cement were used.
Mohd-Ibrahim, M.Y. et al., 2018	[28]	2.5–4.3	0.34	4.75–12.5	Use of nano black rise husk ash and crushed granites.
Tataranni, P. and Sangiorgi, C., 2019	[18]	0.48–0.60	—	6.3–12.5	Use of a polymeric binder, activated with sodium silicate and sodium hydroxide. Synthetic and limestone aggregate.
Alshareedah, O. et al., 2019	[29]	1.1–1.5	0.35	4.75–9.5	Addition of cured carbon fiber composite material.
Elizondo-Martinez, E.J. et al., 2020	[9]	1.62–2.75	0.30	4–8	Combination of superplasticizer, air entraining, and polypropylene fibers for the highest result.

[1] IT: Indirect tensile strength; Agg: coarse aggregate.

In addition, considering some authors obtained high ITS values by employing additives, fibers, or other additions (such as sand and certain types of ash) to improve the properties of the mixture, the results obtained in the present investigation demonstrated that the methodology of design, as well as the compaction method used, can achieve very good results, with a range between 0.65 MPa and 2.83 MPa.

3.3. Optimal Mixtures and Performance Requirements

Figure 7 shows the graph proposed by Bonicelli et al. [26], showing the performance requirements for different urban uses of porous concrete pavements and locating results of the present investigation. As stated by the authors, mid-volume urban roads require ITS values over 1.9 MPa, low-volume urban roads and parking lots require instead ITS values between 1.7 and 1.9 MPa and permeability results over 1 cm/s. ITS values between 1.5 and 1.7 MPa and permeability over 1.5 cm/s are recognized as suitable for bike paths, while permeability values over 2 cm/s work better for pedestrian areas, squares, foot paths, and parks.

As seen in Figure 6, all the cement-base mixtures can be considered suitable for mid-volume urban roads because of the high mechanical properties and relatively low permeability. The geopolymer mixtures are suitable for pedestrian areas, squares, footpaths, and parks because of the high permeability capacity. However, for future research, the reduction in the design AV for geopolymer concrete might improve the final ITS of the mixtures.

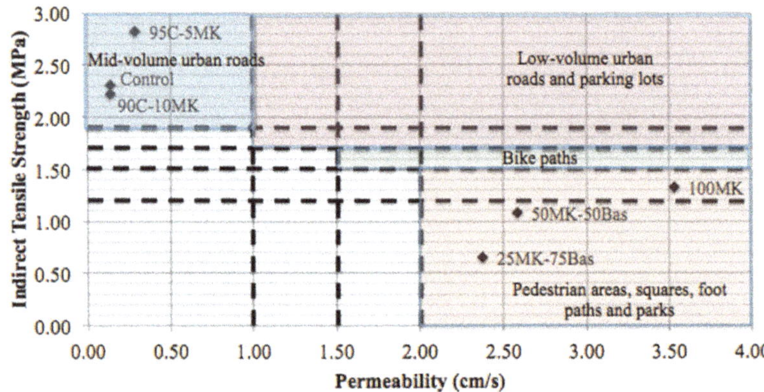

Figure 7. Performance requirements for different urban uses of porous concrete pavements according to Bonicelli et al., 2016 [26].

3.4. Water Susceptibility of Geopolymer Mixtures

Despite the environmental advantages that porous pavements made with geopolymers can present, the exposure of these materials to water can decrease the strength of the mixtures, as Table 5 shows. Once again, an optimization of the design parameters could decrease the porosity and so increase the VMAs and the final mechanical properties. It is worth noting that in the case of the 100MK mixture, the ITS reduction was not significant (16.67%). For the other experimental mixtures the addition of basalt had a negative effect on the ITS reduction, probably because of the already poor cohesion between particles, which was further limited by the presence of water.

Table 5. Indirect tensile results of geopolymers before and after water exposure.

Mixture	Indirect Tensile Strength (MPa)	Indirect Tensile Strength after Exposure to Water (MPa)	Strength Reduction (%)
100MK	1.33	1.10	16.67
50MK-50BAS	1.08	0.65	40.13
25MK-75BAS	0.65	0.45	31.37

4. Conclusions

Cement production has a big environmental impact and, consequently, alternative materials are being studied to replace it in pavements.

The present research shows a comparison of the functional and mechanical properties of different porous concretes produced with different amounts of metakaolin and alternative geopolymer porous mixtures containing metakaolin and basalt powder.

Based on the discussed results, the following conclusions can be stated:

- Replacing 5% of cement with metakaolin increases both the ITS and the permeability, but a substitution of 10% of cement with metakaolin reduces both the properties.
- Cement base mixtures (only with reductions of 5% or 10%) show very high ITS values and acceptable porosity if compared to the geopolymer ones (100% metakaolin).
- A design porosity of 20% is considered low for the geopolymer mixtures, where, because of the behavior of the paste material, the sample tends to clog. Meanwhile, a design porosity of 30% causes an excessive AV in the material that negatively affects the ITS.
- The increase in the amount of basalt powder in the mixture lowers the mechanical properties of the sample, both the compressive strength of the mortar cubes and the ITS of the porous samples.

- According to the results, for geopolymer porous pavements, an amount of 50% or lower of basalt powder in the mixture is recommended to maintain an average mechanical–permeability relation. The amount will depend on the use the pavement is going to have. A high amount can decrease these results considerably.
- The mechanical properties of geopolymer mixtures with basalt are strongly affected when exposed to water. As the main purpose of porous pavements is to infiltrate water through their structure, adjustments in the design parameters (such as lower porosity and higher VMA) are needed to reduce the water susceptibility.
- Considering the results obtained with the dosages evaluated, geopolymer mixtures are suitable for pavements with low load, like pedestrian areas, which can comprise a large area in a city, and cement use can be decreased. They also prevent runoff during rain events.
- Cement-based mixtures, according to the results of the present investigation, can be considered for use in mid-volume urban roads (secondary streets), which represent a high percentage of pavement in a city, decreasing some amount of cement and increasing the permeable capacity of the soil, especially during rain events.
- Both cement mixtures with metakaolin and mixtures with geopolymer paste represent good alternatives for sustainable pavements, reducing the use of cement.

Author Contributions: Conceptualization, E.-J.E.-M. and P.T.; methodology, E.-J.E.-M., P.T., J.R.-H. and D.C.-F.; software, E.-J.E.-M.; validation, P.T., J.R.-H. and D.C.-F.; investigation, E.-J.E.-M.; resources, P.T.; writing—original draft preparation, E.-J.E.-M.; writing—review and editing, P.T., J.R.-H. and D.C.-F.; supervision, P.T. All authors have read and agreed to the published version of the manuscript.

Funding: This research received no external funding.

Acknowledgments: The authors would like to thank the DICAM Department of the University of Bologna for providing the materials and facilities to make this investigation.

Conflicts of Interest: The authors declare no conflict of interest.

References

1. Tataranni, P. Recycled Waste Powders for Alkali-Activated Paving Blocks for Urban Pavements: A Full Laboratory Characterization. *Infrastructures* **2019**, *4*, 73. [CrossRef]
2. Sinha, K.C.; Bullock, D.; Hendrickson, C.T.; Levinson, H.S.; Lyles, R.W.; Radwan, A.E.; Li, Z. Development of Transportation Engineering Research, Education, and Practice in a Changing Civil Engineering World. *J. Transp. Eng.* **2002**, *128*, 301–313. [CrossRef]
3. Rodriguez-Hernandez, J.; Fernández-Barrera, A.H.; Andrés-Valeri, V.C.A.; Vega-Zamanillo, A.; Castro-Fresno, D. Relationship between Urban Runoff Pollutant and Catchment Characteristics. *J. Irrig. Drain. Eng.* **2013**, *139*, 833–840. [CrossRef]
4. Elizondo-Martínez, E.; Andrés-Valeri, V.; Jato-Espino, D.; Rodriguez-Hernandez, J. Review of porous concrete as multifunctional and sustainable pavement. *J. Build. Eng.* **2020**, *27*, 100967. [CrossRef]
5. Rangelov, M.; Nassiri, S.; Haselbach, L.; Englund, K. Using carbon fiber composites for reinforcing pervious concrete. *Constr. Build. Mater.* **2016**, *126*, 875–885. [CrossRef]
6. Brake, N.A.; Allahdadi, H.; Adam, F. Flexural strength and fracture size effects of pervious concrete. *Constr. Build. Mater.* **2016**, *113*, 536–543. [CrossRef]
7. Giustozzi, F. Polymer-modified pervious concrete for durable and sustainable transportation infrastructures. *Constr. Build. Mater.* **2016**, *111*, 502–512. [CrossRef]
8. Khankhaje, E.; Salim, M.R.; Mirza, J.; Salmiati; Hussin, M.W.; Khan, R.; Rafieizonooz, M. Properties of quiet pervious concrete containing oil palm kernel shell and cockleshell. *Appl. Acoust.* **2017**, *122*, 113–120. [CrossRef]

9. Elizondo-Martínez, E.; Andrés-Valeri, V.; Rodríguez-Hernádez, J.; Sangiorgi, C. Selection of Additives and Fibers for Improving the Mechanical and Safety Properties of Porous Concrete Pavements through Multi-Criteria Decision-Making Analysis. *Sustainability* **2020**, *12*, 2392. [CrossRef]
10. Nguyen, L.; Moseson, A.J.; Farnam, Y.; Spatari, S. Effects of composition and transportation logistics on environmental, energy and cost metrics for the production of alternative cementitious binders. *J. Clean. Prod.* **2018**, *185*, 628–645. [CrossRef]
11. Jato-Espino, D.; Lizasoain-Arteaga, E. Life cycle assessment of construction materials: Analysis of environmental impacts and recommendations of eco-efficient management practices. In *Handbook of Environmental Materials Management*; Hussain, C.M., Ed.; Springer: Cham, Switzerland, 2018.
12. Jang, J.G.; Ahn, Y.B.; Souri, H.; Lee, H.K. A novel eco-friendly porous concrete fabricated with coal ash and geopolymeric binder: Heavy metal leaching characteristics and compressive strength. *Constr. Build. Mater.* **2015**, *79*, 173–181. [CrossRef]
13. Chen, X.; Guo, Y.; Ding, S.; Zhang, H.; Xia, F.; Wang, J.; Zhou, M. Utilization of red mud in geopolymer-based pervious concrete with function of adsorption of heavy metal ions. *J. Clean. Prod.* **2018**, *207*, 789–800. [CrossRef]
14. Bouguermouh, K.; Bouzidi, N.; Mahtout, L.; Pérez-Villarejo, L.; Martínez-Cartas, M.L. Effect of acid attack on microstructure and composition of metakaolin-based geopolymers: The role of alkaline activator. *J. Non. Cryst. Solids* **2017**, *463*, 128–137. [CrossRef]
15. Mohammed, A.; Salih, A.; Raof, H. Vipulanandan Constitutive Models to Predict the Rheological Properties and Stress–Strain Behaviour of Cement Grouts Modified with Metakaolin. *J. Test. Eval.* **2020**, *48*. [CrossRef]
16. Busari, A.; Dahunsi, B.; Akinmusuru, J. Sustainable concrete for rigid pavement construction using de-hydroxylated Kaolinitic clay: Mechanical and microstructural properties. *Constr. Build. Mater.* **2019**, *211*, 408–415. [CrossRef]
17. Elizondo-Martinez, E.J.; Andres-Valeri, V.C.; Rodriguez-Hernandez, J.; Castro-Fresno, D. Proposal of a New Porous Concrete Dosage Methodology for Pavements. *Materials* **2019**, *12*, 3100. [CrossRef]
18. Tataranni, P.; Sangiorgi, C. Synthetic aggregates for the production of innovative low impact porous layers for urban pavements. *Infrastructures* **2019**, *4*, 48. [CrossRef]
19. Alvarez, A.E.; Martin, A.E.; Estakhri, C.K.; Button, J.W.; Glover, C.J.; Jung, S.H. *Synthesis of Current Practice on the Design, Construction, and Maintenance of Porous Friction Courses*; Technical Report No. FHWA/TX-06/0-5262-1; Texas Department of Transportation Research and Technology Implementation Office: Austin, TX, USA; Texas Transportation Institute at the Texas A&M University System: College Station, TX, USA, 2006; p. 76.
20. Mallick, R.B.; Kandhal, P.S.; Cooley, L.A.; Watson, D.E. *Design, Construction and Performance of New Generation Open-Graded Friction Courses*; NCAT Report No. 2000-01; National Center for Asphalt Technology (NCAT), Auburn University: Auburn, AL, USA, 2000.
21. Praticòo, F.G.; Moro, A. Permeability and Volumetrics of Porous Asphalt Concrete. A Theoretical and Experimental Investigation. *Road Mater. Pavement Des.* **2007**, *8*, 799–817. [CrossRef]
22. Yasarer, H.; Najjar, Y. Development of void prediction models for Kansas concrete mixes used in PCC pavement. *Procedia Comput. Sci.* **2012**, *8*, 473–478. [CrossRef]
23. Torres, A.; Hu, J.; Ramos, A. The effect of the cementitious paste thickness on the performance of pervious concrete. *Constr. Build. Mater.* **2015**, *95*, 850–859. [CrossRef]
24. Bonicelli, A.; Giustozzi, F.; Crispino, M. Experimental study on the effects of fine sand addition on differentially compacted pervious concrete. *Constr. Build. Mater.* **2015**, *91*, 102–110. [CrossRef]
25. Adewumi, A.A.; Owolabi, T.O.; Alade, I.O.; Olatunji, S.O. Estimation of physical, mechanical and hydrological properties of permeable concrete using computational intelligence approach. *Appl. Soft Comput. J.* **2016**, *42*, 342–350. [CrossRef]
26. Bonicelli, A.; Arguelles, G.M.; Pumarejo, L.G.F. Improving Pervious Concrete Pavements for Achieving More Sustainable Urban Roads. *Procedia Eng.* **2016**, *161*, 1568–1573. [CrossRef]
27. Ho, H.L.; Huang, R.; Hwang, L.C.; Lin, W.T.; Hsu, H.M. Waste-based pervious concrete for climate-resilient pavements. *Materials* **2018**, *11*, 900. [CrossRef]

28. Ibrahim, M.Y.M.; Ramadhansyah, P.J.; Rosli, H.M.; Ibrahim, M.H.W. Mechanical performance of porous concrete pavement containing nano black rice husk ash. *IOP Conf. Ser. Mater. Sci. Eng.* **2018**, *290*, 012050. [CrossRef]
29. AlShareedah, O.; Nassiri, S.; Chen, Z.; Englund, K.; Li, H.; Fakron, O. Field performance evaluation of pervious concrete pavement reinforced with novel discrete reinforcement. *Case Stud. Constr. Mater.* **2019**, *10*, e00231. [CrossRef]

Article

Shear-Torque Fatigue Performance of Geogrid-Reinforced Asphalt Interlayers

Davide Ragni [1,*], Francesco Canestrari [1], Fatima Allou [2], Christophe Petit [2] and Anne Millien [2]

[1] Department of Civil and Building Engineering and Architecture, Università Politecnica delle Marche, via Brecce Bianche, 60131 Ancona, Italy; f.canestrari@univpm.it

[2] Laboratoire GC2D, Université de Limoges, Bd J. Derche, 19300 Egletons, France; fatima.allou@unilim.fr (F.A.); christophe.petit@unilim.fr (C.P.); anne.millien@unilim.fr (A.M.)

* Correspondence: d.ragni@pm.univpm.it

Received: 5 May 2020; Accepted: 25 May 2020; Published: 27 May 2020

Abstract: Interlayer reinforcement systems represent a valid solution to improve performance and extend the service life of asphalt pavements, reducing maintenance costs. The main issue is that the presence of reinforcement may hinder the full transmission of stresses between asphalt layers, reducing the overall pavement bearing capacity. This study aimed at evaluating the mechanical behavior of geogrid-reinforced asphalt interlayers under cyclic shear loading. To this purpose, a trial section, characterized by three types of interface (reinforced with carbon fiber grid, reinforced with glass fiber grid and unreinforced), was built. Cores were taken from the trial section to carry out shear-torque fatigue tests. Static Leutner shear tests were also performed on cored specimens having the same interface configuration. From data gathered in the present study, shear-torque fatigue tests have proved to be a powerful tool for investigating reinforced specimens. Results clearly ranked the investigated materials, showing that the glass fiber grid has the lowest shear fatigue performance in comparison with the other two interfaces at 20 °C. However, the shear fatigue resistance of glass fiber grid increases significantly at 10 °C. Finally, an interesting correlation was found between cyclic and static shear test results that should be better investigated in future studies.

Keywords: maintenance; reinforced asphalt pavement; geogrid; interlayer bonding; static shear test; cyclic shear test; fatigue properties

---

1. Introduction

In recent decades, highway agencies are facing a twofold problem. Pavement construction costs are shooting up due to the scarcity of pavement materials along with strict environmental regulations. The intensification of traffic and the increase in axle loads on road pavements are generating premature failure processes and rapid loss of structural and functional pavement characteristics. This degradation process is drastically accelerated by extreme weather conditions connected to climate changes [1,2]. Therefore, the reduced budgets for pavement rehabilitation coupled with the scarcity of raw materials are leading to the need for adopting maintenance strategies as effective and durable as possible.

The conventional method for pavement rehabilitation is the construction of asphalt overlays usually applied as partial replacement of existing cracked layers. As a consequence, asphalt overlay represents a cost-effective method, but it is rarely durable because of the propagation of pre-existing cracks from the lower old pavement (not replaced) to the upper new asphalt overlay. This mechanism of distress is well-known as reflective cracking.

In recent years, maintenance and rehabilitation processes in the road networks are often performed by inserting reinforcement systems within pavement layers. The idea of introducing interlayer reinforcement systems in road pavements dates to the 1950s and 1960s, when first attempts were carried out placing metal meshes between asphalt layers to improve pavement performance and durability. The results of these initial experiences were not encouraging because the installation system was too rudimentary, and the reinforcing material was not suitable for road pavement applications [3]. The use of new technologies and materials, such as geosynthetics, has provided new incentives for the use of interlayer reinforcement systems in pavement engineering. Contrarily to asphalt overlays, geosynthetics are able to significantly increase the maintenance intervals of road pavements, resulting in a cost-effective and long-lasting pavement rehabilitation method. Less frequently, reinforcement systems can be also used for new pavement construction. The use of geosynthetics also allows a reduction in the thickness of the old layers to be milled and of the new asphalt layers to be built above the reinforcement, leading to a reduction in materials to be disposed of, lower exploitation of raw materials, lower energy consumption (transport, laying and compaction), lower damage and inconvenience to secondary roads. Moreover, certain types of geosynthetics can be also milled and recycled [4]. Therefore, the pavement industry may benefit from adopting these interlayer systems by constructing more sustainable infrastructures.

Geosynthetics can fulfil various functions as separation, filtration, reinforcement, stiffening and drainage [5]. Several types of geosynthetics are available on the market produced by many manufacturers worldwide and can be grouped into four categories: geotextile, geomembrane, geogrid and geocomposite [4]. Among all, geogrids are the most used geosynthetics for reinforcement applications where no waterproofing functions are required.

The primary ability of a pavement reinforced with geosynthetics is to distribute the applied load to a wider area on top of the unbound layers, thus resulting in smaller strain–stress values, as shown in Figure 1. However, the efficiency of the interlayer reinforcement system strongly depends on the proper choice of the geosynthetics, correct installation, and characteristics of the asphalt concrete layers.

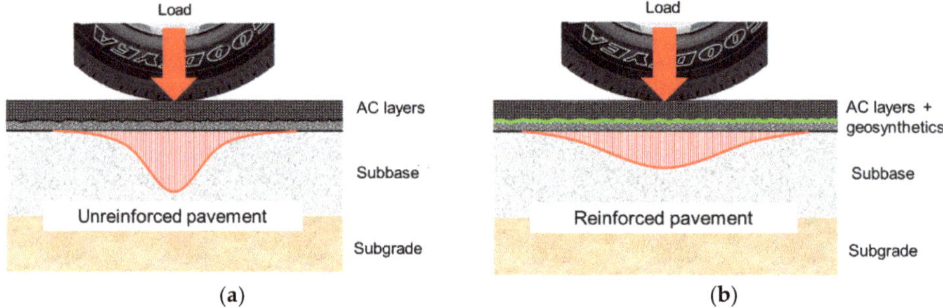

Figure 1. Distribution of vertical stress: (**a**) unreinforced pavement; (**b**) reinforced pavement.

Before the reinforcement installation, the underlying layer must be devoid of structural defects (e.g., rutting, depressions, etc.,) and tack coat can be applied to improve the bonding of the layers, paying attention both to the application rate and curing time [6]. Regarding the reinforcement installation, the reinforcement has to be perfectly laid, avoiding any possible corrugation, and must remain flat during the laying of the upper layer [7–11]. It is good practice to follow the recommendations of the manufacturers when installing the products, otherwise, the application of reinforcements at the interface can be technically and economically ineffective or even harmful.

Although considerable studies have been conducted to investigate the behavior of reinforced asphalt pavement, there are still many open issues to be investigated. Different studies performed both in the laboratory and on real pavements showed that geosynthetics can extend the pavement fatigue life and improve resistance to reflective cracking and rutting [12–21]. Therefore, the extra endeavors and costs associated with the application of geosynthetics are justified by the longer service life and lower lifecycle costs of the pavement.

On the other hand, the presence of geosynthetics inevitably causes a significant reduction in the shear resistance between asphalt layers, this phenomenon is known as the debonding effect [18,22–28].

Since a good bonding between the pavement layers is essential to maintain the structural integrity, the debonding effect considerably influences the pavement response in terms of the stress–strain distribution, and, therefore, negatively impacts the pavement lifespan [9,20,29,30]. Graziani et al. [31] built a reinforced asphalt pavement instrumented with pressure cells and strain gauges. In this study, falling weight deflectometer (FWD) tests along with a layered elastic theory (LET) model analysis showed that certain geogrids cause a noticeable interface slip, and this could lead to an increase in the tensile strain within the pavement due to the debonding effect. Therefore, if the shear resistance excessively decreases due to the presence of the reinforcement, the overall pavement performance would be negatively affected, and slippage could occur at the pavement surface due to shear stresses produced by traffic loads.

In the laboratory, the interlayer bonding of double-layered reinforced specimens is typically evaluated by measuring the interlayer shear strength (ISS or τ_{peak}) by means of static (i.e., monotonic) shear tests [32,33]. Nevertheless, road pavements are subjected to cyclic traffic loads with magnitudes considerably lower than those that cause the interface failure during static shear tests. In this sense, static (i.e., monotonic) shear tests can be used for quality assessment of the interlayer bonding properties at failure, whereas the adoption of cyclic shear tests can offer a more complex evaluation of interlayer bonding [34–37]. Moreover, cyclic shear test results can be used for modeling or pavement design purposes. The first cyclic shear tests used to investigate the shear fatigue performance of asphalt interlayers were conducted in the early 2000s [38–40]. So far, only a few studies were addressed to characterize the reinforced asphalt systems under cyclic shear loading [28,40–42] and, consequently, the shear fatigue behavior of reinforced asphalt pavement is not yet fully understood. This lack of exhaustive scientific knowledge regarding the shear fatigue behavior of reinforcements is also an obstacle for innovation and industrial practice.

Lastly, another crucial aspect that should be considered is that laboratory tests carried out on reinforced specimens fabricated in the laboratory may lead to results that do not occur with in situ cored specimens [18,43]. This may be due to the different compaction methods and reinforcement installation techniques used in the laboratory and in situ. Consequently, the construction of full-scale trial sections is more appropriate for evaluating the effect of reinforcement systems [44,45].

Objective and Scope

Given this background, this study focuses on the analysis of the interlayer bonding between asphalt layers and reinforcements. The main goal of this study was to evaluate the behavior of reinforcement systems and their effects on the interlayer mechanical properties under cyclic shear loading at the interface. To accomplish the objective of this study, a full-scale trial section, characterized by three types of interfaces (two reinforced with different geogrids and one unreinforced for comparison purposes), was built. Shear-torque fatigue tests were performed on in situ cored specimens to evaluate the fatigue performance of the reinforcement. Different fatigue failure criteria were adopted to select the most appropriate fatigue approach in order to determine the failure of each specimen. Besides, static (i.e., monotonic) direct shear tests were also performed on the same cored specimens by using the Leutner device to search for a possible correlation between cyclic and static shear tests in order to get useful insights for the future deepening of such an interesting goal.

2. Experimental Program

2.1. Reinforcing Materials

Two different geosynthetics (coded as CF and FG) were used as reinforcements in this experimental study. The CF geogrid (Figure 2a) was composed of carbon fiber rovings with a square 15 mm mesh pre-coated with bitumen in conjunction with a burn-off film applied on the underside, characterized by a tensile strength of 200 kN/m (in both directions), whereas, the FG geogrid (Figure 2b) was composed of glass fiber yarns with a square 25 mm mesh in conjunction with a light polyester knitted veil applied on the underside, characterized by a tensile strength of 100 kN/m (in both directions). Besides, the CF geogrid was characterized by a lower tensile elongation at failure with respect to the FG geogrid (1.75% in both directions for CF vs. 3% in both directions for FG).

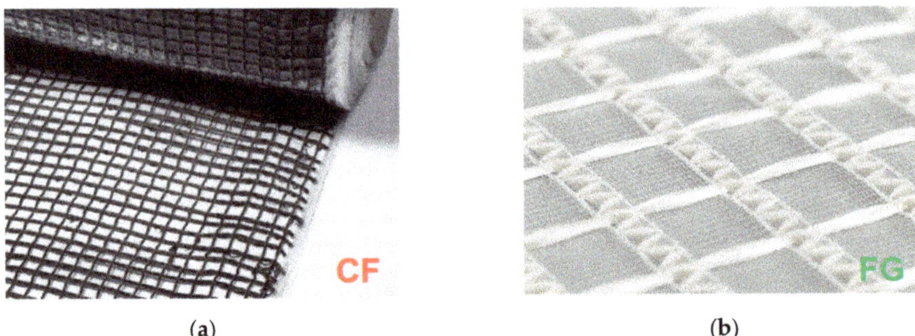

Figure 2. Detail of the installed geosynthetics: (**a**) CF carbon fiber geogrid; (**b**) FG glass fiber geogrid.

2.2. *Trial Section and Specimen Preparation*

A full-scale trial section (8 m long, 3 m wide, and 2 m deep) was built at the Laboratoire GC2D of the University of Limoges (Egletons, France) in July 2017, in a pit installed in a building with the possibility to control several conditions (e.g., temperature, humidity). The trial section was characterized by different interfaces as follows:

- unreinforced with a tack coat interface used as a reference for comparison purposes (coded as UN);
- tack coat and carbon fiber geogrid (coded as CF);
- tack coat and glass fiber geogrid (coded as FG).

The main construction activities of the pavement section are summarized below (Figure 3a–c):

- 20/40 gravel with a thickness of 20 cm on the bottom of the pit;
- subgrade course, having a thickness of 140 cm, prepared with decomposed granite;
- subbase course, having a thickness of 30 cm, prepared with untreated gravel GNT 2 (GNT stands for "Grave Non-Traitée" in French) [46] of maximum diameter 31.5 mm;
- base course, having a thickness of 9 cm, prepared with asphalt concrete GB 3 0/14 (GB stands for "Grave Bitume" in French) [47];
- accurate cleaning and preparation of the upper base course surface and application of the tack coat (bituminous emulsion of pure bitumen) with a residual dosage of 0.5 kg/m^2;
- application of the geogrids directly on the fresh emulsion right after spreading (except for the unreinforced section);
- application of the tack coat with a residual dosage of 0.5 kg/m^2 only above the FG geogrid;
- wearing course, having a thickness of 5 cm, with asphalt concrete BBSG 3 0/10 (BBSG stands for "Béton Bitumineux Semi Grenu" in French) [47], once the tack coat emulsion was fully cured.

Figure 3d shows the cross-section of the full-scale trial section. More details regarding the trial section are available in the references [48,49].

In May 2018, several cores with a nominal diameter of 100 and 150 mm and a thickness of 140 mm were extracted from the experimental pavement section (Figure 4). Each core was marked by an identification code (ID) defining its location in the trial section; for example, UN_2 represents the specimen number 2 taken from the unreinforced section (UN). In the laboratory, each core was sawed in order to obtain a total thickness of 90 mm (both layers of 45 mm). The average bulk density of the specimens, measured according to [50], was 2.23 g/cm^3.

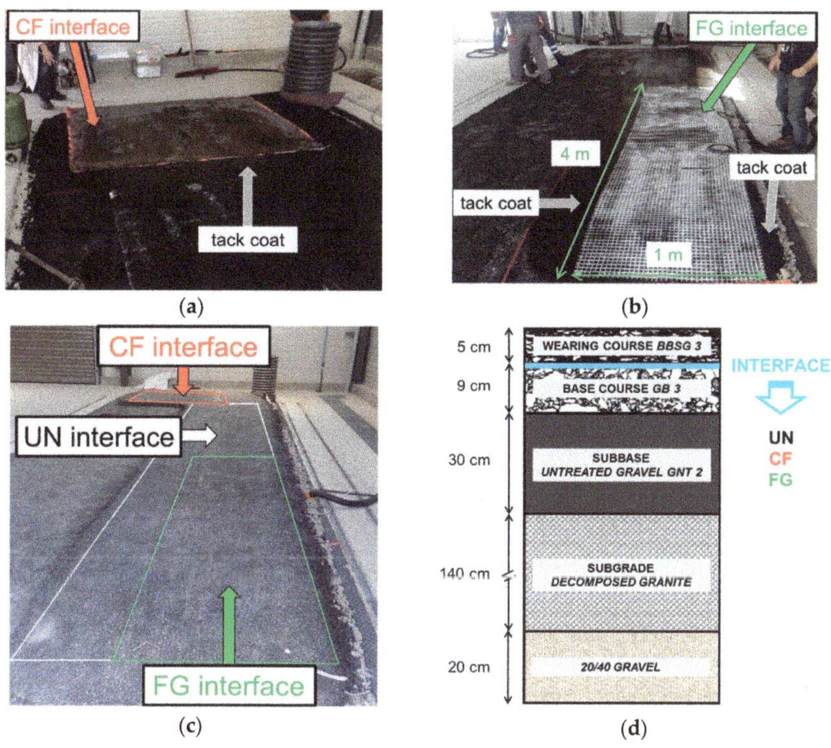

Figure 3. Full-scale trial section: (**a**) installation of CF carbon fiber geogrid; (**b**) installation of FG glass fiber geogrid; (**c**) completed trial section; (**d**) cross-section of the trial section.

Figure 4. (**a**) Detail of the trial section after coring; (**b**) cored specimen.

2.3. Testing Methods

2.3.1. Shear-Torque Fatigue Test

Shear-torque fatigue test carried out in stress-controlled mode consists of measuring the sinusoidal torsional rotation angle (α), when a sinusoidal torque (T) is applied along with a small axial compression load (N) on a cylindrical asphalt concrete specimen through a servo-hydraulic (MTS) device (Figure 5). The torsional rotation angle is measured with a magnetic non-contact angular sensor (accuracy 0.001°), which is located on the upper steel plate (Figure 5). The load cell measures the torque up to ±1 kNm and the axial load up to ±100 kN. During the test, small axial load amplitude is applied to ensure a good alignment of the specimen and steel plates, to guarantee the homogeneity of the stress states in the specimen. The sinusoidal evolution with time of the two measured values is defined by the following equations:

$$T(t) = T_0 \sin(\omega t) \tag{1}$$

$$\alpha(t) = \alpha_0 \sin(\omega t - \varphi) \tag{2}$$

where T_0 is the amplitude of the applied torque, ω is the torque pulsation ($\omega = 2\pi f$ with f the load frequency), t is the time, α_0 is the amplitude of the torsional rotation angle, and φ is the phase angle related to the lag between stress and strain.

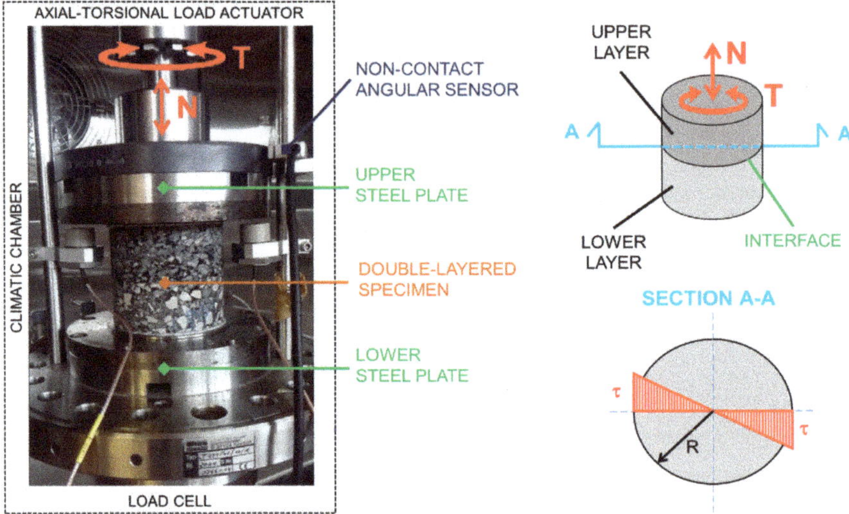

Figure 5. Shear-torque fatigue test.

Considering complex notations, where j is the complex number defined by $j^2 = -1$, the measured values can be written as follows:

$$T^* = T_0 \exp[j\omega t] \tag{3}$$

$$\alpha^* = \alpha_0 \exp[j(\omega t - \varphi)] \tag{4}$$

The correspondence principle allows the application of known solutions for linear elastic structures also for geometrically identical bodies made of linear viscoelastic materials. Therefore, for cylindrical specimens, the applied torque (T) generates shear stress (τ) which varies linearly with the radius of the specimen (R) (Figure 5). From cyclic torque tests, complex shear modulus G^* of materials can be calculated with the following equation:

$$G^* = \frac{H}{I_p} \frac{T_0 \ \exp[j\omega t]}{\alpha_0 \ \exp[j(\omega t - \varphi)]} = |G^*| \exp[j\varphi] \tag{5}$$

where H is the specimen height, $|G^*|$ is the norm (or absolute value) of the complex shear modulus, and I_p is the polar moment of inertia of the circular section.

The apparatus is placed in a climatic chamber to control the temperature during the test. Prior to testing, the specimen is glued between two steel plates using an epoxy resin (Figure 5) and care must be taken to avoid eccentricity of the specimen during gluing which could affect the test results. More details of the shear-torque fatigue test can be found in the references [37,51,52].

By analyzing fatigue test data, the choice of fatigue criterion has paramount importance for the understanding of material behavior. The fatigue life value (N_f) at a selected stress level is defined as the number of cycles corresponding to the failure point calculated by adopting a given specific criterion for the tested specimen. Different approaches for the prediction of fatigue life can be found in the literature. Usually, the traditional approach defines failure as the point at which the decrease in the material modulus reaches a certain value (Figure 6a). The most classical fatigue criterion ($N_f = N_{50}$) uses a threshold value of 50% of the initial modulus values [53]. As an alternative to the traditional approach, Reese [54] suggested a new failure approach based on the evolution of the phase angle (φ) considering the viscoelastic behavior of asphalt materials. During cyclic loading, the measured phase angle of asphalt concrete generally shows a steady increase followed by a sudden decrease (Figure 6b). The cycle corresponding to this sudden decrease is defined as the number of cycles to failure ($N_f = N_{\varphi max}$). Compared to the traditional approach, this approach seems to have a more theoretical underpinning, as the sudden reduction in the phase angle represents a viscoelastic behavior modification of the material probably due to the formation of macro-cracks. However, the real mechanism governing the phase angle evolution (e.g., nonlinear viscoelasticity, fatigue damage) is not yet fully understood. Fatigue failure criterion that can accurately define the effective failure of the double-layered asphalt concrete specimens during cyclic shear-torque tests has yet to be developed. A recent study [51] adopted the acoustic emission (AE) technique to investigate the fatigue behavior of asphalt interlayers in cyclic torque tests, highlighting that the damage evolution phase occurs in the specimen when the norm of its complex shear modulus $|G^*|$ decreases by about 70% (Figure 6a). According to these results, the 70% decrease in the stiffness initial value can be used as fatigue criterion ($N_f = N_{70}$) for this type of test.

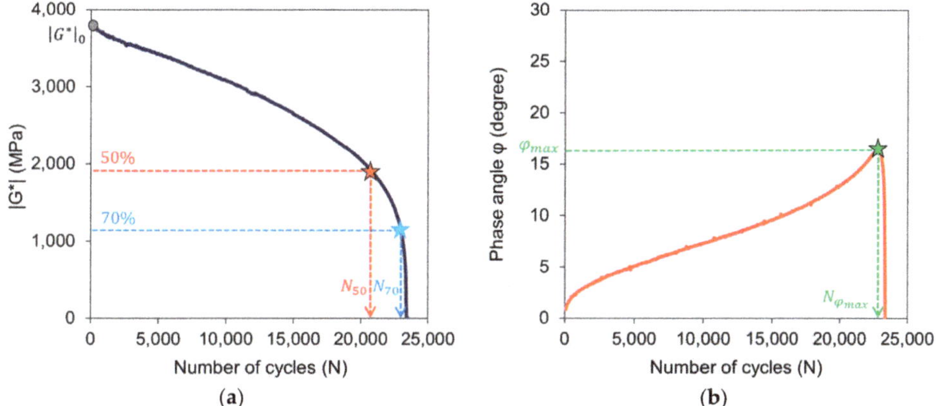

Figure 6. Determined failure point by different criteria: (**a**) material modulus approach; (**b**) phase angle approach.

2.3.2. Leutner Test

The Leutner test [55], which is a static (i.e., monotonic) shear test compliant with prEN 12697-48 [56], consists of measuring the shear force when a constant shear displacement rate is applied across the interface of a double-layered specimen, without applying a normal load perpendicular to the interface. A specimen with a nominal diameter of 100 or 150 mm is clamped in the test equipment between two shear rings, taking care to ensure that the specimen interface is correctly aligned with the shear plane. The Leutner device is installed into a servo-mechanic press frame able to apply displacement rates up to 50 mm/min. An external linear variable displacement transducer (LVDT) is used to measure the shear displacement of the specimen at the interface. The shear force and displacement are recorded during the test. By plotting instant by instant the shear stress at the interface (τ), calculated as the ratio between the shear force and the specimen cross-sectional area, as a function of shear displacement, it is possible to determine the maximum shear stress (i.e., interlayer shear strength ISS or τ_{peak}). Lower τ_{peak} implies lower interlayer bonding.

2.4. Testing Program

The experimental program consisted of evaluating the shear fatigue performance of reinforced and unreinforced cylindrical specimens by performing shear-torque fatigue tests in stress-controlled mode. Static Leutner tests were carried out in strain-controlled mode on the same specimens. A summary of the testing program is reported in Table 1. Prior to testing, all specimens were conditioned at the testing temperature in a climatic chamber for at least 4 h.

Table 1. Testing program.

Interface Type	Diameter (mm)	Shear-Torque Fatigue Test Replicates (#)		Static Leutner Test Replicates (#)	
		20 °C; 10 Hz	10 °C; 10 Hz	20 °C; 50.8 mm/min	10 °C; 50.8 mm/min
UN	100	5	-	-	-
	150	-	-	3	3
CF	100	4	-	-	-
FG	100	5	3	-	-
	150	-	-	3	3
Total		14	3	6	6

As shown in Table 1, shear-torque fatigue tests were carried out only on 100 mm nominal diameter specimens applying a sinusoidal torque at the frequency of 10 Hz. In this study, an alternate cyclic loading (signal centered at zero) was adopted to simulate the stress–strain state induced by a moving wheel in a straight pavement section (without braking and acceleration conditions), whereas the frequency of 10 Hz was chosen to simulate a traffic speed of roughly 80 km/h on a pavement at a depth of 10–20 cm [57]. Shear-torque fatigue tests were conducted at a temperature of 20 °C for each interface type as usually suggested for static shear tests [32]. For FG specimens, tests were also carried out at 10 °C. A preliminary test was conducted on an unreinforced specimen (UN), considering two different torque amplitudes (T_0 = 20 and 40 Nm), to select the suitable loading range to apply during this experimentation. Different torque amplitudes (T_0) ranging from 20 to 80 Nm were chosen to obtain a wide range of the number of cycles to failure (N_f).

Static Leutner tests were carried out only on 150 mm nominal diameter specimens applying the standard displacement rate of 50.8 mm/min. Tests were conducted at 10 and 20 °C on UN and FG interface types. Three repetitions were performed for each test condition.

After each test, the specimen was visually inspected to determine the mode of shear failure: break at the interface, break within the asphalt layer or mixed break (both at the interface and within the asphalt layer).

3. Results

3.1. Shear-Torque Fatigue Test Results

3.1.1. Viscoelastic Properties

The relationship between the applied shear stress amplitude ($\tau_{max,0}$) and the corresponding initial value of the norm of the complex shear modulus ($|G^*|_0$) is depicted in Figure 7. The equation used for calculating the amplitude of the applied shear stress ($\tau_{max,0}$) is shown in the following:

$$\tau_{max,0} = \frac{2T_0}{\pi R^3} \tag{6}$$

The initial value $|G^*|_0$ is assumed as the norm of the complex shear modulus evaluated at the 50th cycle, because at this stage of the test, the double-layered specimen is not damaged yet and, at the same time, the induced stress–strain field can be considered not affected by the initial perturbation (i.e., steady). The results presented in Figure 7 show that the initial norm of the complex shear modulus ($|G^*|_0$) depends on the applied shear stress amplitude ($\tau_{max,0}$), i.e., the interface displays nonlinear viscoelastic behavior within this loading range. In particular, the measured $|G^*|_0$ decreases as the applied shear stress amplitude increases. It can be also observed that the presence of a geogrid at the interface leads to smaller initial values of the norm of the complex shear modulus ($|G^*|_0$). Besides, due to the presence of the asphalt concrete layers, $|G^*|_0$ increases as testing temperature decreases (from 20 to 10 °C) for the FG interface type.

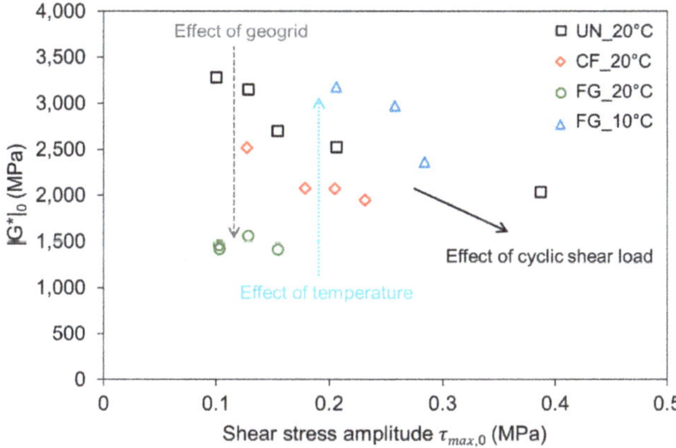

Figure 7. Initial norm of the complex shear modulus $|G^*|_0$ vs. applied shear stress amplitude $\tau_{max,0}$.

The damage of the specimen was analyzed by using the evolution of the phase angle (φ) and the normalized norm of complex shear modulus ($|G^*|_n$). The latter is given by the following equation:

$$|G^*|_n = \frac{|G^*|_N}{|G^*|_0} \tag{7}$$

where $|G^*|_N$ is the norm of the complex shear modulus calculated at any given number of loading cycles (N).

The results of specimen FG_7 tested with a torque amplitude T_0 = 55 Nm at 10 Hz and 10 °C are presented herein as a typical example. In Figure 8, the normalized norm of complex shear modulus ($|G^*|_n$) and the phase angle (φ) are presented as a function of the number of cycles. It is interesting to observe in Figure 8 that $|G^*|_n$ decreases with the number of cycles, indicating a progressive weakening of the interface properties during the test characterized by a typical three-phase fatigue curve [51,52,58,59], whereas the phase angle (φ) increases during the cyclic test and drops suddenly approaching the end of the test. Four phases can be identified for the phase angle curve. The first phase consists of a quick increase in the phase angle; this is attributable to bulk reversible phenomena (e.g., self-heating) that tend to appear during the initial test cycles. The second phase is associated with a quasi-linear increase in the phase angle. In the third phase, irreversible phenomena (e.g., fatigue damage) appear and the phase angle quickly increases until a sudden drop (fourth phase). During the fourth phase, macro-cracks propagate at the interface, generating a not homogeneous distribution of stresses and strains. According to Reese [54], the maximum point of the phase angle defines the point at which the location of the damage begins.

Figure 9 shows the evolution of the normalized norm of complex shear modulus ($|G^*|_n$) of the FG interface type at various torque amplitudes (T_0) at 10 Hz and 10 °C. It is possible to note that $|G^*|_n$ decreases faster with the number of loading cycles as the applied torque amplitude increases.

Figure 10 shows the evolution of the phase angle (φ) of the FG interface type at 10 Hz and two testing temperatures (10 and 20 °C). It is possible to note that, for both temperatures, φ increases faster with the number of loading cycles by applying higher torque amplitude values. Besides, the phase angle values at 20 °C are greater than those at 10 °C because, as expected, asphalt materials are more viscous at higher temperatures. This observation is in agreement with a previous study [60], and the measured values of the phase angle are also comparable.

After the test, the failure occurred exactly at the interface for all the specimens, i.e., a complete detachment between the two layers of the specimen was observed. In particular, the failure for the FG interface type was on the polyester knitted veil side, denoting that the veil could be an obstacle to bonding the two layers in contact (Figure 11).

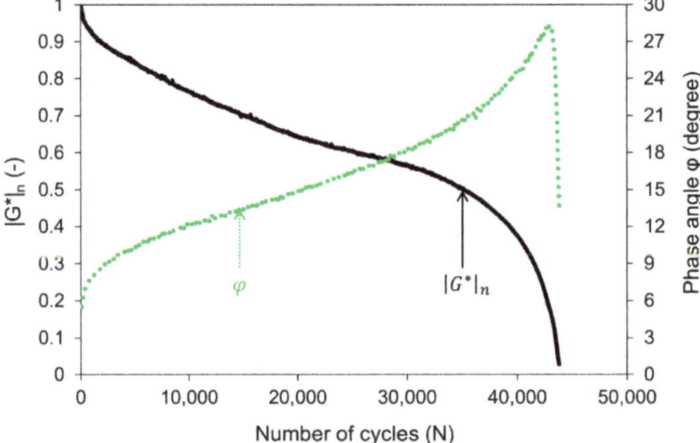

Figure 8. Evolution of normalized norm of complex shear modulus $|G^*|_n$ and phase angle φ of specimen FG_7 during the shear-torque fatigue test at 10 Hz and 10 °C.

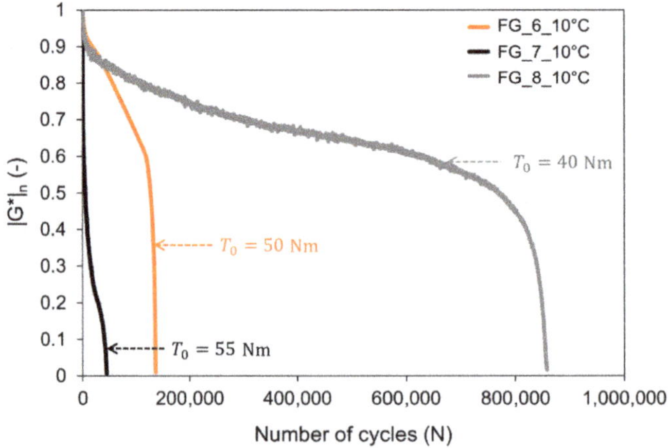

Figure 9. Evolution of normalized norm of complex shear modulus $|G^*|_n$ of FG interface at various torque amplitudes T_0 during the shear-torque fatigue test at 10 Hz and 10 °C.

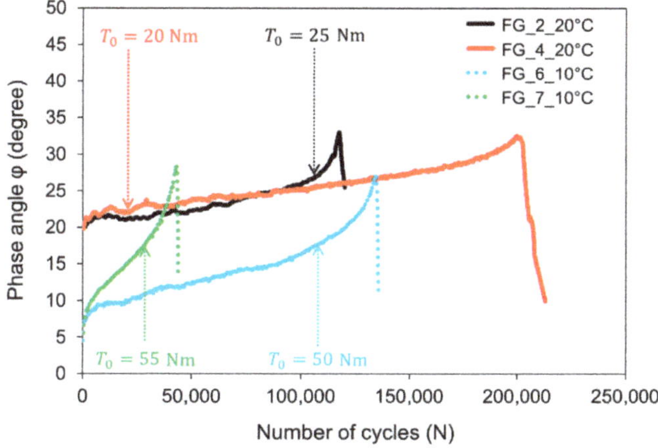

Figure 10. Evolution of phase angle φ of FG interface during the shear-torque fatigue test at 10 Hz and two temperatures (10 and 20 °C).

Figure 11. Failure mechanism of FG reinforced specimens at the end of shear-torque fatigue test.

3.1.2. Interlayer Shear Fatigue Curve

The interlayer shear fatigue curves of the tested interface types are shown in a log–log plane from Figures 12–15. A typical power-law model was used to obtain the relationship between the amplitude of the applied shear stress amplitude ($\tau_{max,0}$) and the number of cycles to failure (N_f) according to the following equation:

$$\tau_{max,0} = a \cdot N_f{}^{-b} \tag{8}$$

where parameters a and b are regression coefficients. In particular, b represents the slope of the linear regression in a log–log plane.

In each plot, interlayer shear fatigue curves obtained by using the classical fatigue criterion (N_{50}) were compared to those established by considering more appropriate failure criteria (N_{70} and $N_{\varphi max}$). The corresponding regression coefficients for the power-law model (a and b) are also presented in Table 2, as well as the coefficient of determination (R^2).

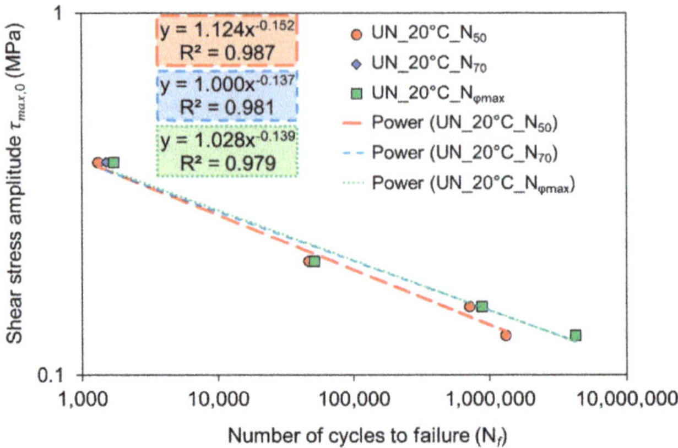

Figure 12. Interlayer shear fatigue curves for UN interface type at 20 °C.

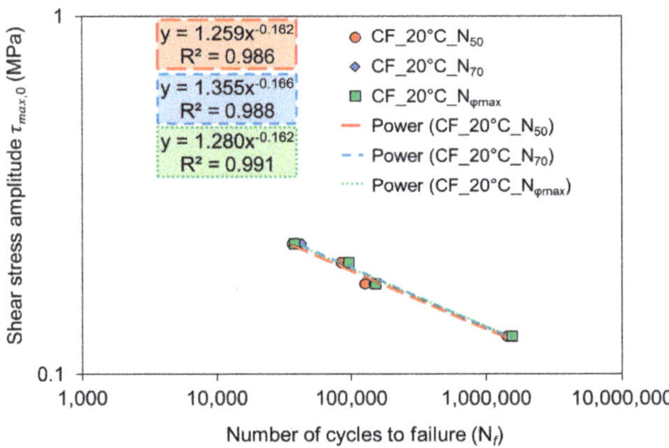

Figure 13. Interlayer shear fatigue curves for CF interface type at 20 °C.

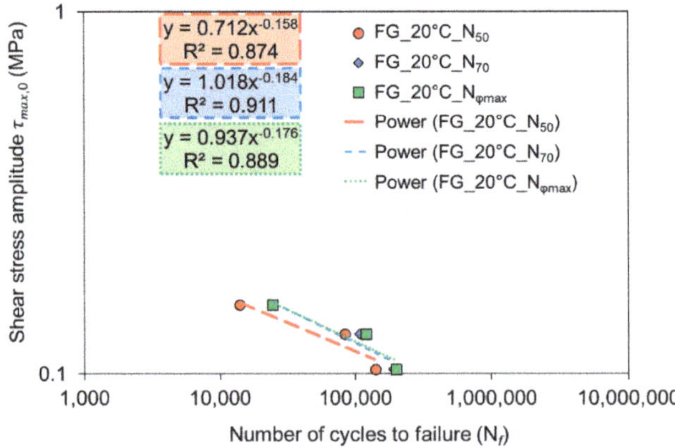

Figure 14. Interlayer shear fatigue curves for FG interface type at 20 °C.

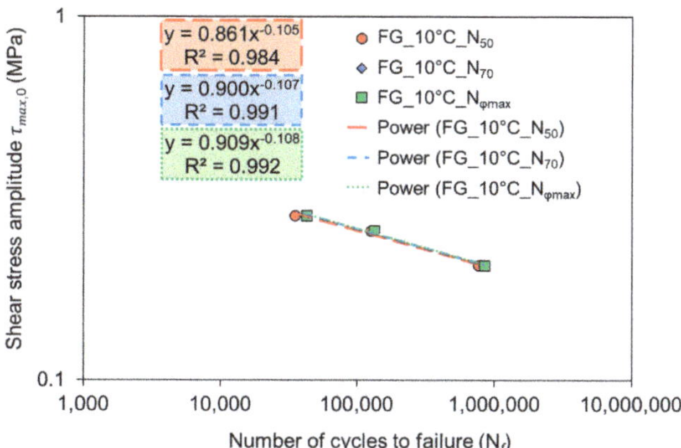

Figure 15. Interlayer shear fatigue curves for FG interface type at 10 °C.

Table 2. Parameters *a* and *b* for all interface types according to Equation (8).

Interface Type	Temperature (°C)	Failure Criterion	*a*	*b*	R^2
UN	20	N_{50}	1.124	−0.152	0.987
		N_{70}	1.000	−0.137	0.981
		$N_{\varphi max}$	1.028	−0.139	0.979
CF	20	N_{50}	1.259	−0.162	0.986
		N_{70}	1.355	−0.166	0.988
		$N_{\varphi max}$	1.280	−0.162	0.991
FG	20	N_{50}	0.712	−0.158	0.874
		N_{70}	1.018	−0.184	0.911
		$N_{\varphi max}$	0.937	−0.176	0.889
FG	10	N_{50}	0.861	−0.105	0.984
		N_{70}	0.900	−0.107	0.991
		$N_{\varphi max}$	0.909	−0.108	0.992

Looking at the experimental results, it can be seen that the obtained interlayer shear fatigue curves are very similar by applying the failure criteria N_{70} and $N_{\varphi max}$, whereas in some cases, the N_{50} failure criterion is not always in agreement with the previous ones (UN and FG interface types at 20 °C, Figures 12 and 14, respectively). As a consequence, the traditional failure criterion (N_{50}) can probably lead to a misleading ranking, since it is not capable of quantifying the damage mechanisms that occur within the interface. Meanwhile, the maximum phase angle ($N_{\varphi max}$) and the 70% failure criterion (N_{70}) can better correlate the number of cycles to failure with the damage process at the interlayer because they are related to a change in the inner behavior of the specimen. For example, once the specimen becomes severely damaged at the interface, the strain response curve in a stress-controlled test varies significantly from an actual sinusoidal function and this distortion is responsible for the drop in phase angle. These results also confirm the effectiveness of the 70% failure criterion already highlighted in a previous study carried out on unreinforced asphalt interlayers [51]. Thus, considering the weakness of the traditional approach, these results illustrate that the maximum phase angle and the 70% failure criterion provide similar results and can offer an accurate shear fatigue life prediction.

Several interesting findings can be drawn also looking at the results listed in Table 2. By comparing the fatigue law parameters at 20 °C for the N_{70} and $N_{\varphi max}$ criteria, it is possible to observe that the FG interface shows the lowest and highest values for *a* and *b*, respectively. In general, coefficients of determination (R^2) are greater than 0.9 for all the interface types, which indicates a very good correlation between measured data and the linear fatigue law. Nevertheless, R^2 values increase as the temperature decreases (greater than 0.99) for the FG interface, indicating that the specimen-to-specimen interlayer shear variability increases at higher temperatures. Meanwhile, the parameter b values decrease as the testing temperature decreases, indicating a clear thermo-dependency for the interlayer shear fatigue properties.

In order to rank the different interface types (UN, CF and FG) and to investigate the influence of testing temperature on the FG interface, interlayer shear fatigue curves are represented in Figure 16 according to 70% norm of the complex shear modulus reduction criterion (N_{70}). Since the asphalt mixture and compaction method of the tested specimens are the same, it can be asserted that the resistance to shear fatigue damage is only a function of the interface type.

Figure 16 shows that UN and CF interfaces provide very similar results in term of interlayer shear fatigue life, although it appears that UN interface guarantees slightly higher performance at a lower shear stress level than the CF interface. Moreover, for a given shear stress amplitude, FG reinforced specimens are characterized by a number of cycles to failure considerably lower than unreinforced and CF reinforced specimens (Figure 16). For example, with $\tau_{max,0}$ = 0.15 MPa (i.e., T_0 = 30 Nm) as input level (orange dotted line in Figure 16), the FG interface requires less than 30,000 cycles to failure at 20 °C, whereas the other CF reinforced interface undergoes more than 700,000 cycles at the same temperature.

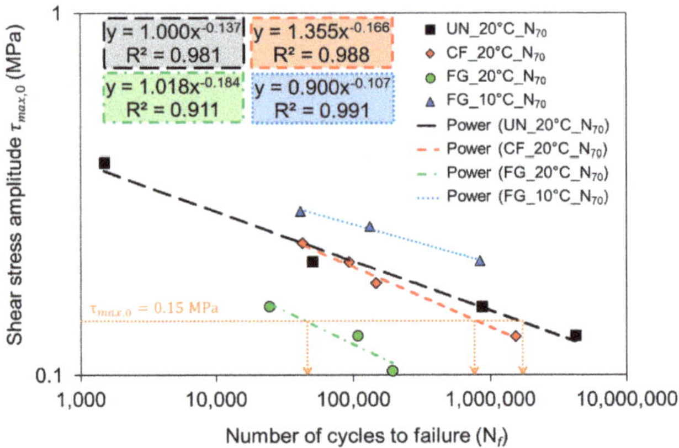

Figure 16. Interlayer shear fatigue curves for all interface types.

Starting from these results, it is expected that the CF geogrid is able to perform well in the field since the debonding effect highlighted by shear-torque fatigue loading is not so evident compared to the unreinforced interface UN. The fairly good performance of this type of geogrid has already been observed in previous studies by performing static shear tests on specimens reinforced with a similar geogrid [13,45]. This could be due to the presence of the pre-coating and the fact that the grid knots are not fixed, which allows the grid structure to move freely during the laying and compaction of the asphalt mixture ensuring the achievement of an optimal interlocking. Besides, the presence of the film applied on the underside of the CF geogrid, which is burned before installation, further improves the bonding properties on the underlying layer. On the contrary, the FG geogrid provides the lowest performance with respect to the other two interface types (UN and CF). This could be due to the presence of the polyester knitted veil and the fixed knots of the FG geogrid (unlike the CF geogrid), which probably hinder the achievement of an optimal bonding and interlocking between the two asphalt layers in contact as already observed in Figure 11.

As far as the testing temperature is concerned, the FG interface at 10 °C provides higher shear fatigue performance compared to those at 20 °C for the same reinforcement (Figure 16). This is in accordance with previous investigations carried out with various shear tests in cyclic modality on unreinforced specimens [34,39,61] and in static modality on reinforced specimens [45], where an improvement of interlayer resistance was measured at low temperatures. Therefore, it can be assumed that as the temperature decreases, since the asphalt concrete is a thermo-dependent material, the interlayer becomes stiffer and more loading cycles of the same stress intensity are needed to cause the failure of the specimen.

To allow a better comparison between the different interface types (UN, CF and FG), it is possible to calculate, from the power-law models reported in Figure 16, the parameter τ_6 shown in Figure 17. The parameter τ_6 is defined as the shear stress level that leads to a fatigue life of 1 million cycles ($N_f = 10^6$) in a cyclic shear test and it is inspired by ε_6, defined as the strain level leading to specimen failure for 1 million cycles, which is used to calculate the admissible strain in asphalt pavement layers in the French pavement design method [34,62]. Lower τ_6 implies lower shear fatigue performance. As shown in Figure 17, the values of τ_6 confirm the outcomes previously discussed in Figure 16, but the comparison of τ_6 allows to easily rank the different interface types (UN, CF and FG), denoting that it can be a useful parameter to characterize the interlayer bonding in cyclic shear tests.

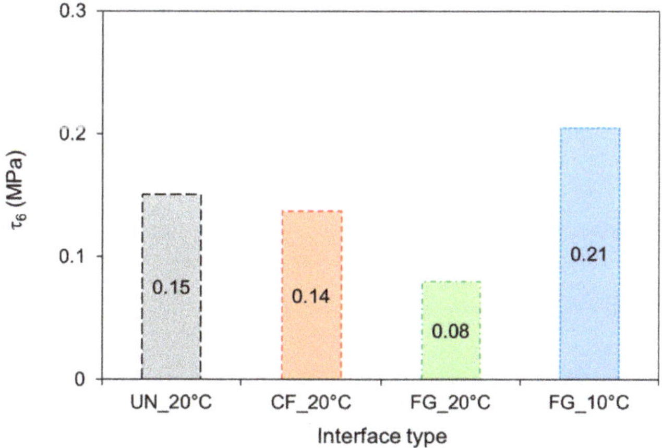

Figure 17. τ_6 values for all interface types.

In synthesis, the obtained results demonstrate that an appropriate choice of the most suitable interlayer reinforcement system could increase the cyclic shear fatigue resistance strictly linked to the debonding effect. Moreover, shear-torque fatigue tests could provide useful guidance for the selection of the most appropriate reinforcement because the results are clearly sensitive to the testing parameters (i.e., type of interface and testing temperature). However, further work is needed to adopt a method for selecting effective torque levels because different reinforcement and/or type of interface experience different levels of sensitivity to changes in stress level. Furthermore, another shortcoming is that shear-torque fatigue tests are highly time-consuming, especially at very low stress–strain levels. On the other hand, the analysis of failure of fatigue curves could help for a better understanding of the experimental results obtained with routine testing protocols such as static (i.e., monotonic) shear tests for the evaluation of the interlayer shear strength (ISS or τ_{peak}).

3.2. Static Leutner Test Results

Figure 18 shows the results of static Leutner test, in terms of average interlayer shear strength (τ_{peak}), for UN and FG specimens with a diameter of 150 mm at 10 and 20 °C. The τ_{peak} value decreases with increasing testing temperature for both interface types but the reduction (in percentage) is greater for the UN system compared to the FG system (i.e., 45% and 22%, respectively).

Figure 18 also shows that the presence of the FG geogrid at the interface leads to lower τ_{peak} compared to the corresponding unreinforced system (UN) for both temperatures according to the results of shear-torque fatigue tests. The reduction (in percentage) of τ_{peak} by comparing UN and FG systems is of 59% and 42% at 10 and 20 °C, respectively. These results allow remarking that the interlayer reinforcement worsens the interlayer properties by decreasing the adhesion between the two asphalt layers [45].

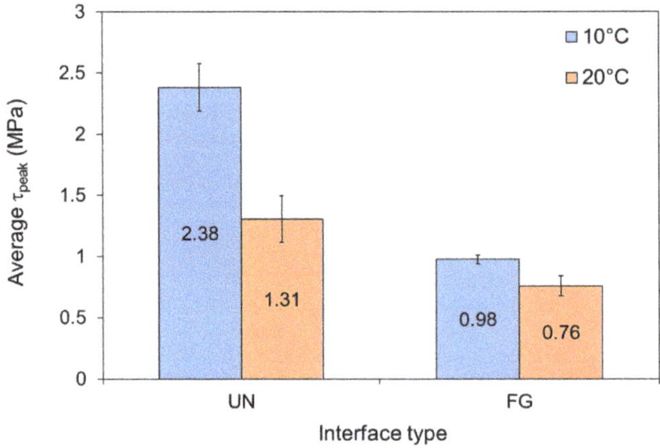

Figure 18. Average interlayer shear strength (τ_{peak}) from static Leutner tests at 10 and 20 °C for UN and FG specimens with a diameter of 150 mm (error bars provide the variability of the results).

3.3. Comparison between Cyclic and Static Shear Tests

As explained in the introduction, cyclic shear tests allow the determining of accurate parameters closely linked to field performance, but they are more time-consuming and require more effort to process the data compared to the static shear tests. In this sense, there is a need to find links between cyclic and static shear test to correlate different laboratory results and to predict interlayer shear fatigue performance from rapid and simple static shear tests. In the wake of this discussion, a possible interrelationship between cyclic and static shear tests can be found by calculating the cyclic–static shear ratio resistance (C2S2R) parameter as follows:

$$\mathrm{C2S2R} = \frac{\tau_6}{\tau_{peak}} \tag{9}$$

where τ_6 is the shear stress level that leads to a fatigue life of 1 million cycles ($N_f = 10^6$) determined in a cyclic shear test, and τ_{peak} is the interlayer shear strength determined in a static shear test.

Based on the shear-torque fatigue tests results reported in Figure 17 and the static Leutner test results reported in Figure 18, C2S2R values were calculated for the UN and FG interface types at 20 °C, as shown in Table 3.

Table 3. Cyclic–static shear ratio resistance (C2S2R) parameter.

Interface Type	Temperature (°C)	τ_6 (MPa)	τ_{peak} (MPa)	C2S2R (-)
UN	20	0.15	1.31	0.12
FG	20	0.08	0.76	0.11

The C2S2R values in Table 3 are roughly the same for UN and FG interfaces at 20 °C, specifically τ_6 is almost 10% of τ_{peak}. From a practical point of view, an empirical correlation between static Leutner test and shear-torque fatigue test results could consist of multiplying the interlayer shear strength

(τ_{peak}) of the static Leutner test by 0.10 to obtain the cyclic shear strength at 1 million cycles (τ_6) for this interface type at 20 °C. This means that an interface type characterized by τ_{peak} of 1 MPa can withstand 1 million cycles at a cyclic shear stress level of 0.10 MPa. Bearing in mind the very limited number of tests, it can be concluded that the C2S2R parameter can be assumed approximately equal to 0.1 for GB/BBSG interface, with pure bitumen emulsion as the tack coat (with and without reinforcement), but a larger number of testing specimens is required to obtain statistically significant results.

However, it is expected that C2S2R values could depend on the interface type (i.e., interlocking effect and tack coat contribution). For example, the higher the interlocking effect, the higher the interlayer shear strength (τ_{peak}) with static shear tests, but the same effect is not yet clear with cyclic shear tests because of their recent development. Therefore, further investigation on different interface types, different asphalt mixtures and with a larger number of repetitions are needed to find correlations between cyclic and static shear tests and confirm these interesting results. In the future, this would allow the use of the static Leutner test to evaluate shear fatigue performance, applying simple empirical correlations to the test results.

4. Conclusions

This study was performed to investigate the shear fatigue performance of geogrid-reinforced asphalt interlayers. To this end, a full-scale trial section was built with three different types of interface: unreinforced (UN), reinforced with a carbon fiber geogrid (CF) and reinforced with a glass fiber geogrid (FG). Cores were taken directly from the trial section to carry out shear-torque fatigue tests. Three different failure criteria (50% and 70% stiffness modulus value reduction and maximum phase angle) were used to analyze shear fatigue life of test data. Besides, static (i.e., monotonic) shear tests were carried out with the Leutner device on the same specimens in an attempt to find a relationship between cyclic and static shear tests.

Based on the experimental results, the following main conclusions can be drawn:

- Shear-torque fatigue test results clearly ranked the studied materials, showing that the carbon fiber geogrid (CF) reinforced interface provides similar shear fatigue behavior to the unreinforced interface (UN). In contrast, a significant reduction in shear fatigue behavior is evident with the glass fiber geogrid (FG) reinforced interface.
- As far as the temperature effect is concerned, it was observed that shear fatigue resistance significantly increases with decreasing temperature for the FG interface. Further research is needed to investigate the influence of temperature also for the CF interface.
- Good correlations were found between maximum phase angle and 70% stiffness modulus value reduction failure criteria. The results indicate that even though these fatigue failure criteria were not originally developed to be used with double-layered reinforced specimens, they may still be useful in ranking the different reinforced interfaces and appear to be able to predict the actual interlayer shear fatigue life.
- Static Leutner test results showed that the interlayer shear strength (τ_{peak}) decreases with increasing temperature and with the presence of the FG geogrid.
- A promising correlation was found between the shear-torque fatigue test and the static Leutner test results. Such an approach is worthy of further investigation but needs to be validated through extensive research activity.

In conclusion, these findings showed that a correct choice of geogrid could reduce the debonding effect that inevitably occurs by introducing a reinforcement system within asphalt pavement. Shear-torque fatigue tests have proved to be powerful tools for investigating the damage progress in double-layered reinforced asphalt specimens. Considering the crucial importance to properly select and assess the reinforcement system to be inserted in asphalt pavements, this test method could provide useful information on the interlayer bonding condition and interlayer fatigue failure of reinforced systems under stresses and strains similar to those existing in a real pavement. However, the presented evaluation is very limited and needs to be deepened with an extended investigation.

Author Contributions: Conceptualization, F.C. and C.P.; formal analysis, D.R., F.C., F.A., C.P. and A.M.; investigation, D.R., C.P. and A.M.; resources, F.A., C.P. and A.M.; writing—original draft preparation, D.R.; writing—review and editing, F.C., C.P. and A.M.; supervision, C.P. All authors have read and agreed to the published version of the manuscript.

Funding: This research received no external funding.

Acknowledgments: The authors truly acknowledge S&P Clever Reinforcement Company and AfiTexinov for providing the reinforcing materials.

Conflicts of Interest: The authors declare no conflict of interest.

References

1. Gudipudi, P.; Underwood, B.S.; Zalghout, A. Impact of climate change on pavement structural performance in the United States. *Transp. Res. D Trans. Environ.* **2017**, *57*, 172–184. [CrossRef]
2. Underwood, B.S.; Guido, Z.; Gudipudi, P.; Feinburg, Y. Increased costs to US pavement infrastructure from future temperature rise. *Nat. Clim. Change* **2017**, *7*, 704–707. [CrossRef]
3. Al-Qadi, I.L.; Elseifi, M.A. Field installation and design considerations of steel reinforcing netting to reduce reflection of cracks. In Proceedings of the 5th International RILEM Conference on Reflective Cracking in Pavements, Limoges, France, 5–8 May 2004; pp. 97–104.
4. Button, J.W.; Lytton, R.L. Guidelines for using geosynthetics with hot-mix asphalt overlays to reduce reflective cracking. *Transp. Res. Rec. J. Transp. Res. Board* **2007**, *2004*, 111–119. [CrossRef]
5. Zornberg, J.G. Functions and applications of geosynthetics in roadways. *Procedia Eng.* **2017**, *189*, 298–306. [CrossRef]
6. Ferrotti, G.; Canestrari, F.; Virgili, A.; Grilli, A. A strategic laboratory approach for the performance investigation of geogrids in flexible pavements. *Constr. Build. Mater.* **2011**, *25*, 2343–2348. [CrossRef]
7. Bocci, M.; Grilli, A.; Santagata, F.A.; Virgili, A. Influence of reinforcement geosynthetics on flexion behaviour of double-layer bituminous systems. In Proceedings of the International Conference on Advanced Characterisation of Pavement and Soil Engineering Materials, Athens, Greece, 20–22 June 2007; pp. 1415–1424.
8. Francken, L. Prevention of cracks in pavements: Achievements and open questions. *Road Mater. Pavement Des.* **2005**, *6*, 407–425. [CrossRef]
9. Shukla, S.K.; Yin, J.-H. Functions and installation of paving geosynthetics. In Proceedings of the 3rd Asian Regional Conference on Geosynthetics, Seoul, Korea, 21–23 June 2004.
10. Uijting, B.G.J.; Jenner, C.G.; Gilchrist, A.J.T. Evaluation of 20 years experience with asphalt reinforcement using geogrids. In Proceedings of the 3rd International Conference Bituminous Mixtures and Pavements, Thessaloniki, Greece, 21–22 November 2002; pp. 869–877.
11. Vanelstraete, A.; De Visscher, J. Long term performance on site of interface systems. In Proceedings of the 5th International RILEM Conference on Reflective Cracking in Pavements, Limoges, France, 5–8 May 2004; pp. 699–706.
12. Brown, S.F.; Thom, N.H.; Sanders, P.J. A study of grid reinforced asphalt to combat reflection cracking. *J. Assoc. Asphalt Paving Technol.* **2001**, *70*, 543–569.
13. Canestrari, F.; Belogi, L.; Ferrotti, G.; Graziani, A. Shear and flexural characterization of grid-reinforced asphalt pavements and relation with field distress evolution. *Mater. Struct.* **2015**, *48*, 959–975. [CrossRef]
14. Correia, N.S.; Zornberg, J.G. Strain distribution along geogrid-reinforced asphalt overlays under traffic loading. *Geotex. Geomembr.* **2018**, *46*, 111–120. [CrossRef]
15. Ferrotti, G.; Canestrari, F.; Pasquini, E.; Virgili, A. Experimental evaluation of the influence of surface coating on fiberglass geogrid performance in asphalt pavements. *Geotex. Geomembr.* **2012**, *34*, 11–18. [CrossRef]
16. Ingrassia, L.P.; Virgili, A.; Canestrari, F. Investigating the effect of geocomposite reinforcement on the performance of thin asphalt pavements through accelerated pavement testing and laboratory analysis. *Case Stud. Constr. Mater.* **2020**, *12*, e00342. [CrossRef]
17. Nejad, F.M.; Asadi, S.; Fallah, S.; Vadood, M. Statistical-experimental study of geosynthetics performance on reflection cracking phenomenon. *Geotex. Geomembr.* **2016**, *44*, 178–187. [CrossRef]

18. Ragni, D.; Montillo, T.; Marradi, A.; Canestrari, F. Fast falling weight accelerated pavement testing and laboratory analysis of asphalt pavements reinforced with geocomposites. *Lect. Notes Civil Eng.* **2020**, *48*, 417–430. [CrossRef]
19. Saride, S.; Kumar, V.V. Influence of geosynthetic-interlayers on the performance of asphalt overlays on pre-cracked pavements. *Geotex. Geomembr.* **2017**, *45*, 184–196. [CrossRef]
20. Sobhan, K.; Tandon, V. Mitigating reflection cracking in asphalt overlay using geosynthetic reinforcements. *Road Mater. Pavement Des.* **2008**, *9*, 367–387. [CrossRef]
21. Zofka, A.; Maliszewski, M.; Maliszewska, D. Glass and carbon geogrid reinforcement of asphalt mixtures. *Road Mater. Pavement Des.* **2017**, *18*, 471–490. [CrossRef]
22. Caltabiano, M.A.; Brunton, J.M. Reflection cracking in asphalt overlays. *J. Assoc. Asphalt Paving Technol.* **1991**, *60*, 310–330.
23. Canestrari, F.; Grilli, A.; Santagata, F.A.; Virgili, A. Interlayer shear effect of geosynthetic reinforcements. In Proceedings of the 10th International Conference on Asphalt Pavements, Québec City, QC, Canada, 12–17 August 2006.
24. Canestrari, F.; Pasquini, E.; Belogi, L. Optimization of geocomposite for double layer bituminous system. In *7th RILEM International Conference on Cracking in Pavements*; Scarpas, A., Kringos, N., Al-Qadi, I.A.L., Eds.; Springer: Dordrecht, The Netherlands, 2012; pp. 1229–1239. [CrossRef]
25. Pasquini, E.; Bocci, M.; Ferrotti, G.; Canestrari, F. Laboratory characterisation and field validation of geogrid-reinforced asphalt pavements. *Road Mater. Pavement Des.* **2013**, *14*, 17–35. [CrossRef]
26. Pasquini, E.; Bocci, M.; Canestrari, F. Laboratory characterisation of optimised geocomposites for asphalt pavement reinforcement. *Geosynth. Int.* **2014**, *21*, 24–36. [CrossRef]
27. Raab, C.; Partl, M.N. Interlayer shear performance: Experience with different pavement structures. In Proceedings of the 3rd Eurasphalt and Eurobitume Congress, Vienna, Austria, 12–14 May 2004; pp. 535–545.
28. Zamora-Barraza, D.; Calzada-Peréz, M.; Castro-Fresno, D.; Vega-Zamanillo, A. New procedure for measuring adherence between a geosynthetic material and a bituminous mixture. *Geotex. Geomembr.* **2010**, *28*, 483–489. [CrossRef]
29. Canestrari, F.; Santagata, E. Temperature effects on the shear behaviour of tack coat emulsions used in flexible pavements. *Int. J. Pavement Eng.* **2005**, *6*, 39–46. [CrossRef]
30. Jaskula, P.; Rys, D. Effect of interlayer bonding quality of asphalt layers on pavement performance. *IOP Conf. Ser. Mater. Sci. Eng.* **2017**, *236*, 012005. [CrossRef]
31. Graziani, A.; Pasquini, E.; Ferrotti, G.; Virgili, A.; Canestrari, F. Structural response of grid-reinforced bituminous pavements. *Mater. Struct.* **2014**, *47*, 1391–1408. [CrossRef]
32. Canestrari, F.; Ferrotti, G.; Lu, X.; Millien, A.; Partl, M.N.; Petit, C.; Phelipot-Mardelé, A.; Piber, H.; Raab, C. Mechanical testing of interlayer bonding in asphalt pavements. In *Advances in Interlaboratory Testing and Evaluation of Bituminous Materials, RILEM State-of-the-Art Reports*; Partl, M., Bahia, H.U., Canestrari, F., De la Roche, C., Di Benedetto, H., Piber, H., Sybilski, D., Eds.; Springer: Dordrecht, The Netherlands, 2013; Volume 9, pp. 303–360. [CrossRef]
33. Petit, C.; Chabot, A.; Destrée, A.; Raab, C. Interface debonding behavior. In *Mechanisms of Cracking and Debonding in Asphalt and Composite Pavements: State-of-the-Art of the RILEM TC241-MCD, RILEM State-of-the-Art Reports*; Buttlar, W.G., Chabot, A., Dave, E.V., Petit, C., Tebaldi, G., Eds.; Springer International Publishing: Cham, Switzerland, 2018; pp. 103–153. [CrossRef]
34. Petit, C.; Diakhaté, M.; Millien, A.; Phelipot-Mardelé, A.; Pouteau, B. Pavement design for curved road sections: Fatigue performance of interfaces and longitudinal top-down cracking in multilayered pavements. *Road Mater. Pavement Des.* **2009**, *10*, 609–624. [CrossRef]
35. Ktari, R.; Millien, A.; Fouchal, F.; Pop, I.-O.; Petit, C. Pavement interface damage behavior in tension monotonic loading. *Constr. Build. Mater.* **2016**, *106*, 430–442. [CrossRef]
36. Ragni, D.; Graziani, A.; Canestrari, F. Cyclic interlayer testing in bituminous pavements. In Proceedings of the 7th International Conference Bituminous Mixtures and Pavements, Thessaloniki, Greece, 12–14 June 2019; pp. 207–212. [CrossRef]

37. Canestrari, F.; Attia, T.; Di Benedetto, H.; Graziani, A.; Jaskula, P.; Kim, Y.R.; Maliszewski, M.; Pais, J.; Petit, C.; Raab, C.; et al. Interlaboratory test to characterize the cyclic behavior of bituminous interlayers: An overview of testing equipment and protocols. In Proceedings of the RILEM International Symposium on Bituminous Materials, Lyon, France, 14–16 December 2020.
38. Romanoschi, S.A.; Metcalf, J.B. Characterization of asphalt concrete layer interfaces. *Transp. Res. Rec. J. Transp. Res. Board* **2001**, *1778*, 132–139. [CrossRef]
39. Diakhaté, M.; Phelipot, A.; Millien, A.; Petit, C. Shear fatigue behaviour of tack coats in pavements. *Road Mater. Pavement Des.* **2006**, *7*, 201–222. [CrossRef]
40. Donovan, E.P.; Al-Qadi, I.L.; Loulizi, A. Optimization of tack coat application rate for geocomposite membrane on bridge decks. *Transp. Res. Rec. J. Transp. Res. Board* **2000**, *1740*, 143–150. [CrossRef]
41. Cho, S.H.; Safavizadeh, S.A.; Kim, Y.R. Verification of the applicability of the time–temperature superposition principle to interface shear stiffness and strength of GlasGrid-reinforced asphalt mixtures. *Road Mater. Pavement Des.* **2017**, *18*, 766–784. [CrossRef]
42. Safavizadeh, S.A.; Kim, Y.R. DIC technique to investigate crack propagation in grid-reinforced asphalt specimens. *J. Mater. Civil Eng.* **2017**, *29*. [CrossRef]
43. Raab, C.; Arraigada, M.; Partl, M.N.; Schiffmann, F. Cracking and interlayer bonding performance of reinforced asphalt pavements. *Eur. J. Environ. Civ. Eng.* **2017**, *21*, 14–26. [CrossRef]
44. Arraigada, M.; Perrotta, F.; Raab, C.; Tebaldi, G.; Partl, M.N. Use of APT for validating the efficiency of reinforcement grids in asphalt pavements. In *The Roles of Accelerated Pavement Testing in Pavement Sustainability*; Aguiar-Moya, J., Vargas-Nordcbeck, A., Leiva-Villacorta, F., Loría-Salazar, L., Eds.; Springer International Publishing: Cham, Switzerland, 2016; pp. 509–521. [CrossRef]
45. Canestrari, F.; D'Andrea, A.; Ferrotti, G.; Graziani, A.; Partl, M.N.; Petit, C.; Raab, C.; Sangiorgi, C. Advanced interface testing of grids in asphalt pavements. In *Testing and Characterization of Sustainable Innovative Bituminous Materials and Systems, RILEM State-of-the-Art Reports*; Partl, M., Porot, L., Di Benedetto, H., Canestrari, F., Marsac, P., Tebaldi, G., Eds.; Springer International Publishing: Cham, Switzerland, 2018; pp. 127–202. [CrossRef]
46. NF EN 13285. *French Standards for Unbound Mixtures–Material Specifications.*
47. NF EN 13108-1. *French Standards for Bituminous Mixtures–Material Specifications–Part 1: Asphalt Concrete.*
48. Petit, C.; Lesueur, D.; Millien, A.; Leguernevel, G.; Dopeux, J.; Picoux, B.; Allou, F.; Terhani, F. Smart geosynthetics for strain measurements in asphalt pavements. In Proceedings of the 13th International Conference on Asphalt Pavements, Fortaleza, Brazil, 19–21 June 2018.
49. Petit, C.; Lesueur, D.; Millien, A.; Leguernevel, G.; Dopeux, J.; Picoux, B.; Allou, F.; Terhani, F. Des géosynthétiques intelligents pour renforcer et suivre les déformations de chaussées bitumineuses. In Proceedings of the 12th Rencontres Géosynthétiques, Nancy, France, 11–13 March 2019.
50. EN 12697-6. *European Standards for Bituminous Mixtures–Test Methods for Hot Mix Asphalt–Part 6: Determination of Bulk Density of Bituminous Specimens.*
51. Ragni, D.; Takarli, M.; Petit, C.; Graziani, A.; Canestrari, F. Use of acoustic techniques to analyse interlayer shear-torque fatigue test in asphalt mixtures. *Int. J. Fatigue* **2020**, *131*, 105356. [CrossRef]
52. Ragni, D.; Ferrotti, G.; Petit, C.; Canestrari, F. Analysis of shear-torque fatigue test for bituminous pavement interlayers. *Constr. Build. Mater.* **2020**, *254*, 119309. [CrossRef]
53. Shen, S.; Lu, Z. Energy based laboratory fatigue failure criteria for asphalt materials. *J. Test. Eval.* **2011**, *39*, 313–320. [CrossRef]
54. Reese, R. Properties of aged asphalt binder related to asphalt concrete fatigue life. *J. Assoc. Asphalt Paving Technol.* **1997**, *66*, 604–632.
55. Leutner, R. Untersuchung des schichtenverbundes beim bituminosen oberbau. *Bitumen* **1979**, *41*, 84–91.
56. prEN 12697-48. *European Pre-Standard for Bituminous Mixtures–Test Methods for Hot Mix Asphalt–Part 48: Interlayer Bonding.*
57. Boudabbous, M.; Millien, A.; Petit, C.; Neji, J. Energy approach for the fatigue of thermoviscoelastic materials: Application to asphalt materials in pavement surface layers. *Int. J. Fatigue* **2013**, *47*, 308–318. [CrossRef]
58. Di Benedetto, H.; De la Roche, C.; Baaj, H.; Pronk, A.; Lundström, R. Fatigue of bituminous mixtures. *Mater. Struct.* **2004**, *37*, 202–216. [CrossRef]
59. Pérez-Jiménez, F.; Botella, R.; López-Montero, T.; Miró, R.; Martínez, A.H. Complexity of the behaviour of asphalt materials in cyclic testing. *Int. J. Fatigue* **2017**, *98*, 111–120. [CrossRef]

60. Diakhaté, M.; Millien, A.; Petit, C.; Phelipot-Mardelé, A.; Pouteau, B. Experimental investigation of tack coat fatigue performance: Towards an improved lifetime assessment of pavement structure interfaces. *Constr. Build. Mater.* **2011**, *25*, 1123–1133. [CrossRef]
61. Collop, A.C.; Sutanto, M.H.; Airey, G.D.; Elliott, R.C. Development of an automatic torque test to measure the shear bond strength between asphalt. *Constr. Build. Mater.* **2011**, *25*, 623–629. [CrossRef]
62. Corte, J.-F.; Goux, M.-T. Design of pavement structures: The French technical guide. *Transp. Res. Rec. J. Transp. Res. Board* **1996**, *1539*, 116–124. [CrossRef]

Article

Evaluation of Gas Emissions, Energy Consumption and Production Costs of Warm Mix Asphalt (WMA) Involving Natural Zeolite and Reclaimed Asphalt Pavement (RAP)

Alejandra T. Calabi-Floody *, Gonzalo A. Valdés-Vidal, Elsa Sanchez-Alonso and Luis A. Mardones-Parra

Department of Civil Engineering, Universidad de La Frontera, Temuco 1145, Chile; gonzalo.valdes@ufrontera.cl (G.A.V.-V.); elsa.sanchez@ufrontera.cl (E.S.-A.); luis.mardones.p@ufrontera.cl (L.A.M.-P.)

* Correspondence: alejandra.calabi@ufrontera.cl; Tel.: +56-(45)-232-56-80

Received: 3 June 2020; Accepted: 22 July 2020; Published: 10 August 2020

Abstract: Asphalt mixture is the most widely used material in road construction, and the industry is developing more sustainable technologies. Warm mix asphalt (WMA) is a promising alternative as it saves energy, reduces fuel consumption and generates fewer gas and fume emissions, while maintaining a similar performance to hot mix asphalt (HMA). This paper presents an evaluation of the gas emissions at laboratory scale, as well as the energy consumption and production costs, of five types of WMA with the addition of natural zeolite. The control mixture was a HMA manufactured at 155 °C. The mixtures evaluated were two WMA manufactured at 135 °C with 0.3% and 0.6% natural zeolite, and three WMA with partial replacement of raw materials by 10%, 20% and 30% of reclaimed asphalt pavement (RAP); these mixtures, called WMA–RAP, were manufactured at 125 °C, 135 °C and 145 °C, respectively. The results indicated that all the mixtures evaluated reduced CO and CO_2 emissions by 2–6% and 17–37%, respectively. The energy consumption presented a 13% decrease. In the current situation, the production costs for WMA with 0.3 and 0.6% natural zeolite are slightly higher than the control mixture, because the saving achieved in fuel consumption is lower than the current cost of the additive. On the other hand, WMA manufactured with the addition of natural zeolite and RAP could produce cost savings of up to 25%, depending on the amounts of RAP and natural zeolite used.

Keywords: warm mix asphalt; natural zeolite; gas emissions; energy consumption; production costs

1. Introduction

Worldwide, environmental awareness has forced governments and leaders to take action to protect the planet from global warming and climate chaos. This trend was formalized in the Kyoto protocol (1997), where the main goal was to reduce greenhouse gas emissions [1], and reaffirmed in the Paris Agreement. In this context, the asphalt industry is currently developing more sustainable technologies.

Hot mix asphalt (HMA) is the most widely used construction material for paving roads. Nearly 90% of paved roads in the world are made of asphalt mixtures [2]. In Europe, more than 90% of paved roads are made of asphalt materials. Similar levels are found in the United States, Canada and Mexico, where the use of asphalt for paved roads exceeds 92%, 90% and 96%, respectively [3]. Asphalt mixtures use asphalt binder, aggregates (sand and crushed rock) and mineral filler. This mixture is produced at high temperatures, between 150 °C and 190 °C, depending of the type of asphalt binder used. In the field, the mixture is spread and compacted on the road at high temperatures (120–140 °C). During these processes (production, storage and handling at high temperatures), a complex mixture of gases is released into the atmosphere [4] and a large amount of energy is consumed [4,5]. At the same time, large volumes of raw materials are required due to the high demand for HMA [3]. As mentioned above, technologies are needed to produce sustainable pavements, focusing on reducing emissions, saving energy and conserving natural resources [6], while keeping costs as low as possible. Some researchers have identified and quantified asphalt mixture emissions, both in the plant and in the laboratory [7,8], resulting in different techniques for reducing emissions. Quian et al. [9] proposed adding different fume suppressor agents, observing that by adding 3% of SBS, PE, melamine, nano-calcium carbonate and activated carbon in separate tests, asphalt fumes were reduced by 19.2%, 6.5%, 45.1, 4.8% and 41.6%, respectively. Autelitano et al. [4] concluded that a 30 °C reduction in the manufacturing temperature of asphalt mixture, achieved by the addition of wax, cut emissions by half. Croteau and Tessier [10] showed that a reduction of 20 °C in the manufacturing temperature decreased greenhouse gas emissions by between 20% and 35%, with a fuel saving of 2 l/ton of mixture. One of the main greenhouse gases is CO_2, which in asphalt paving is generated by two principal mechanisms: energy consumption during mixing and compaction processes; and binder oxidation at the high temperatures required for mixing [11]. These processes involve CO_2 generation through molecular reaction by hydrocarbon oxidation [12]. Some authors have indicated that naphthenic aromatic and polar aromatic compounds are the principal components responsible for CO_2 emissions [13,14]. This agrees with the study by Brandt and De Groot [15], who concluded that in the temperature range between 140 °C and 190 °C, the fume-emission rate increased by a factor of 2 for each 12 °C increase in temperature. Therefore, reducing the manufacturing temperature of asphalt mixtures would appear to be the most effective way of reducing CO_2 emissions during asphalt mixture production and pavement construction [12].

In the report presented by West et al., an average fuel saving of 22.1% was obtained by a temperature reduction of 9 °C in the asphalt mixture [16]; thus, reducing manufacturing temperatures helps to reduce energy consumption [17]. Many of the sustainable technologies reported relate to methods for producing warm mix asphalt (WMA) because these technologies help to decrease both energy consumption and emissions, and in some cases production costs (the reduction in fuel consumption may help compensate for the cost of WMA additive or equipment) [10,16]. WMA technologies allow the reduction in manufacturing and compaction temperatures without affecting mixture performance [18]. Mohd et al. indicated that energy savings between 23% and 29% are possible through the use of WMA rather than HMA [19].

WMA technologies are classified in three major categories: asphalt foaming technologies, organic additives and chemical additives [18]. The benefits of these technologies include reducing emissions, saving energy, reducing fuel consumption, extending the paving season, reducing workers' exposure to gas and temperature, and improving field compaction, as well as the potential for including a higher proportion of reclaimed asphalt pavement (RAP) [20]. Foaming processes are the most common WMA technology [21]. One of these, the indirect foaming technique, includes the addition of synthetic zeolite, one of the most common additives for WMA production [22]. Synthetic zeolite is a porous mineral with crystallized water in its structure (ca. 20%) composed of aluminosilicates of alkali metals [23]. This water is released for 6 or 7 h when the synthetic zeolite comes into contact with the preheated asphalt binder; it generates micro-foam, leading to an increase in the volume of the binder and reducing its viscosity [20], thereby increasing aggregate coating and the workability of the mixture [24].

Finally, it is necessary to solve the high consumption of raw materials and the waste generated by the replacement of deteriorated pavement. In this context, the replacement of raw materials by RAP in asphalt mixtures (WMA–RAP) achieves important benefits, because the use of virgin materials, both asphalt binder and aggregates, is reduced [25]. The use of RAP is an opportunity to reduce the impacts associated with extraction of the used material as well as the transportation of raw materials (energy consumption, carbon footprint and cost). Furthermore, the quantity of waste produced is reduced, helping to solve the final disposal problems of highway construction materials in landfills [26,27]. Some authors indicate that WMA technologies may help to increase the proportion of RAP in the asphalt mixture [28,29]. A larger proportion of RAP can be included in foamed WMA [27] than in other

WMA technologies. In this context, the Road Pavement Research Group at Universidad de La Frontera has carried out a project to design and develop asphalt mixtures with greater energy efficiency and lower environmental impact. These are WMA and WMA–RAP mixtures which use a local natural zeolite (clinoptilolite-modernite type), extracted from the central zone of Chile, as the additive. In the present work, we studied five WMA types: two WMA with 0.3% and 0.6% natural zeolite content, and three WMA–RAP with 0.6% natural zeolite and different RAP contents (10%, 20% and 30%). The control was a standard HMA. The first part of this study, published recently, concluded that these WMA and WMA–RAP performed well in rutting, cracking resistance, fatigue and moisture damage tests [30]. The present paper presents the second part of the study, the object of which is to assess emissions and energy consumption in the manufacturing process, including a cost analysis of the different mixtures.

2. Materials and Methods

2.1. Materials and Mix Design

One type of aggregate and one type of conventional asphalt binder (CA-24) were used in this research. The aggregates, obtained from fluvial sand deposits, are composed mainly of dolomite, basalt, dacite, andesite, rhyolite, sandstone, quartz and quartzite. A semi-dense asphalt mixture type IV-A-12 was used, which complies with Chilean specifications [31]. The physical properties of the aggregates according to Chilean standards are shown in Table 1. The aggregate gradation is presented in Figure 1. The properties of asphalt binder are shown in Table 2. The specific gravity of asphalt binder is 1040 kg/m^3, and the optimum mix temperature is 155 °C. The Marshall method was used to design the mixtures [32], appropriate for surface layers. This design method is the standard in Chile for asphalt pavements. As the objective is to promote massive use of this environmentally friendly alternative asphalt mixture, it was important follow the standards. The analyses of air voids (%), flow (0.25mm), stability (kN) and voids in mineral aggregates (%) were performed according to Chilean specifications [31]. The optimum asphalt binder content was 5.4% of aggregate weight for the reference HMA mixture. This optimal binder content was used in the manufacture of all asphalt mixtures studied in this research.

Table 1. Physical properties of aggregates.

Tests	Specifications	Results
Coarse Aggregate		
Los Angeles abrasion loss (%)	Max 25	18.4
Disintegration in sodium sulphate (%)	Max 12	2.4
Crushed aggregates (%)	Min 90	97.3
Flakiness index (%)	Max 10	0.1
Static method adhesion (%)	Min 95	>95
Dynamic method adhesion (%)	Min 95	>95
Specific gravity (kg/m3)	-	2661
Absorption (%)	-	1.54
Fine Aggregate		
Plasticity index	Nonplastic	Nonplastic
Riedel–Weber adhesion	Min 0–5	0–9
Disintegration in sodium sulphate (%)	Max 15	1.4
Specific gravity (kg/m3)	-	2629
Absorption (%)	-	1.09
Combined Aggregate		
Soluble salts (%)	Max 2	0.5
Sand equivalent (%)	Min 50	81

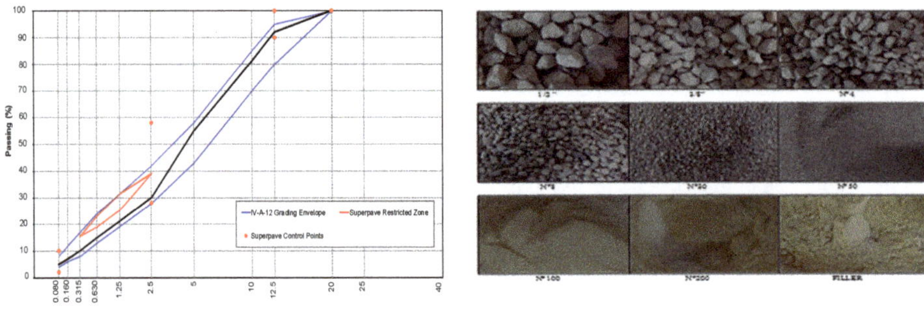

Figure 1. Grading curve asphalt mixture (IV-A-12).

Table 2. Properties of asphalt binder.

Tests	Results	Specifications.
Absolute viscosity at 60 °C, 300 mm Hg (*P*)	3730	Min 2400
Penetration at 25 °C, 100 g. 5 s. (0.1 mm)	60	Min 40
Ductility at 25 °C, (cm)	>100	Min 100
Spot test hep./xyl., (%xylene)	<30	Max 30
Cleveland open cup flash point, (°C)	>232	Min 232
Softening point R&B, (°C)	52	To be reported
Trichloroethylene solubility (%)	99.8	Min 99
Penetration index	−0.2	−1.5 a + 1.0
RTFOT		
Mass loss, (%).	0.03	Max 0.8
Absolute viscosity at 60 °C, 300 mm Hg (*P*)	8800	To be reported
Ductility at 25 °C, 5 cm/min, (cm)	>100	Min 100
Durability index	2.4	Max 4.0

The natural zeolite used in this research is a clinoptilolite-modernite type, obtained from the Quinamávida quarry in the VII Region of Chile. This mineral is found mainly in areas of volcanic activity, and Chile has large deposits. This type of natural zeolite is composed of aluminosilicates that have a three-dimensional arrangement with an open, porous structure [33], and can release crystallized water at 100 °C without affecting its internal structure. When the zeolite comes into contact with hot asphalt binder, the water is released and generates a micro-foam that allows the asphalt binder to involve the aggregates at a lower temperature.

Table 3 shows the characterization of the natural zeolite used in this study. Figure 2 shows the images of different contents of natural zeolite distributed in the asphalt binder obtained through a confocal microscope at two magnification levels. Epifluorescence images allow us to verify the dispersion of the natural zeolite in the binder. These images show that the zeolite-binder mixture is quite homogeneous for 1%, 5% and 10% additions (over weight of binder).

Table 3. Properties of Chilean natural zeolite.

Description	Chilean Natural Zeolite
Chemical name	Calcium magnesium aluminosilicate hydrated
Grain size	0–0.173 mm
Main zeolitic component	Clinoptilolite—Modernite
Other components	Plagioclase, Smectite and Quartz
Colour	Ivory

Table 3. *Cont.*

Description	Chilean Natural Zeolite
Chemical composition (%)	SiO_2: 64.19 TiO_2: 0.51 Al_2O_3: 11.65 Fe_2O_3: 2.53 MnO: 0.03 MgO: 0.66 CaO: 3,42 Na_2O: 0.75 K_2O: 1.60 P_2O_5: 0.03 PxC: 14.64
Cation exchange capacity	86.82 a 112.88 cmol/kg
Specific surface (Bet method)	446 a 480 m^2g^{-1}
Thermal stability	<450 °C
Chemical stability	8.9
Density	0.661 g/cm^3
Natural moisture	12.6%

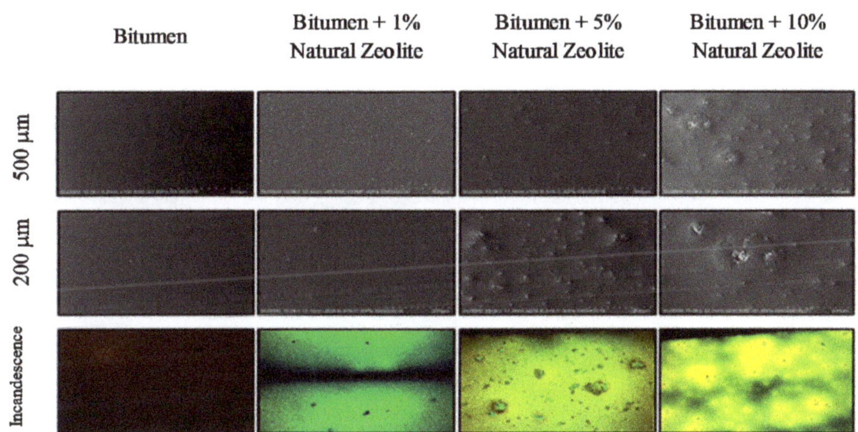

Figure 2. Homogeneity and incandescence of the bitumen with addition of natural zeolite.

The RAP used comes from milling works on Route 5, IX Region of Chile. In order to reduce the heterogeneity of the WMA-RAP mixtures, they were manufactured with two fractions of RAP, as recommended by Solaimanian and Tahmoressi [34]. The RAP fractions used were 0/5 mm and 5/20 mm. Table 4 shows the RAP gradation and the bitumen content for both RAP fractions. The characteristics of binder recovered from the RAP before fractioning are shown in Table 5.

Table 4. Reclaimed asphalt pavement (RAP) gradation and bitumen content (after extraction).

RAP Fraction (mm)	0/5	5/20
Bitumen Content (% by Weight of Mix)	**7.5**	**3.3**
Sieve Size (mm)	**Gradation (% Passing)**	
20	100	100
12.5	100	73
10	100	56
5	100	29
2.5	70	18
0.63	35	11
0.315	25	8
0.16	19	5
0.08	15	4

Table 5. Characteristics of binder recovered from the RAP.

Tests	Results
Penetration at 25 °C, 100 g. 5 s. (0.1 mm)	10
Fraas Breaking Point (°C)	+3
Softening Point R&B, (°C)	55.6
Brookfield Viscosity at 60 °C (Pa s)	30,883

The asphalt mixtures evaluated are shown in Table 6 All these mixtures were evaluated for their resistance to moisture damage, low temperature cracking, fatigue and rutting, showing similar or better performance than the control HMA [30].

Table 6. Description of mixtures evaluated.

Nomenclature	Description
P-155	Control HMA, manufacturing temperature: 155 °C
Z0.3-135	WMA; Natural Zeolite content: 0.3% by weight of aggregates; manufacturing temperature 135 °C
Z0.6-135	WMA; Natural Zeolite content: 0.6% by weight of aggregates; manufacturing temperature 135 °C
PR10-155	Control HMA–RAP, RAP content 10%, manufacture production temperature: 155 °C
R10-Z0.6-125	WMA-RAP, RAP content 10%, Natural Zeolite content: 0.6% manufacturing temperature 125 °C
PR20-155	Control HMA–RAP, RAP content 20%, manufacturing temperature: 155 °C
R20-Z0.6-135	WMA-RAP, RAP content 20%, Natural Zeolite content: 0.6% manufacturing temperature 135 °C
PR-30-155	Control HMA–RAP, RAP content 30%, manufacturing temperature: 155 °C
R30-Z0.6-145	WMA–RAP, RAP content 30%, Natural Zeolite content: 0.6% manufacturing temperature 145 °C

2.2. Test Methods for Gas Emission Evaluation

To manufacture the WMA, aggregates and asphalt binder were heated to the mixing temperature according to Table 3. In the case of WMA–RAP, the RAP temperature was 15 °C and the raw aggregates had to be overheated to achieve the final manufacturing temperature (Table 7). To obtain each gas sample, 4 kg of asphalt mixture were manufactured, maintaining the dosage of each mixture. The mixtures were manufactured in a hermetic chamber for 5 min, followed by 2 min resting time before gas capture. The gases (500 mL) were captured using a gas syringe and stored in hermetic Tedlar gas sample bags with a polypropylene valve (Figure 3). Three gas samples were taken for each asphalt mixture. It is important to note that the gas samples did not include the gases emitted during the combustion process, and only considered the gases generated by heating asphalt binder, RAP, aggregates and natural zeolite because the heating process was carried out without combustion in the laboratory.

Table 7. Overheating temperature of natural aggregates and binder dosage as a function of RAP content and final mixture temperature.

Mixture of Aggregates		Asphalt Binder Content (%)		Aggregate Temperature (°C)			
Natural (%)	RAP (%)	New Binder	RAP Binder Contribution	WMA			HMA (References)
100	0	5.4	0		135		155
90	10	5.19	0.21	142			180
80	20	4.98	0.43		169		200
70	30	4.77	0.54			202	227
Manufacture temperature (°C)				125	135	145	155

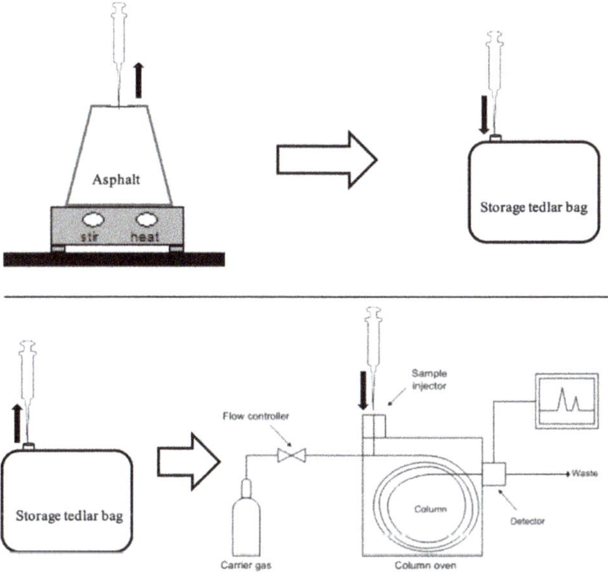

Figure 3. Diagram of fume capture and GEIS sampling systems.

A gas chromatograph with TCD (thermal conductivity detector), Clarus 580 Perkin Elmer, with a packed column 60/80 carboxeen 1000 15ft × 1/8″ (2.1 mm ID, internal diameter) was used to evaluate gas emissions. The gas samples were extracted from the gas storage bags and injected manually at a volume of 250 μL. The temperature of the detector was 250 °C and the injector temperature was 200 °C. Helium carrier gas was used, with a flow of 30 mL/min. The temperature ramp was 180 °C for 4.50 min. The working range was 0.05 to 1%, Figure 3. The gases evaluated were CO_2, CO and SO_2.

The methodology was divided into four phases (Figure 4). The first phase consisted of a review of studies related to the project, combined with visits to different asphalt mixing plants to become familiar with each stage involved in the manufacturing process in order to determine the most significant variables.

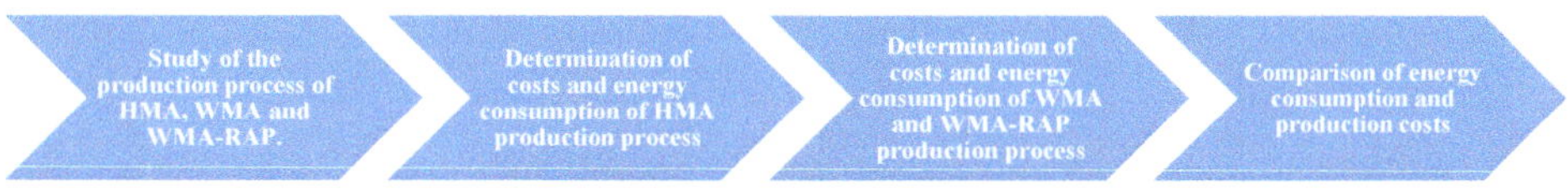

Figure 4. Four-phase model.

To complete the second phase, a cost table was generated by collecting data in the field provided by a local asphalt mixing company, incorporating the main costs related to the production of conventional asphalt mixtures. Additionally, energy consumption equations defined by Romier et al. [35] were used to determine the energy consumption during the production stage and to relate this variable to temperature changes (Equations (1)–(4)).

$$C_{agr} \times m_{agr} \times \left(t_{agr} - t_{amb}\right) \quad (1)$$

energy needed to heat aggregates (cal) (1)

$$C_a \times \left(\frac{m_{agr}}{m_{agr} - \frac{u}{100} \times m_{agr}} - 1\right) \times m_{agr} \times (100 - t_{amb}) \quad (2)$$

energy needed to heat the water present in the aggregates (cal) (2)

$$clv \times \left(\frac{m_{agr}}{m_{agr} - \frac{u}{100} \times m_{agr}} - 1 \right) \times m_{agr} \tag{3}$$

energy needed to vaporize water from aggregates (cal) (3)

$$C_v \times \left(\frac{m_{agr}}{m_{agr} - \frac{u}{100} \times m_{agr}} - 1 \right) \times m_{agr} \times \left(t_{agr} - 100 \right) \tag{4}$$

energy needed to eliminate water vapor (cal) (4)

where: C_{agr} is the specific heat of aggregates (cal/kg^{-1}°C^{-1}); m_{agr} is the mass of aggregates (kg); t_{agr} is the aggregate heating temperature (°C); t_{amb} is the ambient temperature (°C); C_a is the specific heat of water; (cal/kg^{-1*}°C^{-1}); h is the aggregate moisture content (%); clv is the latent heat of vaporization of the water (cal/kg); C_v is the vapor-specific heat (cal/kg^{-1*}°C^{-1}); and t_{agr} is the aggregate heating temperature (°C).

The parameters involved in calculating the energy and fuel consumption are shown in Tables 7 and 8. The moisture content was fixed according to the average moisture content reported by the production plant. To determine the cost and energy consumption of WMA and WMA–RAP, the cost table of the conventional asphalt mixture (previously generated and validated with information provided by the mixing plant) was modified by the inclusion of the new costs according to the different amounts of zeolites and RAP. In this stage, the energy consumption was calculated by the Romier equations, varied according to the changes in the manufacturing temperature. These modifications were caused by the variation in fuel consumption, and in the amount of virgin raw materials (binder and aggregates) used for each WMA–RAP mixture since these were reduced as a function of the amount of RAP included. Finally, the cost tables were compared with the reference mixture. The fuel consumption was calculated by converting the energy consumption to fuel, based on the heating power of fuel oil. The production costs of the different proposed mixtures were obtained by assessing the energy consumption and the costs associated with the raw materials and additive (natural zeolite).

Table 8. Parameters involved in energy and cost consumption.

Parameter	Value	Unit
[1] Specific heat of aggregates (C_{agr})	850	J/(kg * °C)
Specific heat of water (C_a)	4200	J/(kg * °C)
Specific heat of steam (C_v)	1850	J/(kg * °C)
Latent heat of water vaporization (clv)	2,250,000	J/(kg * °C)
Moisture content of aggregates (h)	3	%
Ambient temperature (t_{amb})	15	°C
Heating temperature of aggregates (t_{agr})	155	°C
Heating temperature of bitumen	155	°C
[2] Heating power of fuel oil	9,762,000	Cal/kg
[2] Density of fuel oil	0.9994	kg/L
Assessed quantity of mixture	1000	kg

[1] The Specific heat of granitic aggregates [36]; [2] The information about fuel oil was extracted from the fuel supplier's certificate.

3. Analysis of Results

3.1. Gas Emissions

Before we analyze the results, it is important to remember that this study considers only the emissions generated by heating the asphalt binder in contact with the raw aggregates, and, in some cases, natural zeolite and varying percentages of RAP (i.e., combustion emissions produced by aggregate heating, such as those produced during the production process in asphalt plants, were isolated). In this context, it is observed that the SO_2 emission level was outside the detection ranges for all the mixtures studied. This implies that the SO_2 emissions were less than 0.05%, considered as null for all the mixtures evaluated. These results may be because SO_2 is generated by the combustion process [16] and is highly dependent on fuel type, with fuel oil and recycled fuel oil causing the most pollution. On the other hand, natural gas seems to be an environmentally friendly fuel alternative, since it has a lower sulphur content [16]. This agrees with the results obtained by Lecomte et al. [37], who compared the SO_2 emissions between HMA (180 °C) and WMA foam (125–130 °C). They observed a 35% reduction in fuel consumption, obtaining a reduction in SO_2 emissions between 25% and 30%. On the other hand, Hurley et al. [38] analyzed the natural gas consumption in three WMA mixtures (Evotherm®, Aspha-min® and Sasobit®). The fuel consumption results were + 15.4%, – 8.8% and – 17.9% (Control, HMA, WMA), respectively, while the SO_2 emissions were + 54%, – 83% and – 83% (Control, HMA, WMA), respectively. These results suggest that fuel consumption is the greatest contributing factor to SO_2 emissions.

Similar levels of CO emissions were observed between HMA and WMA. The reduction percentage of CO concentration was 2% for Z-0.3-135, while for Z-0.6-135 the CO emission level was similar to that of P-155 (Figure 5). When HMA–RAP is compared to WMA–RAP (Figure 5), reductions in CO emissions of 6%, 5% and 4% were detected (R10-Z0.6-125, R20-Z0.6-135 and R30-Z0.6-145 respectively). Some researchers suggest that CO emissions are associated with fuel consumption and depend on the type of fuel, reporting that WMA manufactured in a plant powered with natural gas emitted eighteen times less CO than a plant with fuel oil under the same conditions [3]. However, Anderson et al. [39] concluded that CO emissions are more closely related to burner maintenance. Lecomte et al. [37] obtained CO reductions of around 8% when comparing measurements made in the chimney of the mixing plant during the manufacture of HMA (180 °C) and WMA foam mixture (125–130 °C). These findings are consistent with this study, as the effects of fuel burning or burner maintenance are not considered for any of the mixtures evaluated. However, it is remarkable that CO emissions are not correlated exclusively with fuel burning, since CO emissions were detected in this study.

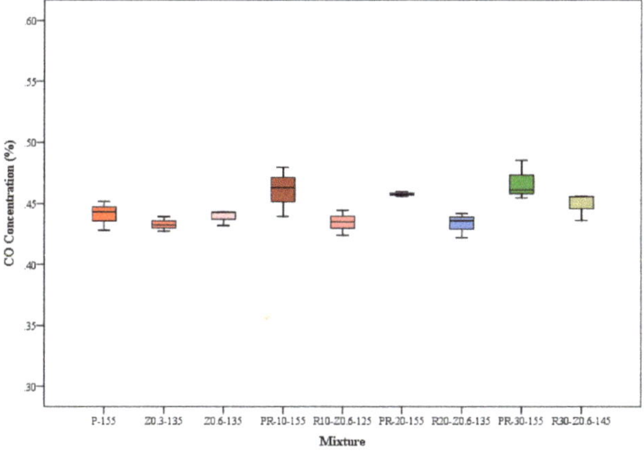

Figure 5. CO concentration (%).

Additionally, the reference mixtures with RAP (PR-10-155, PR-20-155, PR-30-155) showed more CO emissions than the reference mixture without RAP (P-155). However, the reduction in CO emissions in WMA–RAP manufactured at a lower temperature than the reference mixtures with RAP was larger than the difference between WMA and the reference mixture without RAP (Figure 5).

In the analysis of CO_2 emissions, 90% were found to occur directly during the asphalt mixing process [14]. The complete production process includes aggregate heating, asphalt heating, and mixing. In these stages, the carbon emissions observed reached values of 67%, 14% and 12%, respectively [14]. This study focuses on the third stage, the mixing process. The results for CO_2 emissions showed a reduction for all mixtures evaluated; emissions were higher for WMA–RAP (Figure 6). These results are similar to those presented by Sharma et al. [40], who observed CO_2 emission reductions of between 2.5% and 35%, respectively, in a comparison of HMA (180 °C) with WMA (140 °C) with 2%–6% addition of synthetic zeolite (over binder content). Lecomte et al. [37] also observed CO_2 emission reductions of around 35% in a comparison between HMA manufactured at 180 °C and WMA foam at 125–130 °C, both measured in the chimney of the mixing plant. Comparable results were obtained by Mallik et al. [12], who reported a direct reduction in CO_2 emissions of around 32% by decreasing the mixing temperatures by between 10 and 30 °C.

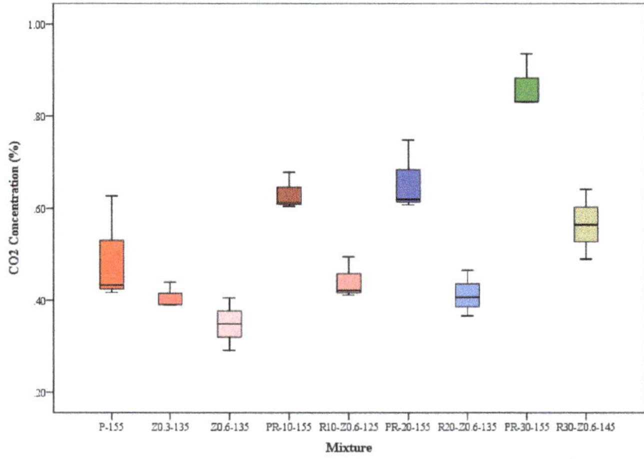

Figure 6. CO_2 concentration (%).

In this study, the CO_2 emission reductions were obtained by comparing HMA manufactured at 155 °C with WMA at 135 °C. The emissions were reduced by 17% and 23%, respectively, with the addition of 0.3% and 0.6% natural zeolite (Figure 6). Larger reductions were obtained in WMA–RAP with the addition of 0.6% natural zeolite (R10-Z0.6-125, R20-Z0.6-135, R30-Z0.6-145), which were compared with the reference mixtures manufactured with RAP (PR-10-155, PR-20-155, PR-30-155). These WMA–RAP mixtures showed reductions in CO_2 emissions in the range of 30% to 37%, the largest reduction being obtained with R20-Z0.6-135 (Figure 6). This may be related to the results obtained by Giani et al. [41], who indicated that WMA produced using RAP instead of virgin raw materials may mean a reduction in CO_2 emissions of approximately 12% over the whole life cycle.

3.2. Energy Consumption

Energy consumption assessment of the production processes of different pavement types may constitute an important tool for engineers and legislators in selecting the type of pavement and technology to use in infrastructure design, in terms of both environmental impact and economic investment [2]. In this context, the manufacture of HMA requires higher energy consumption during asphalt mixing and aggregate heating; the latter process accounts for 53% of energy consumption [2]. Several researchers have concluded that the lower production temperature of WMA technologies may reduce the energy consumption compared to HMA [42,43]. This is consistent with the results shown in Figure 7, which show a reduction in energy consumption due to the lower production temperature. The reference HMA was produced at 155 °C, while the study mixtures were produced at lower temperatures (125 °C, 135 °C and 145 °C). These temperature reductions account for decreases in energy consumption of 13%, 9% and 5%, respectively. These findings are similar to those reported by Anderson et al. [39], who described an energy saving of about 10% using WMA with water-injection foaming systems. Croteau and Tessier [10], in their study of the WMA state-of-practice, indicated energy savings ranging from 20 to 35% according to the reports of WMA trials in an asphalt plant. Similarly, Harder et al. [44] observed reductions in energy consumption of between 10% and 30%. These variations are dependent on the WMA system, moisture content of the aggregates and the type/efficiency of the plant. Expressed in terms of kcal, a reduction of 2051 kcal/ton per 10 °C decrease in temperature is observed. A similar result was reported by Peinado et al. [42], who noted an extra energy demand of about 2253 Kcal for an increase of 10 °C in the mixture temperature.

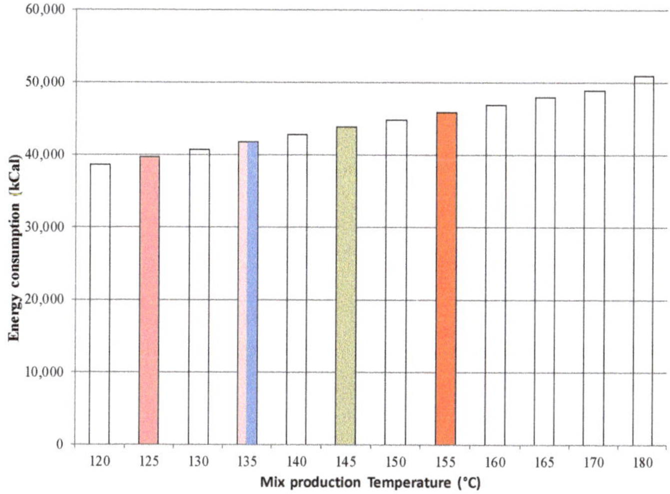

Figure 7. Energy consumption in reference to production temperature.

3.3. Evaluation of Production Cost

Energy savings are related to cost reductions because they imply a reduction in fuel consumption. Several researchers have reported economic benefits from the use of WMA. The different quantities of fuel required to produce all mixtures evaluated in this study were calculated, based on the energy consumption results of the asphalt mixtures produced at different temperatures and the heating power of the fuel. Parameters such as ambient temperature (15 °C), aggregate moisture content (3%) and RAP moisture content (0.3%) were fixed (Table 8), since the focus of the present study was on the energy and cost savings achievable by reducing the manufacturing temperature during the production stage. A reduction in fuel consumption of about 10% was observed for WMA at 135 °C, while for WMA–RAP the reduction in fuel consumption reached 15% for R10-Z0.6-125, R20-Z0.6-135 and R30-Z0.6-145 (Figure 8). These results are similar to those reported by Anderson et al., who indicated that a temperature reduction of 27 °C resulted in an average fuel saving of 22%, with data drawn from 13 projects [39].

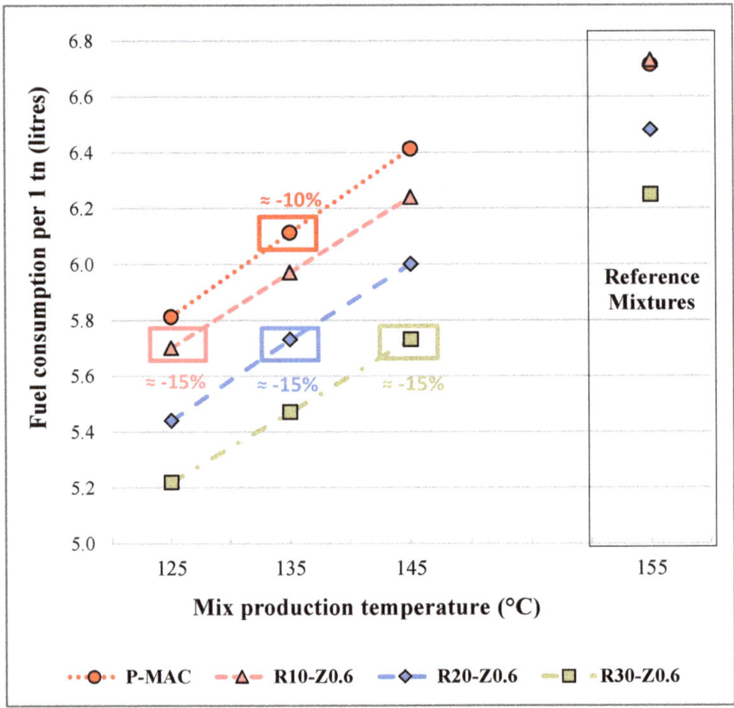

Figure 8. Comparison of fuel consumption per 1 ton of mixture production.

The cost evaluation of WMA with the addition of natural zeolite included the cost of the additive and the savings related to reduced fuel consumption. For the WMA–RAP with natural zeolite, the cost evaluation considered the cost of the additive and the savings due to reduced fuel consumption and raw materials (aggregates and asphalt binder). The cost evaluations consider the production temperature for all the mixtures evaluated, including the aggregate overheating temperature required for the WMA with RAP, as well as the reduced consumption of virgin asphalt binder due to the incorporation of RAP (Table 7).

The current cost of natural zeolite reported by the supplier is US$1.03 per kg. This is a high cost, much higher than the cost of lime (US$0.11 per kg), probably due to the lack of development of this industry. The production process of natural zeolite includes the same steps as the production of lime: extraction of natural stone, grinding, sieving and finally packaging for commercialization; however, lime production requires an important extra step: mineral calcination at 900 °C. This process implies a major additional cost that is not required to obtain natural zeolite. Nevertheless, the cost of lime continues to be considerably lower than that of natural zeolite. In consideration of the above, two cost evaluations were carried out based on two scenarios: the current situation, when the cost of natural zeolite is US$1.03 per kg; and a hypothetical future situation, in the expectation that the natural zeolite industry is likely to grow to meet demand, when the cost of natural zeolite will fall to a similar level to that of lime (US$0.11 per kg). The results shown in Figure 9 indicate that in the current situation, the production cost of manufacturing WMA is higher than the production cost of HMA by a factor ranging from 5% for WMA with 0.3% of natural zeolite to 12% for WMA with 0.6% of natural zeolite. These results were higher than the cost of the reference mixture (P-155) because the savings in fuel costs were lower than the extra cost of the additive. Similar findings were reported by EAPA, which described the production cost of WMA as being similar to or slightly higher than that of HMA [45].

Almeida et al. [46] obtained similar results, comparing rough asphalt concrete with WMA (Sasobit®); they reported little difference in the total costs of the two mixtures. When the hypothetical situation projected above is analyzed, the cost of WMA with 0.3% and 0.6% natural zeolite may be similar to that of HMA.

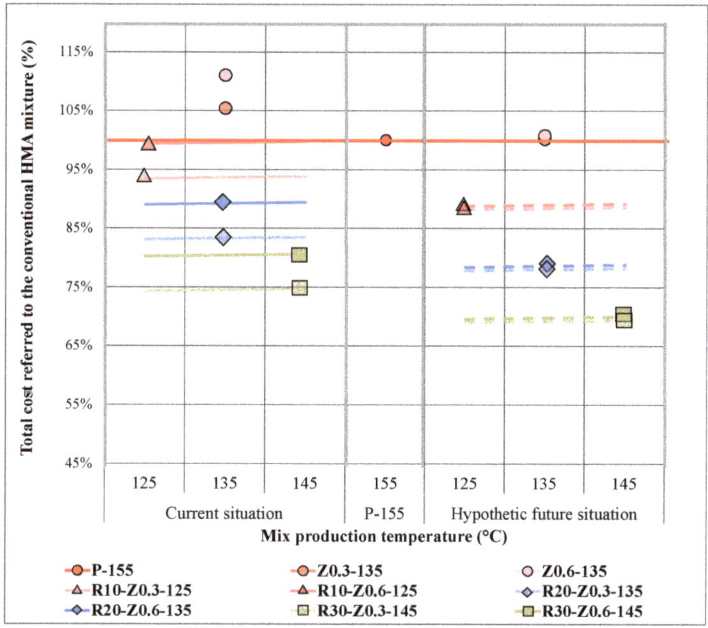

Figure 9. Production cost evaluation for warm mix asphalt (WMA) with natural zeolite addition considering 1 ton of mixture, in the current situation and a hypothetical future situation.

When the cost of WMA with the addition of natural zeolite and RAP is analyzed (Figure 9), all the asphalt mixtures evaluated in this study present reduced costs. In the current situation, WMA with 30% RAP and 0.3% natural zeolite presented cost savings of about 25%, while the cost savings for WMA with 30% RAP and 0.6% natural zeolite were 19% (both manufactured at 145 °C). For WMA with 10% RAP produced at 125 °C, the cost savings were 7% and 1% for the addition of 0.3% and 0.6% of natural zeolite respectively. Finally, if a balance is sought between incorporating RAP and reducing temperature, the best alternative might be WMA with 20% RAP produced at 135 °C, which gives cost savings of 17% and 11% for the addition of 0.3% and 0.6% natural zeolite respectively. Analyzing the WMA-RAP mixtures in the hypothetical future situation, the cost savings may be higher, achieving a cost reduction of 30% for R30-Z0.6-145, 21% for R20-Z0.6-135 and 11% for R10-Z0.6-125 (Figure 9). These results are consistent with the report by Kristjánsdóttir et al. [43], who showed that RAP incorporation may be a good approach to cost savings; they concluded that in the project evaluated, increasing the RAP content from 25% to 40% would result in a net saving of US$4.55/ton.

4. Conclusions

The present study focused on the gas emissions, energy consumption and production costs of WMA and WMA-RAP, in comparison with a reference mixture. In terms of gas emissions, all the WMA showed reductions in direct CO_2 emissions of 17% and 23%, respectively with the addition of 0.3% and 0.6% natural zeolite. The CO_2 reductions for WMA-RAP were higher, reaching 37% for WMA-RAP with 20% RAP manufactured at 135 °C and 35% for WMA-RAP with 30% RAP manufactured at 145 °C.

The reductions in CO emissions were smaller than those achieved for CO_2: for WMA, the CO reduction was not statistically significant, whereas for WMA-RAP the CO emissions decreased by nearly 6%. On this point, it is remarkable that the CO emissions are not only correlated to fuel burning; in this research CO emissions were detected although the fuel burning process was excluded. For all asphalt mixtures evaluated, SO_2 emissions were out of the detection ranges (0.05%); this could be because SO_2 is generated by the combustion process and is highly dependent on fuel type, which was not considered here.

Reductions in energy consumption were found in the range of 5% to 13%, similar to the reductions reported by other researchers.

Finally, in the cost evaluation, the cost reduction for WMA achieved by reduced fuel consumption was not enough to compensate the current additional cost of the additive; these mixtures are still slightly more expensive than the reference mixture. The situation for WMA-RAP was different: production cost savings close to 25% were achieved for the mixtures with 30% RAP and 0.3% natural zeolite.

Our results show that all the asphalt mixtures evaluated presented reduced gas emissions and energy/fuel consumption, with a cost similar to the control HMA. In addition, if the mixtures are manufactured with RAP, the benefits will be greater thanks to the energy and cost savings; mixtures containing RAP also offer environmental benefits, which include lower emissions and the use of recycled asphalt pavement materials.

Author Contributions: L.A.M.-P.: experimental work and data analysis; E.S.-A.: methodology, investigation, data analysis, review and editing, G.A.V.-V.: project administration, investigation, data analysis, review and editing, A.T.C.-F.: methodology, investigation, data analysis, writing—original draft preparation, review and editing. All authors have read and agreed to the published version of the manuscript.

Funding: This research was funded by Comisión Nacional de Investigación Ciencia y Tecnología (CONICYT) conducted within the framework of the FONDEF IDEA Grant Nº ID15I10235; and by Universidad de La Frontera through DIUFRO Grant N° DI18-0053.

Conflicts of Interest: The authors declare no conflict of interest. The funders had no role in the design of the study; in the collection, analyses, or interpretation of data; in the writing of the manuscript, or in the decision to publish the results.

References

1. Chappat, M.; Bilal, J. *The Environmental Road of the Future: Life Cycle Analysis*; Colas SA: Paris, France, 2003; Volume 9, pp. 1–34.
2. Zapata, P.; Gambatese, J.A. Energy Consumption of Asphalt and Reinforced Concrete Pavement Materials and Construction. *J. Infrastruct. Syst.* **2005**, *11*, 9–20. [CrossRef]
3. Thives, L.P.; Ghisi, E. Asphalt mixtures emission and energy consumption: A review. *Renew. Sustain. Energy Rev.* **2016**, *72*, 473–484. [CrossRef]
4. Autelitano, F.; Bianchi, F.; Giuliani, F. Airborne emissions of asphalt/wax blends for warm mix asphalt production. *J. Clean. Prod.* **2017**, *164*, 749–756. [CrossRef]
5. Woszuk, A.; Zofka, A.; Bandura, L.; Franus, W. Effect of zeolite properties on asphalt foaming. *Constr. Build. Mater.* **2017**, *139*, 247–255. [CrossRef]
6. Miller, T.D.; Bahia, H.U. *Sustainable Asphalt Pavements: Technologies, Knowledge Gaps and Opportunities*; University of Wisconsin Madison: Madison, WI, USA, 2009.
7. Gasthauer, E.; Mazé, M.; Marchand, J.P.; Amouroux, J. Characterization of asphalt fume composition by GC/MS and effect of temperature. *Fuel* **2008**, *87*, 1428–1434. [CrossRef]
8. Jullien, A.; Gaudefroy, V.; Ventura, A.; de La Roche, C.; Paranhos, R.; Monéron, P. Airborne Emissions Assessment of Hot Asphalt Mixing: Methods and limitations. *Road Mater. Pavement Des.* **2010**, *11*, 149–169. [CrossRef]
9. Qian, S.L.; Wang, F. The Experimental Study on the Road Asphalt Fumes Inhibitors. *Adv. Mater. Res.* **2012**, *413*, 472–476. [CrossRef]
10. Croteau, J.; Tessier, B. Warm mix asphalt paving technologies: A road builder's perspective. In Proceedings of the 2008 Annual Conference of the Transportation Association of Canada, Whistler, BC, Canada, 21–24 September 2008; pp. 1–12.

11. Jamshidi, M.; Hamzah, O.; You, Z. Performance of Warm Mix Asphalt containing Sasobit®: State-of-the-art. *Constr. Build. Mater.* **2013**, *38*, 530–553. [CrossRef]
12. Mallick, R.B.; Bergendahl, J. A laboratory study on CO_2 emission from asphalt binder and its reduction with the use of warm mix asphalt. *Int. J. Sustain. Eng.* **2009**, *2*, 275–283. [CrossRef]
13. Ahmad, I.; Shakirullah, M.; ur Rehman, H.; Ishaq, M.; Khan, M.A.; Shah, A.A. NMR analysis of cracking products of asphalt and assessment of catalyst performance. *Energy* **2009**, *34*, 127–133. [CrossRef]
14. Peng, B.; Cai, C.; Yin, G.; Li, W.; Zhan, Y. Evaluation system for CO_2 emission of hot asphalt mixture. *J. Traffic Transp. Eng. (Engl. Ed.)* **2015**, *2*, 116–124. [CrossRef]
15. Brandt, H.C.A.; de Groot, P.C. A laboratory rig for studying aspects of worker exposure to bitumen fumes. *Am. Ind. Hyg. Assoc. J.* **1999**, *60*, 182–190. [CrossRef] [PubMed]
16. West, R.; Rodezno, C.; Julian, G.; Prowell, B.; Frank, B.; Osborn, L.V.; Kriech, T. *NCHRP Report 779. Field Performance of Warm Mix Asphalt Technologies*; Transportation Research Board of the National Academies: Washington, DC, USA, 2014.
17. Autelitano, F.; Giuliani, F. Analytical assessment of asphalt odor patterns in hot mix asphalt production. *J. Clean. Prod.* **2018**, *172*, 1212–1223. [CrossRef]
18. Adbullah, M.E.; Zamhari, K.A.; Hainin, M.R.; Oluwasola, E.A.; Yusoff, N.I.M.; Hassan, N.A. High temperature characteristics of warm mix asphalt mixtures with nanoclay and chemical warm mix asphalt modified binders. *J. Clean. Prod.* **2016**, *122*, 326–334.
19. Hasan, M.R.M.; You, Z.; Yang, X. A comprehensive review of theory, development, and implementation of warm mix asphalt using foaming techniques. *Constr. Build. Mater.* **2017**, *152*, 115–133. [CrossRef]
20. Prowell, B.D. *Warm Mix Asphalt, the International Technology Scanning Program Summary Report*; American Trade Initiatives: Alexandria, Egypt, 2007; p. 21.
21. Iwański, M.; Chomicz-Kowalska, A.; Maciejewski, K. Application of synthetic wax for improvement of foamed bitumen parameters. *Constr. Build. Mater.* **2015**, *83*, 62–69. [CrossRef]
22. Sengoz, B.; Topal, A.; Gorkem, C. Evaluation of natural zeolite as warm mix asphalt additive and its comparison with other warm mix additives. *Constr. Build. Mater.* **2013**, *43*, 242–252. [CrossRef]
23. Rubio, M.C.; Martínez, G.; Baena, L.; Moreno, F. Warm Mix Asphalt: An overview. *J. Clean. Prod.* **2012**, *24*, 76–84. [CrossRef]
24. Barthel, W.; Marchand, J.-P.; von Devivere, M. Warm asphalt mixes by adding a synthetic zeolite. In Proceedings of the 3rd Eurasphalt & Eurobitume Congress, Vienna, Austria, 12–14 May 2004; Volume 1.
25. Adnan, M.; Shafi, M.; Sharma, A. Laboratory study on use of RAP in WMA pavements using rejuvenator. *Constr. Build. Mater.* **2018**, *168*, 61–72.
26. Horvath, A. *Life-Cycle Environmental and Economic Assessment of Using Recycled Materials for Asphalt Pavements*; University of California Transportation Center: Berkeley, CA, USA, 2003.
27. Mallick, R.B.; Kandhal, P.S.; Bradbury, R.L. Using warm mix asphalt technology to incorporate high percentage of reclaimed asphalt pavement (RAP) material in asphalt mixtures. *Transp. Res. Rec. J. Transp. Res. Board* **2008**, *2051*, 71–79. [CrossRef]
28. Al-Qadi, I.L.; Elseifi, M.; Carpenter, S.H. *Reclaimed Asphalt Pavement-A Literature Review*; Illinois Center for Transportation: Rantoul, IL, USA, 2007.
29. Farooq, M.A.; Mir, M.S. Use of reclaimed asphalt pavement (RAP) in warm mix asphalt (WMA) pavements: A review. *Innov. Infrastruct. Solut.* **2017**, *2*, 1–9. [CrossRef]
30. Valdes-Vidal, G.; Calabi-Floody, A.; Sanchez-Alonso, E. Performance evaluation of warm mix asphalt involving natural zeolite and reclaimed asphalt pavement (RAP) for sustainable pavement construction. *Constr. Build. Mater.* **2018**, *174*, 576–585. [CrossRef]
31. Públicas, M.D.; de Vialidad, D. *Manual de Carreteras Volumen 5 Especificaciones Técnicas Generales de Construcción*; Ministerio de Obras Públicas: Santiago, Chile, 2017.
32. Ministerio de Obras; de Vialidad, D. Manual de Carreteras Volumen 8 Especificaciones y métodos de muestreo, ensaye y control, Test method 8.302.40. In *Asfaltos Método Para Determinar La Resistencia a La Leformación Plástica de de Mezclas Asfálticas Utilizando el Aparato Marshall*; Ministerio de Obras Públicas: Santiago, Chile, 2017.
33. Woszuk, A.; Franus, W. A Review of the Application of Zeolite Materials in Warm Mix Asphalt Technologies. *Appl. Sci.* **2017**, *7*, 293. [CrossRef]

34. Solaimanian, M.; Savory, E. Variability Analysis of Hot-Mix Asphalt Concrete Containing High Percentage of Reclaimed Asphalt Pavement. *Transp. Res. Rec. J. Transp. Res. Board* **1996**, *1543*, 13–20. [CrossRef]
35. Romier, A.; Audeon, M.; David, J.; Martineau, Y.; Olard, F. Low-Energy Asphalt with Performance of Hot-Mix Asphalt. *Transp. Res. Rec.* **2006**, *1962*, 101–112. [CrossRef]
36. Motta, R.S. *Estudo de Misturas Asfálticas Mornas em Revestimentos de Pavimentos para Redução de Emissão de Poluentes e de Consumo Energético*; Escola Politécnica da Universidade de São Paulo: São Paulo, Brazil, 2011.
37. Lecomte, M.; Deygout, F.; Menetti, A. *Emission and Occupational Exposure at Lower Asphalt Production and Laying Temperatures*; Shell Bitumen: London, UK, 2007.
38. Hurley, G.C.; Prowell, B.D.; Kvasnak, A.N. *Ohio Field Trial of Warm Mix Asphalt Technologies: Construction Summary*; National Center for Asphalt Technology: Auburn, AL, USA, 2009.
39. Anderson, R.; Baumgardner, G.; May, R.; Reinke, G. Engineering properties, emissions, and field performance of warm mix asphalt technologies. In *Transportation Research Board of the National Academies Privileged Document*; National Cooperative Highway Research Program: Washington, DC, USA, 2008.
40. Sharma, A.; Lee, B.K. Energy savings and reduction of CO2emission using Ca(OH)2incorporated zeolite as an additive for warm and hot mix asphalt production. *Energy* **2017**, *136*, 142–150. [CrossRef]
41. Giani, M.I.; Dotelli, G.; Brandini, N.; Zampori, L. Comparative life cycle assessment of asphalt pavements using reclaimed asphalt, warm mix technology and cold in-place recycling. *Resour. Conserv. Recycl.* **2015**, *104*, 224–238. [CrossRef]
42. Peinado, D.; de Vega, M.; García-Hernando, N.; Marugán-Cruz, C. Energy and exergy analysis in an asphalt plant's rotary dryer. *Appl. Therm. Eng.* **2011**, *31*, 1039–1049. [CrossRef]
43. Kristjánsdóttir, Ó.; Muench, S.; Michael, L.; Burke, G. Assessing Potential for Warm-Mix Asphalt Technology Adoption. *Transp. Res. Rec. J. Transp. Res. Board* **2007**, *2040*, 91–99. [CrossRef]
44. Harder, G.A.; LeGoff, Y.; Loustau, A.; Martineau, Y.; Heritier, B.; Romier, A. Energy and environmental gains of warm and half-warm asphalt mix: Quantitative approach. In Proceedings of Transportation Research Board 87th Annual Meeting, Washington, DC, USA, 13–17 January 2008.
45. EAPA. *The Use of Warm Mix Asphalt. EAPA–Position Pap*; EAPA: Belgium, Brussels, 2010; pp. 1–13.
46. Almeida-Costa, A.; Benta, A. Economic and environmental impact study of warm mix asphalt compared to hot mix asphalt. *J. Clean. Prod.* **2016**, *112*, 2308–2317. [CrossRef]

 *sustainability*

Review

Behavior Evaluation of Bituminous Mixtures Reinforced with Nano-Sized Additives: A Review

Raffaella Fusco *, Laura Moretti, Nicola Fiore and Antonio D'Andrea

Department of Civil, Construction and Environmental Engineering, Sapienza University of Rome, 00184 Rome, Italy; laura.moretti@uniroma1.it (L.M.); nicola.fiore@uniroma1.it (N.F.); antonio.dandrea@uniroma1.it (A.D.)

* Correspondence: raffaella.fusco@uniroma1.it; Tel.: +39-06-4458-5124

Received: 2 September 2020; Accepted: 28 September 2020; Published: 29 September 2020

Abstract: This article presents a comparative review of the most commonly used nano-additives for bituminous mixtures: nanoclays (NC), nanosilicates, carbon nanotubes (CNTs), graphene nanoplatelets (GNPs), nano-calcium oxide (CaO), and nano-titanium dioxide (TiO_2). In this study, the mechanical behavior of the obtained additive mixture is evaluated. According to the revised literature, the results strongly depend on type, concentration, and dispersal of used nano-additive. In fact, it has been seen that simple shear mixing followed by sonication homogenizes the distribution of the nanoparticles within the bituminous matrix and favors the bonds' formation. The viscosity of the mixture of bitumen with nanoparticles improves with the increase of the percentage of additive added: it indicates a potential improvement to permanent deformation and rutting. Another benefit is an increased resistance of the binder to aging. Furthermore, it has been shown that the nanoparticles are able to prolong the service life of a bituminous mixture by means of various interdependent chemical–physical mechanisms that can influence the resistance to fatigue failure or the ability to self-heal. However, the effectiveness of these improvements depends on the particle type, added quantity and mixing technique, and the tests carried out.

Keywords: bituminous mixtures; nano-additives; nanoclay; carbon nanotubes; graphene nanoplatelets; nano-calcium oxide; nano-titanium dioxide; sonication; fatigue performance; self-healing

---

1. Introduction

Bitumens are natural or artificial mixtures of solid or semi-solid hydrocarbons, obtained from asphaltic rocks or natural oils; they are used to provide waterproofing and protective coating and as binders in road construction [1]. Bitumen's components are usually grouped into two categories: asphaltenes and maltens. These two can be subdivided into saturates, aromatics, and resins [1]. Bitumen is modeled as a colloid composed of asphaltene micelles covered by a stabilizing phase of polar resins; this phase forms the interface with a continuous oily maltenic medium [1,2].

Different factors affect the physical and mechanical behavior of bituminous mastics, including temperature and loading time. The compound is liquid or solid at high or low temperatures, respectively. Therefore, when asphalt is used for road pavements, cracks at low temperatures or rutting at high temperatures may occur. Moreover, oxygen, ultraviolet (UV) light of sun, and heat affect both the physical properties and chemical structure of asphalt, and cause a phenomenon called aging [3]. Another enemy of the asphalt binder is moisture: it causes the progressive loss of functionality of the material due to loss of the adhesive bond between the asphalt binder and the aggregate surface [4]. Penetration of moisture in asphalt mixtures reduces strength and stiffness of asphalt mixtures and makes the mixtures prone to develop premature pavement distresses (e.g., stripping, raveling, and hydraulic scour [5,6]; rutting, alligator cracking, and potholes [4]). The presence of water in pavements can be detrimental if combined with other environmental factors such as freeze–thaw cycling over extended periods: chemical and physical interactions between bitumen and aggregate at the interface influence the adhesive strength [5]. Moreover, repetitive vehicular loads cause fatigue cracking distress in asphalt pavements. Fatigue in asphalt pavements consists of two consecutive phases. At first (pre-localization), micro-cracks are generated; these grow followed by the formation of macro-cracks during post-localization due to critical stresses or strains [7]. Finally, permanent deformations (i.e., rutting) affect asphalt pavements; they are caused by deformation or consolidation of pavement layers, especially if they are thermosusceptible as asphalt ones are. Several factors influence rutting of asphalt pavement: overloading, low-speed trucks, substance properties, and climate conditions (i.e., high-temperature areas). As traffic loading and tire pressure increase, permanent deformation at the top layer of pavement surface also increases [4]. Service life of the pavement drastically declines because of rutting: when in ruts, asphalt becomes a hydroplaning hazard.

In recent years, increased traffic levels (both volume and load) entailed the need to enhance performance of used asphaltic materials. In addition, both a better understanding of behavior and characteristics of binders and the greater development of technology have encouraged and enabled researchers to examine the benefits of introducing additives and modifiers into the asphalt [8,9]. Available modifiers fit into various categories (e.g., naturally occurring materials, industrial by-products and waste materials, and engineered products). Some of the most common categories include reclaimed rubber products, fillers, fibers, catalysts, polymers (natural and synthetic), and extenders. Among them, a blend of asphalt with polymer is the most often currently used to improve performances of asphalt (and asphalt mixes) [10,11]. Polymer-modified asphalt has been used for many years with mixes and its usage will probably increase in the immediate future. Polymer-modified asphalt improves resistance to rutting, abrasion, cracking, fatigue, stripping, bleeding, and aging at high temperatures, and flexibility at low temperatures [10]. In addition, the structural thickness of asphalt pavement could be reduced. According to some research [12], different types of polymers can be mixed with asphalt: there is no universally superior polymer type and therefore its selection depends on specific needs. Moreover, its effectiveness depends on polymer characteristics, polymer content, and the nature of the asphalt.

In addition to traditional modifiers such as polymers, in recent years various alternative materials have been considered. Particularly, the emergence of nano-technologies has motivated a number of researchers in evaluating the use of nano-materials for such a purpose [13]. Nanotechnology is the study of the control of matter on an atomic and molecular scale: it deals with structures of the size 100 nm or smaller and involves developing materials or devices within that size. The application of nanomaterial technology in asphalt mixtures is a relatively new topic; it has rapidly evolved since the buckminsterfullerene discovery [14]. Nanotechnology allows us to create new materials and devices to be used in many fields of science and applied technology [15]. In the road pavement sector, due to their mechanical properties and large surface area to volume ratio, carbon nanotubes (CNTs) [16] and nanoclays (NCs) are some of the most promising modifiers [17–19]. According to Steyn [20], nanotechnology can play a role in improving existing and available materials to enhance the mechanical characteristics of asphalt binders. Several studies have been carried out on the capability of nano-sized particles to improve rheological characteristics of bitumens, while limited works have focused on the effects on fatigue and healing properties. Khattak et al. [21] demonstrated that carbon nano-fibers could enhance the fatigue resistance of bituminous materials by means of crack bridging and pull-out mechanisms. Santagata et al. [22] showed that the adoption of a proper dispersion of carbon nanotubes in bituminous mastics could have effect against cracking. According to Liu et al. and Wu et al. [23,24], both chemical characteristics of the nanoclay surfactant and an adequate interfacial interaction between bitumen and nanoclay particles can improve fatigue resistance.

However, aspects must be clarified concerning the costs–benefits of nano-reinforced materials and the industrial-scale implementation of bituminous mixtures with nano-additives' production. According to preliminary investigations, it is envisioned that nanotechnologies in the road sector may open undiscovered scenarios in the development of new smart, multifunctional, and high-performance products.

2. Research Methodology

In this study, the systematic literature review (SLR) defined by Kitchenham was performed [25]. It allows identification, analysis, and interpretation of available data about a research question, area, or investigated phenomenon. According to the evidence of reference materials, the goal of the study is to identify and analyze research into the use of nanoparticles as additives in bituminous conglomerates: manuscripts that contribute to SLR are primary studies, while this manuscript is a secondary study.

Peer-reviewed articles on the use of nanotechnology in road pavement, published between 2007 and 2018, were included. The main search strategy was automatic and involved the peer-reviewed databases Scopus, Web of Science, and Google Scholar. Furthermore, a manual search activity was performed as consequence of the results obtained in the automatic process and involved papers published since 2003. This activity involved both peer-reviewed, and not, documents. Finally, classical sources, standards, and regulations were considered, whatever their publication year.

Planning, conduction, and reporting results are the three main phases of the systematic literature review. They include:

1. Planning: Identification of the need which justifies the systematic literature review. Particularly, the research questions are:
    - What are the types of nano-additives considered, and the methods and technological solutions investigated at international level?
    - What are the performance improvements of bituminous mixtures additives with nanoparticles?
    - What is the extent of the self-repairing capacity of bituminous mixtures with the addition of nanoparticles compared to those without additives?
2. Conduction: Implementation of a search strategy compliant with the protocol defined in the previous phase;
3. Reporting results: Description of the results, answers to the goal of the study, and discussion of the results.

After the final application of the work selection strategy, 81 documents were identified: 69 are primary studies (i.e., peer-reviewed indexed research papers), 4 are secondary studies (i.e., reviews), and 8 are classical sources, standards, and regulations.

These works allowed the authors to make a critical assessment of the state of the art and answer the research questions.

3. Materials and Methods

3.1. Materials

3.1.1. Nanoclay

Nanoclays are hydrated aluminosilicates belonging to the class of phyllosilicates such as montmorillonite or caolinite. The nanoclays are composed of nanometric flakes with high specific surface area and aspect ratio: their dispersion in a polymeric matrix gives the nanocomposite increased barrier properties and greater resistance [26]. In general, nanoclays are able to modify profoundly the rheological properties of different materials. Extensive research has been dedicated to the use of nanoclay to reinforce asphalt binders. Although some types of nanoclay did not affect the stiffness or viscosity of the bitumen, other types of nanoclay showed encouraging results. Various physical properties such as stiffness, tensile strength, tension module, flexural strength, and bitumen thermal stability modulus can be improved when a small amount of nanoclay is microscopically dispersed. Generally, the elastic modulus increases for the modified bitumen with the added nanoclay, while the dissipation of the mechanical energy with respect to the unmodified bitumen is lower [26]. Jahromi and Khodaii [26] used bentonite clay (BT) and a chemically modified bentonite (OBT) to modify asphalt binders using the mixing process "sonication and shearing stresses". The modified asphalts had a higher resistance to rutting as well as a significant improvement in behavior at low temperatures with an increase in cracking resistance [27,28]. The improvement in terms of stiffness and resistance of the modified bitumen depends on the temperature and the percentage of added nanoclay. Nanoclay is also used as a second additive to improve the performance of bitumen modified with styrene–butiadene–styrene (SBS) [29–31].

Recent studies have shown an increase in the rutting trigger factor according to Superpave (SUperior PERforming asphalt PAVEments) standard [32], while the rotational viscosity tests indicated a significant increase in viscosity [33,34]. Among other benefits, nanoclays have also shown the ability to improve the resistance to aging of asphalt mixtures [35,36].

3.1.2. Nanosilica

Silica nanoparticles are used in medicine and pharmaceutical industries [37] for diagnostics and treatment of many pathologies, in industry to reinforce elastomers as a rheological solute [38], and in cement mixtures [39]. Similar to nanoclay, they have low production costs and high performance. Indeed, asphalt binders modified with nanosilicates have a lower viscosity value that reduces the compaction temperature or the energy dispersion during the construction process. The addition of nanosilica improves the recovery capacity of asphalt binders and enhances anti-aging capabilities, cracking resistance, resistance against rutting, and anti-stripping properties [27]. On the other hand, the addition of nanosilica does not influence the low-temperature properties of the modified bitumen [31].

According to Yusoff et al. [11], nanosilica could reduce susceptibility to moisture damage and increase resistance to fatigue and rutting of bituminous binders [11].

3.1.3. Carbon NanoTube

Under specific conditions, carbon atoms make up spherical structures, the fellurenes whose structure, after a subsequent relaxation, tends to roll up on itself, resulting in the typical cylindrical structure of carbon nanotubes [16]. Nanotubes are commonly divided into two types:

- single-walled nanotubes or SWCNTs (single-walled carbon nanotubes): consisting of a single graphite sheet wrapped around itself;
- multi-walled nanotubes or MWCNTs (multi-walled carbon nanotubes): formed by multiple sheets coaxially wound one on the other.

The high ratio between the length and diameter of carbon nanotubes gives them mechanical properties that are superior to those of other construction materials. When CNTs are added to bituminous binder mixtures with a sufficient percentage (>1% by bitumen weight), they can have significant effects on their rheological properties [21,40]. With the addition of CNTs to asphalt, an increase in adhesive strength was noted as well as increased susceptibility to moisture [41]. Moreover, the addition of CNTS produces positive effects on fatigue and rutting resistance compared to the not-additivated bituminous binders. [42,43]. Furthermore, the susceptibility to thermal cracking [35] and oxidative aging is reduced in bituminous mixtures [22].

3.1.4. Graphene Nanoplatelets

Graphene nanoplatelets (GNPs) consist of stacks of graphene sheets (Figure 1) (material consisting of a monoatomic layer of hexagonally arranged carbon atoms) that can be characterized by diameter of sub-micrometers and thickness of the order of nanometers. Graphene nanoplatelets exhibit superior mechanical properties if compared to nanoclay and nanosilica. They have the advantage of having a lower cost compared to CNTs. Having regard to cement mortar, GNP could effectively improve the electrical conductivity and reduce the critical pore diameter of cement mortars [44] with a significant impact on its durability [44]. Experiments have also shown that the addition of GNP would increase mechanical performances of cement mortars and contribute to the self-healing properties of bituminous mixtures [44,45].

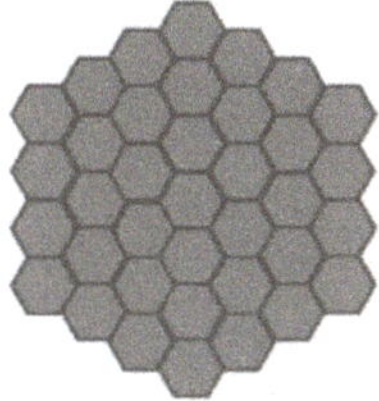

Figure 1. Graphene nanoplatelet (GNP) sheet.

3.1.5. Nano-Oxides

This category includes a wide variety of spherical or aggregate forms of nanoparticles. It includes metal oxides, semiconductors, and ultrafine inorganic compounds, produced through various chemical processes or through pyrolytic processes and the recovery of combustion waste in industrial processes. It is a type of particle whose dimensions are quite variable, particularly suitable for production in large quantities and on an industrial scale compared to other types [46].

Nano-oxides have greater thermal and chemical resistance and mechanical stability; greater resistance to atmospheric agents and greater resistance to aging.

In the infrastructure sector, mainly calcium oxide and titanium dioxide are used.

Nano-Calcium Oxide (CaO)

According to recent studies [47,48], the addition of nano-calcium oxide (CaO) in bituminous binder produces:

- A reduction of the penetration value by 7%, which directly relates to resistance to high temperature [47];
- An increase of softening point's value by 45% with resilience modulus value increased by 1.7 times that of neat bitumen [47];
- An improvement of bitumen's characteristics in colder regions to avoid thermal cracking when added at 4% and 6% of weight of bitumen [48].

Nano-Titanium Dioxide (TiO_2)

According to recent studies [49–51], the addition of titanium dioxide (TiO_2) to a bituminous binder can:

- improve fatigue resistance, permanent deformation, and oxidative aging of the binder [49–51];
- in association with other modifiers, such as polymers, it can improve the softening point and the ductility of the binder [52];
- have the ability to remove air pollutants [53];
- degrade most pollutants from automobile exhaust [54];
- improve creep comportment and prevent vertical cracks (added in 5% of bitumen mass) [49];
- improve fatigue life and flexural stiffness [55].

3.2. Methods

In the literature, the presented additives have been added to a bituminous base compliant with AASHTO M 320 [56]. The added percentage by weight of bitumen varies between 0.1% and 1% for CNTs and GNP, and between 3% and 6% for nanoclays and nanosilica; in any case dosage of nanoclays and nanosilica is higher than that of CNTs [22,57–60]. The added percentage by weight of bitumen varies between 4% and 6% for nano-calcium oxide [47] and between 1% and 7% for nano-titanium dioxide. In the latter case, for many authors the optimum content of TiO_2 is 5% of binder mass [61].

Both the dosage of added nano-additive and the mixing techniques used for preparing the mixture influence the bitumen's behavior [62,63].

3.2.1. Mixing Techniques

Usually two mixing techniques are used:

- The first is based on a simple shear mixing procedure [22,57,64,65] and consists of two phases. In the first phase, the nano-additives are manually added to the bitumen. A second phase follows this (pre-mixing) one; a mechanical stirrer (Figure 2) with heating mixes at 1550 rpm and a constant temperature of 150 °C the additivated bitumen for 90 min. This technique is not only the most convenient to use in laboratory, but is easily transferable on industrial scale in hot-mix asphalt plants.

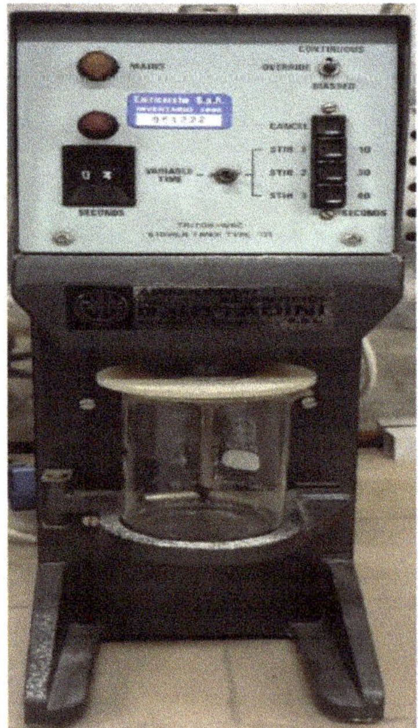

Figure 2. Mechanical stirrer.

- The second mixing procedure consists of three phases: a third phase (i.e., ultrasonic sonication) is added to those of the first procedure. In one research project [63], a UP 200S ultrasonic homogenizer (200 W and 24 kHz) equipped with a titanium cylindrical sonotrode (7 mm diameter) was used. When it was immersed in the fluid mixture at a constant temperature of 150 °C, the generated ultrasounds propagated inside the material. The transmitted compression waves allowed separation of individual nanoparticles from the existing agglomerations and ensured a greater homogeneity of dispersion [60]. In order to improve the dispersion, several researchers tested duration and amplitude of the waves during sonication [27,66,67]. Finally, the effects of sonication on the distribution of nanoparticles were evaluated directly by microscope or indirectly through rheometric methods. Particularly, the storage modulus was sensitive to dispersion of nanoparticles: it increases with increasing energy spent during homogenization (both sonication duration and wave amplitude [68]).

The rotation rate and the power of the mixer were varied for high shear, mechanical, and ultrasonic mixers (Figure 3) [69].

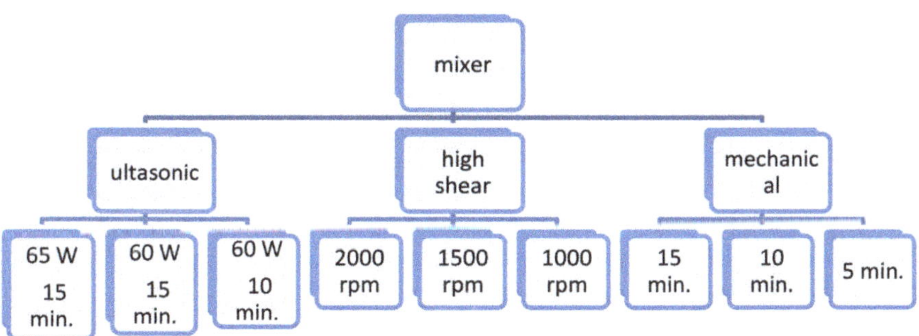

Figure 3. Alternative mixing techniques reproduced from [69].

3.2.2. Testing Program

To study the distribution of the various nanocomposites in the bituminous matrix, researchers have commonly used atomic force microscopy (AFM), X-ray diffraction spectroscopy (XRD), scanning electron microscopy (SEM), and tunneling electron microscopy (TEM). These tests are also useful to understand the effect of nanocomposites on the mechanical and rheological properties of the modified bitumen [70].

Having regard to the three critical phases in bitumen life (i.e., storage and transport; mixture production and laying, and aging), laboratory tests analyzed non-aged short-lived and long-lasting specimens [71,72]. The carried out laboratory tests concerned:

1. Viscosity tests according to AASHTO T316-04 [73].
2. Through the use of the DSR (dynamic shear rheometer), OSL (oscillatory shear loading) and MSCR (multiple stress creep recovering) tests were carried out, increasing stresses from 0.1 to 3.2 kPa and different temperature values according to AASHTO TP 70-10 [74]. By applying a shear stress, the equipment measures the binder's response in shear-cut terminations. Through the relationship between the applied cutting force and the obtained deformation, the rheometer allows us to obtain the following parameters:
    - G^* (complex modulus);
    - δ (phase angle);
    - $G^*/\sin\delta$ (parameter correlated to rutting);
    - $G^* \sin\delta$ (parameter correlated to fatigue).
3. Fatigue tests according to AASHTO T 315 [75].

4. Self-repair tests performed by means of cyclic load tests of the time sweep type interrupted by multiple rest periods; the test ends when a reduction in initial dissipation energy of 5%, 10%, 30%, and 50% is reached.

4. Conclusions

This document presents a review of the nano-additives most commonly used to modify bitumen: it analyzes and compares their mixing conditions and their influence on the mechanical characteristics of the binder. Performances of bitumen additivated with nanoclays, nanosilica, carbon nanotubes, graphene nanoplatelets, and nano-oxides were considered having regard to different distresses (i.e., fatigue resistance, rutting, and self-healing processes).

Given the results in the literature, simple shear mixing technique is potentially more easily transferable to the industrial production of hot bituminous mixes, while the dispersion of nanoparticles in the bituminous matrix improves through the ultrasonic mixing technique.

In fact, with reference to storage modulus that is highly sensitive to nanoparticle dispersion and interfacial interaction, by increasing the energy input for homogenization, either by extending sonication duration or by expanding wave amplitude, the storage modulus increases. It can be seen that similar trends were recorded regardless of the considered additive type. This is also supported by the results obtained under the microscope and by other characterization techniques such as AFM, TEM, SEM [68–70].

The addition of a sufficient amount of CNTs (>0.5% by weight of bitumen) significantly increases the stiffness and elasticity of the bituminous base at low frequencies and high temperatures: it could result in a potential improvement in resistance to bending. Binders containing CNT have revealed a high sensitivity to the level of damage. In addition, there is a different fatigue behavior depending on the mixing technique adopted to disperse the CNTs in the bituminous matrix; an improvement is noted with the increase in sonication times or with the modification of the amplitude of the ultrasonic waves.

As indicated by the upward displacement of the corresponding curves τ–NDERmax [59], the mixtures with NC have shown better performance in terms of fatigue than net bitumen in the entire spectrum of loading and damage conditions simulated in laboratory [76–79].

CNT helps in improving tensile strength, flexural strength, and rutting resistance, and reduces thermal cracking.

In terms of self-repair, the recoverable damage component depends on both the load history and the type of bitumen considered. By increasing the degree of damage suffered by the sample, the repair potential tends to decrease; on the other hand, materials containing nano-additives show a higher recovery component than that of net bitumen.

In particular, the higher viscosity of the admixed mixtures can delay the process of formation of the surface cracks which represents the initial step for the self-repair process [64,65,80].

While NCs improve the self-repairing ability of net bitumen when limited damage occurs, CNTs make an even more significant contribution after high-load levels.

Nano-TiO_2/SiO_2 have also been reported to be used as an additive to improve the rheological properties of conventional bitumen [4]. The nano-modified bitumen showed improved adhesive bonding of aggregate, better interlock between aggregates, reduced deformation, improved fatigue life of pavement, low phase angle value, high complex modulus value at low temperatures, and high rutting resistance [4].

The test results indicate that modified asphalt binders show an increase in the complex modulus and a decrease in the phase angle compared to unmodified asphalt binders.

The phase angles of both unmodified and nano-modified asphalt increase as the temperature increases.

In comparison, the nano-modified asphalt was less sensitive to temperature changes. In other words, the modified asphalt binders demonstrate a higher ability to maintain elastic/viscous capability than the unmodified asphalt [4].

Bitumen composites synthesized with nano-additives exhibit improved properties in hotter and colder regions [70].

In conclusion, the efficacy as bitumen modifiers of additives of nanometric dimensions strongly depends both on the volume within the mixtures (due to a simple filling effect) and on the interactions that can arise with the continuous bituminous matrix (depending on the surface specification and compatibility).

The use of nano-additives for bituminous binders promises a series of advantages, with particular attention to bitumen durability. But their use in bituminous mixes still has many aspects to be clarified and optimized. The first problem is the lack of univocal procedures that makes it difficult to compare data coming from different laboratories [81].

Moreover, according to the opinion of several researchers, the interactions at nanoscale can lead to a new generation of bituminous nanocomposites with tailored chemical–physical properties.

Other specific aspects of binder behavior should be subjected to analysis, possibly by introducing in the evaluation a cost–benefit analysis in order to stimulate applications at the industrial scale. In addition, the possibility of creating a bituminous binder with nano-additives and 100% recycled aggregates should be considered in order to obtain high-performance mixtures and reduce environmental impacts and maintenance costs.

Author Contributions: Conceptualization, A.D. and N.F.; methodology, N.F. and R.F.; investigation, R.F. and N.F.; visualization, L.M.; supervision, L.M.; A.D. and N.F. All authors have read and agreed to the published version of the manuscript.

Funding: This research received no external funding.

Acknowledgments: The authors thank DICEA's road materials laboratory (Sapienza University of Rome) for Figure 2.

Conflicts of Interest: The authors declare no conflict of interest.

References

1. Jahromi, S.G.; Khodaii, A. Effects of nanoclay on rheological properties of bitumen binder. *Constr. Build. Mater.* **2009**, *23*, 2894–2904. [CrossRef]
2. Whiteoak, D.; Read, J.; Hunter, R. *The Shell Bitumen Handbook*, 5th ed.; Thomas Telford Publishing: London, UK, 2003.
3. Canestrari, F.; Cardone, F.; Graziani, A.; Santagata, F.A.; Bahia, H.U. Adhesive and cohesive properties of asphalt-aggregate system subjected to moisture damage. *Road Mater. Pavement* **2010**, *11*, 11–32. [CrossRef]
4. Shafabakhsh, G.H.; Ani, O.J. Experimental investigation of effect of Nano TiO_2/SiO_2 modified bitumen on the rutting and fatigue performance of asphalt mixtures containing steel slag aggregates. *Constr. Build. Mater.* **2015**, *98*, 692–702. [CrossRef]
5. Ameri, M.; Kouchaki, S.; Roshani, H. Laboratory evaluation of the effect of nano-organosilane anti-stripping additive on the moisture susceptibility of HMA mixtures under freeze–thaw cycles. *Constr. Build. Mater.* **2013**, *48*, 1009–1016. [CrossRef]
6. Kiggundu, B.M.; Roberts, F.L. Stripping in HMA mixtures: State-of-the-art and critical review of test methods. In *National Center for Asphalt Technology*; Report No. 88–02; Auburn University: Auburn, GA, USA, 1988.
7. Ameri, M.; Nowbakht, S.H.; Molayem, M.; Aliha, M.R.M. Investigation of fatigue and fracture properties of asphalt mixtures modified with carbon nanotubes. *Fatigue Fract. Eng. Mater. Struct.* **2016**, *39*, 896–906. [CrossRef]
8. Bahia, H.U.; Hanson, D.I.; Zeng, M.; Zhai, H.; Khatri, M.A.; Anderson, R.M. Characterization of modified asphalt binders in Superpave mix design. In *NCHRP, Report 459*; National Cooperative Highway Research Program: Washington, DC, USA, 2001.
9. Santagata, E.; Baglieri, O. Experimental evaluation of modified bituminous binders for heavy duty applications. In Proceedings of the 3rd International SIIV Congress, Bari, Italy, 22–24 September 2005.
10. Steyn, W.J.; Bosman, T.E.; Galle, S.; Van Heerden, J. Evaluating the properties of bitumen stabilized with carbon nanotubes. *Open J. Adv. Mater. Res.* **2013**, *723*, 312–319. [CrossRef]

11. Yusoff, N.I.M.; Breem, A.A.S.; Alattug, H.N.M.; Hamima, A.; Ahmad, J. The effects of moisture susceptibility and ageing conditions on nano-silica/polymer-modified asphalt mixtures. *Constr. Build. Mater.* **2014**, *72*, 139–147. [CrossRef]
12. Isacsson, U.; Lu, X. Testing and appraisal of polymer modified road bitumen-state of art. *Mater. Struct.* **1995**, *28*, 139–159. [CrossRef]
13. Bergman, C.P.; de Andrade, M.J. *Nanostructured Materials for Engineering Applications;* Springer: Berlin/Heidelberg, Germany, 2011.
14. Kroto, H.; Heath, J.; O'Brien, S.C.; Curl, R.F.; Smalley, R.E. C_{60}: Buckminsterfullerene. *Nature* **1985**, *318*, 162–163. [CrossRef]
15. Buzea, C.; Pacheco, I.I.; Robbie, K. Nanomaterials and nanoparticles: Sources and toxicity. *Biointerphases* **2007**, *2*, 17–71. [CrossRef]
16. Iijima, S. Helical microtubules of graphitic carbon. *Nature* **1991**, *354*, 56–58. [CrossRef]
17. Chong, K.P.; Larsen-Basse, J. Challenges in mechanics and materials research in the twenty-first century. *J. Mater. Civ. Eng.* **2005**, *17*, 241–245. [CrossRef]
18. Sobolev, K.; Gutierrez, M.F. How nanotechnology can change the concrete world. *Am. Ceram. Soc. Bull.* **2005**, *84*, 14–18.
19. Buehler, M.J.; Ackbarow, T. Fracture mechanics of protein materials. *Mater. Today Commun.* **2007**, *10*, 46–58. [CrossRef]
20. Steyn, W.J. Potential application of nanotechnology in pavement engineering. *J. Transp. Eng. ASCE* **2009**, *135*, 764–772. [CrossRef]
21. Khattak, M.J.; Khattab, A.; Rizvi, H.R.; Zhang, P. The impact of carbon nano-fiber modification on asphalt binder rheology. *Constr. Build. Mater.* **2012**, *30*, 257–264. [CrossRef]
22. Santagata, E.; Baglieri, O.; Tsantilis, L.; Dalmazzo, D. Rheological characterization of bituminous binders modified with carbon nanotubes. *Procedia Soc. Behav. Sci.* **2012**, *53*, 546–555. [CrossRef]
23. Liu, G.; Wu, S.; Van de Ven, M.F.C.; Molenaar, A.A.A.; Besamusca, J. Modification of bitumen with organic montmorillonite nanoclay. In Proceedings of the Third International Conference on Advances and Trends in Engineering Materials and Their Applications, Ottawa, ON, Canada, 6–10 July 2009.
24. Liu, G.; Wu, S.; Van de Ven, M.F.C.; Yu, J.; Molenaar, A.A.A. Influence of sodium monymorillonites on the properties of bitumen. *Appl. Clay Sci.* **2010**, *49*, 69–73. [CrossRef]
25. Kitchenham, B. *Procedures for Performing Systematic Reviews;* Keele University: Keele, UK, 2004; Volume 33, pp. 1–26.
26. Paul, D.R.; Robeson, L.M. Polymer nanotechnology: Nanocomposites. *Polymer* **2008**, *49*, 3187–3204. [CrossRef]
27. Zare-Shahabadi, A.; Shokuhfar, A.; Ebrahimi-Nejad, S. Preparation and rheological characterization of asphalt binders reinforced with layered silicate nanoparticles. *Constr. Build. Mater.* **2010**, *24*, 1239–1244. [CrossRef]
28. Ganesh, V.K. Nanotechnology in Civil Engineering. *Eur. Sci. J.* **2012**, *8*, 96–109.
29. Yu, J.; Zeng, X.; Wu, S.; Wang, L.; Liu, G. Preparation and properties of montmorillonite modified asphalts. *Mater. Sci. Eng. A* **2007**, *447*, 233–238. [CrossRef]
30. Polacco, G.; Krix, P.; Filippi, S.; Stastna, J.; Biondi, D.; Zanzotto, L. Rheological properties of asphalt/SBS/clay blends. *Eur. Polym. J.* **2008**, *44*, 3512–3521. [CrossRef]
31. Yao, H.; You, Z.; Li, L.; Shi, X.; Goh, S.W.; Mill-Beale, J.; Wingard, D. Performance of asphalt binder blended with non-modified and polymer-modified nanoclay. *Constr. Build. Mater.* **2012**, *35*, 159–170. [CrossRef]
32. Kennedy, T.W.; Huber, G.A.; Harrigan, E.T.; Cominsky, R.J.; Hughes, C.S.; Von Quintus, H.; Moulthrop, J.S. *Superior Performing Asphalt Pavements (Superpave);* The Product of the S.H.R.P; Asphalt Research Program; National Research Counciul: Washinghton, DC, USA, 1994.
33. Hossain, Z.; Zaman, M.; Saha, M.C.; Hawa, T. Evaluation of viscosity and rutting properties of nanoclay-modified asphalt binders. In Proceedings of the Geo-Congress 2014, Atlanta, GA, USA, 23–26 February 2014.
34. Abdelrahman, M.; Katti, D.R.; Ghavibazoo, A.; Upadhyay, H.B.; Katti, K.S. Engineering physical properties of asphalt binders through nanoclay-asphalt interactions. *J. Mater. Civ. Eng.* **2014**, *26*, 04014099. [CrossRef]

35. Yang, J.; Tighe, S. A review of advances of nanotechnology in asphalt mixtures. 13th COTA International Conference of Transportation Professionals (CICTP 2013). *Procedia Soc. Behav. Sci.* **2013**, *96*, 1269–1276. [CrossRef]
36. Mun, S.; Lee, H. Modeling viscoelastic crack growth in hot-mix asphalt concrete mixtures using a disk-shaped compact tension test. *J. Eng. Mech.* **2011**, *137*, 431–438. [CrossRef]
37. Barik, T.K.; Sahu, B.; Swain, V. Nanosilica-from medicine to pest control. *Parasitol. Res.* **2008**, *103*, 253–258. [CrossRef]
38. Chrissafis, K.; Paraskevopoulos, K.M.; Papageorgiou, G.Z.; Bikiaris, D.N. Thermal and dynamic mechanical behavior of bionanocomposities: Fumed silica nanoparticles dispersed in poly(vinyl pyrrolidone), chitosan, and poly(vinyl alcohol). *J. Appl. Polym.* **2008**, *110*, 1739–1749. [CrossRef]
39. Quercia Bianchi, G.; Brouwers, H.J.H. Application of nano-silica (nS) in concrete mixtures. In Proceedings of the 8th FIB PhD Symposium in Civil Engineering, Lyngby, Denmark, 20–23 June 2010.
40. Xiao, F.; Amirkhanian, A.N.; Amirkhanian, S.N. Influence on rheological characteristics of asphalt binders containing carbon nanoparticles. *J. Mater. Civ. Eng.* **2011**, *23*, 423–431. [CrossRef]
41. Al-Adham, K.; Arifuzzaman, M. Moisture damage evaluation in carbon nanotubes reinforced asphalts. In *Sustainability, Eco-Efficiency, and Conservation in Transportation Infrastructure Asset Management*, 1st ed.; Taylor and Francis Group: London, UK, 2014; pp. 103–109.
42. Amirkhanian, A.N.; Xiao, F.; Amirkhanian, S.N. Characterization of unaged asphalt binder modified with carbon nano particles. *Int. J. Pavement Res. Technol.* **2011**, *4*, 281–286.
43. Amirkhanian, A.N.; Xiao, F.; Amirkhanian, S.N. Evaluation of high temperature rheological characteristics of asphalt binder with carbon nano particles. *J. Test. Eval.* **2011**, *39*, 583–591.
44. Le, J.L.; Du, H.; Pang, S.D. Use of 2D graphene nanoplatelets (GNP) in cement composites for structural health evaluation. *Compos. Part B Eng.* **2014**, *67*, 555–563. [CrossRef]
45. Le, J.; Marasteanu, M.; Turos, M. Graphene nanoplatelet (GNP) reonforced asphalt mixtures: A novel multifunctional pavement material. In *NCHRP IDEA, Final Report Project 173*; TRB, University of Minnesota: Minneapolis, MN, USA, 2016.
46. Leone, M.F. *Alte Prestazioni ed Ecoefficienza: Nanotecnologie per L'evoluzione dei Materiali Cementizi*; Dottorato di Ricerca in Tecnologia dell'Architettura XXI Ciclo Scuola di Dottorato in Architettura [in Italian]; University of Naples "FedericoII": Naples, Italy, 2008.
47. Farag, K.; Abd-El-Sadek, M.S.; Hamdy, S.E. Mechanical properties of modified asphalt concrete mixtures using CA(OH)2 nanoparticles. *Int. J. Civ. Eng.* **2014**, *5*, 61–68.
48. Javad, T.; Atousa, K. Decreasing thermal cracking on asphalt pavement by nano calcium carbonate (CCN) Modified bitumen. *J. Des. Built Environ.* **2016**, 64–70. [CrossRef]
49. Shafabakhsh, G.H.; Mirabdolazimi, S.M.; Sadeghnejad, M. Evaluation the effect of nano-TiO_2 on the rutting and fatigue behavior of asphalt mixtures. *Constr. Build. Mater.* **2014**, *54*, 566–571. [CrossRef]
50. Tanzadeh, J.; Vahedi, F.; Kheiry, P.T.; Tanzadeh, R. Laboratory study on the effect of nano TiO_2 on rutting performance of asphalt pavements. *Adv. Mater. Res.* **2013**, *622*, 990–994.
51. Sun, Z.; Yi, J.; Huang, Y.; Feng, D.; Guo, C. Properties of asphalt binder modified by bio-oil derived from waste cooking oil. *Constr. Build. Mater.* **2016**, *102*, 496–504. [CrossRef]
52. Zhang, H.; Su, M.; Zhao, S.; Zhang, Y.; Zhang, Z. High and low temperature properties of nano-particles/polymer modified. *Constr. Build. Mater.* **2016**, *114*, 323–332. [CrossRef]
53. Hassan, M.; Mohammad, L.; Cooper, S.; Dylla, H. Evaluation of nano-titanium dioxide additive on asphalt binder aging properties. *Transp. Res. Rec. J. Transp. Res. Board* **2011**, *2207*, 11–15. [CrossRef]
54. Hu, C.; Ma, J.; Jiang, H.; Chen, Z.; Zhao, J. Evaluation of nano-TiO_2 modified waterborne epoxy resin as fog seal and exhaust degradation material in asphalt pavement. *J. Test. Eval.* **2017**, *45*, 260–267. [CrossRef]
55. Tanzadeh, J.; Tanzadeh, R.; Nazari, H. Fatigue evaluation of hot mix asphalt (HMA) mixtures modified by optimum percent of TiO_2 nanoparticles. *Adv. Eng. Forum* **2017**, *24*, 55–62. [CrossRef]
56. American Association of State Highway and Transportation Officials (AASHTO). AASHTO M 320. In *Standard Specification for Performance-Graded Asphalt Binder*; AASHTO: Washington, DC, USA, 2010.
57. Santagata, E.; Baglieri, O.; Tsantilis, L.; Chiappinelli, G. Effects of nano sized additives on the high temperature properties of bituminous binders: A comparative study. In *International RILEM Symposium on Multi-Scale Modeling and Characterization of Infrastructure Materials*; RILEM Bookseries; Springer: Dordrecht, The Netherlands; Holland, MI, USA, 2013; pp. 297–309.

58. Santagata, E.; Baglieri, O.; Tsantilis, L.; Chiappinelli, G. Fatigue and healing properties of nano-reinforced bituminous binders. *Int. J. Fatigue* **2015**, *80*, 30–39. [CrossRef]
59. Santagata, E.; Baglieri, O.; Tsantilis, L.; Chiappinelli, G. Fatigue properties of bituminous binders reinforced with carbon nanotubes. *Int. J. Pavement Eng.* **2015**, *16*, 80–90. [CrossRef]
60. Santagata, E.; Baglieri, O.; Tsantilis, L.; Chiappinelli, G. Storage stability of bituminous binders reinforced with nano-additives. In *8th RILEM International Symposium on Testing and Characterization of Sustainable and Innovative Bituminous Materials*; RILEM Bookseries; Springer: Basel, Switzerland, 2015; Volume 11, pp. 75–87.
61. Filho, P.G.T.M.; Rodrigues dos Santos, A.T.; Lucena, L.C.; Tenorio, E.A.G. Rheological evaluation of asphalt binder modified with nanoparticles of titanium dioxide. *Int. J. Civ. Eng.* **2020**, *18*, 1195–1207. [CrossRef]
62. Ma, P.C.; Siddiqui, N.A.; Marom, G.; Kim, J.K. Dispersion and functionalization of carbon nanotubes for polymer-based nanpcomposities: A review. *Compos. Part A* **2010**, *41*, 1345–1367. [CrossRef]
63. Planellas, M.; Sacristan, M.; Rey, L.; Olmo, C.; Aymamì, J.; Casa, M.T.; del Valle, L.J.; Franco, L.; Puiggalì, J. Micro-molding with ultrasonic vibration energy: New method to disperse nanoclays in polymer matrices. *Ultrason. Sonochem.* **2014**, *21*, 1557–1569. [CrossRef]
64. Santagata, E.; Baglieri, O.; Dalmazzo, D.; Tsantilis, L. Damage and healing test protocols for the evaluation of bituminous binders. In Proceedings of the 5th Eurasphalt & Eurobitume Congress 2012a, Istanbul, Turkey, 13–15 June 2012.
65. Santagata, E.; Baglieri, O.; Tsantilis, L.; Dalmazzo, D. Evaluation of self healing properties of bituminous binders taking into account steric hardening effects. *Constr. Build. Mater.* **2013**, *41*, 60–67. [CrossRef]
66. Khattak, M.J.; Khattab, A.; Rizvi, H.R. Characterization of carbon nano-fiber modified hot mix asphalt mixtures. *Constr. Build. Mater.* **2013**, *40*, 738–745. [CrossRef]
67. Khattak, M.J.; Khattab, A.; Zhang, P.; Rizvi, H.R.; Pesacreta, T. Microstructure and fracture morphology of carbon nano-fiber modified asphalt and hot mix asphalt mixtures. *Mater. Struct.* **2013**, *46*, 2045–2057. [CrossRef]
68. Santagata, E.; Baglieri, O.; Tsantilis, L.; Chiappinelli, G.; Brignone Aimonetto, I. Effect of sonication on the high temperature properties of bituminous binders reinforced with nano-additives. *Constr. Build. Mater.* **2015**, *75*, 395–403. [CrossRef]
69. Ziari, H.; Rahim-of, K.; Fazilati, M.; Goli, A.; Farahani, H. Evaluation of different conditions on the mixing bitumen and carbon nano-tubes. *Int. J. Civ. Environ. Eng.* **2012**, *12*, 53–59.
70. Deepa, P.; Laad, M.; Singh, R. An overview of use of nanoadditives in enhancing the properties of pavement construction binder bitumen. *Word J. Eng.* **2019**, *16*, 132–137.
71. American Association of State Highway and Transportation Officials (AASHTO). AASHTO T240-09-UL. In *Test for Effect of Heat and Air on a Moving Film of Asphalt (Rolling Thin-Film Oven Test)*; AASHTO: Washington, DC, USA, 2009.
72. American Association of State Highway and Transportation Officials (AASHTO). AASHTO R028-09-UL. In *Standard Practice for Accelerated Aging of Asphalt Binder Using a Pressurized Aging Vessel (PAV)*; AASHTO: Washington, DC, USA, 2009.
73. American Association of State Highway and Transportation Officials (AASHTO). AASHTO T316-04. In *Standard Method of Test for Viscosity Determination of Asphalt Binder Using Rotational Viscometer*; AASHTO: Washington, DC, USA, 2010.
74. American Association of State Highway and Transportation Officials (AASHTO). AASHTO TP 70-10. In *Standard Method of Test for Multiple Stress Creep Recovery (MSCR) Test of Asphalt Binder Using a Dynamic Shear Rheometer (DSR)*; AASHTO: Washington, DC, USA, 2010.
75. American Association of State Highway and Transportation Officials (AASHTO). AASHTO T 315. In *Standard Method of Test for Determining the Rheological Properties of Asphalt Binder Using a Dynamic Shear Rheometer (DSR)*; AASHTO: Washington, DC, USA, 2010.
76. Santagata, E.; Baglieri, O.; Tsantilis, L.; Dalmazzo, D.; Chiappinelli, G. Fatigue and healing properties of bituminous mastics reinforced with nano-sized additives. *Mech. Time-Depend. Mater.* **2016**, *20*, 367–387. [CrossRef]
77. Airey, G.D.; Thom, N.H.; Osman, H.; Collop, A.C. A comparison of bitumen/mastic fatigue data from different test methods. In Proceedings of the 5th International RILEM Conference on Reflective Cracking in Pavements, Limoges, France, 5–8 May 2004; pp. 383–390.

78. Santagata, E.; Baglieri, O.; Dalmazzo, D. Experimental investigation on the fatigue damage behaviour of modified bituminous binders and mastics. *J. Assoc. Asph. Paving* **2008**, *77*, 851–883.
79. Planche, J.P.; Anderson, D.A.; Gauthier, G.; Le Hir, Y.M.; Martin, D. Evaluation of fatigue properties of bituminous binders. *Mater. Struct. Mater. Constr.* **2004**, *37*, 356–359. [CrossRef]
80. Santagata, E.; Baglieri, O.; Dalmazzo, D.; Tsantilis, L. Rheological and chemical investigation on the damage and healing properties of bituminous binders. *J. Assoc. Asph. Paving* **2009**, *78*, 567–595.
81. Filippi, S.; Cappello, M.; Merce, M.; Polacco, G. Effects of nanoadditives on bitumen aging resistance: A critical review. *J. Nanomater.* **2018**, 1–17. [CrossRef]

Article

Ageing and Cooling of Hot-Mix-Asphalt during Hauling and Paving—A Laboratory and Site Study

Edoardo Bocci [1,*], Emiliano Prosperi [2], Volkmar Mair [3] and Maurizio Bocci [2]

1 Faculty of Engineering, Università degli Studi eCampus, 22060 Novedrate, Italy
2 Department of Construction, Civil Engineering and Architecture, Università Politecnica delle Marche, 60131 Ancona, Italy; e.prosperi@pm.univpm.it (E.P.); m.bocci@univpm.it (M.B.)
3 Ufficio Geologia e Prove Materiali, Autonomous Province of Bolzano, 39053 Cardano, Italy; volkmar.mair@provincia.bz.it
* Correspondence: edoardo.bocci@uniecampus.it

Received: 12 August 2020; Accepted: 15 October 2020; Published: 17 October 2020

Abstract: In road construction, it can happen that, for different reasons, the time between hot-mix asphalt (HMA) production and paving is extended to some hours. This can be reflected in several problems such as mix cooling and temperature segregation, but also in an extremely severe bitumen ageing due to its prolonged exposure to high temperatures. This paper deals with the investigation of these phenomena both in the laboratory and on site. In particular, the first part of the research aimed at observing the influence of the conditioning time, when the loose HMA is kept in the oven at a high temperature, on the mix properties. The second part focused on the ageing/cooling that happens on site during HMA hauling, as a function of time and type of truck. Temperatures were monitored using a thermal camera and different probes, and gyratory compactor specimens were produced by sampling some HMA from the trucks every 1 h for 3 h. The results showed that HMA stiffness rises if the time when the loose mix stays in the laboratory oven before compaction increases. However, on site, the HMA volumetric and mechanical properties do not change with hauling time up to 3 h, probably because the external material in the truck bed protects the HMA core from the access of oxygen, hindering bitumen oxidation and loss of volatiles. The temperature monitoring highlighted that temperature segregation, after 3 h hauling, can be higher than 30 °C but it can be reduced using insulated truck beds.

Keywords: hot-mix asphalt; ageing; cooling; temperature segregation; hauling; insulated truck; re-heating

1. Introduction

(HMA) is a composite material used extensively worldwide for the construction of pavements and it consists of aggregates and bitumen. The nature of the components and the production procedure of HMA, as well as the construction techniques and the environmental conditions during hauling and paving, influence the volumetric and mechanical properties of flexible layers and, hence, the pavement service life [1,2].

One of the factors that mainly affect HMA performance is bitumen ageing. At a chemical level, two mechanisms mainly govern the ageing process of bitumen: oxidation reaction and loss of volatiles [3–5]. The oxidation reaction refers to the introduction of oxygen-rich functional groups on the asphaltene molecules, determining the generation of strong intermolecular interactions and the formation of asphaltene clusters [6–8]. The loss of volatiles entails the decrease in the bitumen aromatic content, which causes the destabilization of the asphaltenes and intensifies their agglomeration [9–11]. In the components saturates, aromatics, resins and asphaltenes (SARA), these phenomena determine, in percentage, the reduction in the aromatics and the increase in the resins (which in turn generate asphaltenes), whereas saturates basically do not change, because of their poor reactivity [12]. As bitumen viscosity is strictly related to the asphaltene content, the effect of oxidation clearly reflects on the stiff and hard behavior, the low adhesive properties and the poor coating ability [13–15]. From an engineering point of view, two steps can be identified in the ageing process, namely the short-term ageing, which happens during the HMA production and paving, and long-term ageing, which gradually happens during pavement service life.

Temperature plays a fundamental role in bitumen ageing, as it can accelerate the chemical modifications. This is particularly important in the short-term step [16]. For this reason, during HMA manufacturing, it is fundamental to avoid bitumen overheating, which means containing the temperatures in the bitumen storage silo and inside the drum.

Moreover, both in the laboratory and in real road construction, HMA is sometimes kept at a high temperature for a long time. On site, this issue is typically associated with long hauling distances or to logistical problems during paving operations (paver running and HMA supply can be badly synchronized). In the laboratory, this happens, for example, when big HMA batches need to be heated (large volumes require a long time for the heat to reach the inner part of the material) or when many specimens have to be compacted from the same HMA batch (the material for the last specimen stays in the oven for a longer time compared to the first). If the HMA stays at a high temperature for an extended time, this can be reflected in a more severe short-term ageing [17,18]. The phenomenon is even more magnified if the HMA is left cooling and later reheated [19]. Kidd et al. [20] observed that the reheating of loose asphalt mixture results in changes to mix properties, particularly in the linear viscoelastic field (resilient and flexural modulus). Daniel [21] compared HMA specimens compacted at the plant or in the laboratory (after reheating) and noted that the latter have a significantly higher stiffness and brittleness, as a result of the double short-term ageing.

For HMA hauling, two types of truck can be used, equipped with ordinary or insulated beds [22]. In the ordinary truck, heat is transferred to the surrounding environment by convection and radiation, thus the HMA temperature decreases, particularly in correspondence with the material surface (generating a cool thin crust) and on the bed edges and corners. Conversely, insulated trucks have a protected bed that prevents heat loss and are typically used when HMA has to be delivered to far away paving locations. The most common insulating materials used for the protection of the truck bed are rock wool, plywood or wood of a different nature [22,23].

Therefore, in the case of ordinary trucks, the main concern that must be managed is represented by HMA cooling, which can interest the entire material volume in the truck bed or only some parts (the so-called temperature segregation) [24]. Several researchers demonstrated that temperature decreases determine a significant reduction in HMA compactability during paving operations [25–28]. In particular, it is essential to complete the compaction process before the temperature of HMA falls below a certain point known as cessation temperature, or a further increase in HMA density cannot be achieved despite repetitive passes of rollers [29,30]. This is reflected in a higher risk of moisture damage, raveling, rutting and cracking [24,31,32].

In the case of insulated trucks, the main issue is the extended time during which the material is kept at a high temperature, which can determine an extra short-term ageing. This can also cause a lack of HMA compactability, because of the increased bitumen viscosity, but in particular it determines high stiffness and brittleness, leading to premature fatigue and thermal cracking [33].

In light of these matters, the present research investigates the influence of the time during which HMA is kept at high temperature, both in the truck bed or in the laboratory oven, and the influence of the truck type on HMA ageing (or cooling). The aim is to help the scientific community to understand the beneficial (temperature preservation and reduction in segregation?) and detrimental (extra-ageing?) effects related to the use of insulated trucks, in comparison with the ordinary trucks and the lab-simulated conditions.

2. Objectives and Experimental Program

2.1. Research Purposes

The purposes of this research program were multiple. The first part of the experimentation was carried out in the laboratory and aimed to study the evolution of the HMA volumetric and mechanical characteristics with a growing conditioning time of the loose mixture in the oven. Four HMAs were investigated in particular, including different amounts of Reclaimed Asphalt Pavement—RAP (0%, 25% and 40%) and produced through different rejuvenating techniques (indicated with the codes R1 and R2). After mixing, the HMAs were left in the oven at 180 °C (which is the typical temperature of HMA with a polymer-modified binder when exiting from the production plant) for different times (from 30 to 180 min) before specimen compaction in order to observe the influence of conditioning time on air voids content and indirect tensile stiffness modulus. Table 1 summarizes the tested mixes.

Table 1. Experimental program for the evaluation of the thermal conditioning in the oven.

Mix	RAP Content [%]	Rejuvenating Technique	Conditioning Time [min]	Conditioning Temperature [°C]
0RAP	0	-	30, 60, 90, 120, 180	180
25RAP-R1	25	R1	30, 60, 90	180
25RAP-R2	25	R2	30, 60, 90	180
40RAP-R2	40	R2	30, 60, 90	180

The second part of the experimental program, including field and laboratory tests, had a two-fold objective:

(1) investigating the ageing that happens during HMA hauling as a function of time and type of truck and
(2) highlighting the issues related to temperature segregation of HMA during paving operations.

Four different batch plants (A, B, C and D) operating in the Autonomous Province of Bolzano (northern Italy) were selected to produce HMA with polymer-modified bitumen (PMB). The loose mixtures were dumped into two different kinds of trucks (normal and insulated) which had travelled around the nearby roads for 3 h, in order to simulate the hauling phase. Every 1 h, they came back to the plant and the temperature of the HMA was evaluated using probe and infrared thermometers. In addition, an infrared thermal camera was used to check the temperature segregation on the surface of the loose mixture inside the truck bed (Figure 1a). During each stop, some HMA was sampled from the truck bed at a depth of 50 cm and compacted using a shear gyratory compactor (SGC) providing a 100 mm diameter and 100 revolutions according to EN 12697-31 [34]. Finally, the HMA was laid down using a paving machine (Figure 1b). During this phase, temperatures were checked in order to evaluate any temperature segregation in the road surface. After a few days, cores were taken out from the pavement for laboratory testing. For plant A, the loose mix was also sampled during paving, left cooling, re-heated in the laboratory at 170 °C for 150 min (heating the loose mix for 60 min, separating it into 1200 g samples and further heating for 90 min) and compacted using SGC. Table 2 reports the dataset available for each batch plant. On the specimens and cores, air voids content, indirect tensile stiffness modulus, indirect tensile strength and cracking tolerance index were determined.

Table 2. Dataset available for each hot-mix asphalt (HMA) plant.

		Plant A	Plant B	Plant C	Plant D
Truck	Normal	X	X	X	X
	Insulated	X	X		
Specimens compacted on site	After 0 h hauling	X	X	X	X
	After 1 h hauling	X	X	X	X
	After 2 h hauling	X	X	X	X
	After 3 h hauling	X	X	X	X
Cores		X		X	X
Specimens compacted in the lab after reheating		X			

(a)

(b)

Figure 1. Pictures from the field investigation: (**a**) temperature measurement; (**b**) HMA delivery to the paver.

2.2. *Test Methods*

2.2.1. Volumetric Analysis

The volumetric characteristics of the HMA mixtures were evaluated in terms of air voids content according to EN 12697-8 [35]. In particular, the air voids content v_m was calculated through Equation (1).

$$v_m[\%] = 100 \times \frac{\rho_m - \rho_b}{\rho_m} \quad (1)$$

where ρ_m is the maximum density and ρ_b is the bulk density. The maximum density ρ_m was calculated according to the EN 12697-5—mathematic method [36], considering the real aggregate and bitumen percentage contents (measured through bitumen extraction) and assuming the densities of 2.70 mg/m^3 and 1.02 mg/m^3 for aggregate and bitumen, respectively. The bulk density ρ_b was measured according to EN 12697-6, Procedure C—sealed specimen [37], by weighting the sealed specimen in air and in water.

2.2.2. Indirect Tensile Stiffness Modulus Test

The indirect tensile stiffness modulus (ITSM) is the most popular form of non-destructive stress-strain measurement used to evaluate the elastic properties of bituminous mixtures and it is considered a very important performance parameter in pavement design [38]. European Norm EN 12697-26—Annex C [39] defines the ITSM (MPa) according to Equation (2).

$$ISTM = \frac{F \times (\nu + 0.27)}{(z + h)} \quad (2)$$

where F (N) is the peak value of the applied vertical load, z (mm) is the amplitude of the horizontal deformation obtained during the load cycle, h (mm) is the mean thickness of the test specimen and v is the Poisson's ratio, here assumed to be 0.35 [40]. The test was performed in control-horizontal displacement configuration using a servo-pneumatic device. The system adjusted the magnitude of the vertical force during the first ten conditioning pulses, such that the specified target-peak transient horizontal displacement of 3 μm was achieved. During testing, the rise time, which is the time when the applied load increases from zero to the maximum value, was set at 124 ms. The test was performed at 20 °C.

2.2.3. Temperature Monitoring

On site, the HMA temperature was determined by using both infrared and probe thermometers. The first uses the amount of infrared energy emitted by the HMA and its emissivity to evaluate, within a certain range, its actual temperature. They are sometimes called non-contact thermometers or temperature guns, to describe the device's ability to measure temperature from a distance. The second is a thermometer that has a pointy metal stem that can be inserted into the loose mixture. The temperature probe was introduced into the loose mixture, orthogonally to the surface, to a depth of about 2 cm. This kind of measurement requires approximately 2 min to reach the thermal equilibrium and obtain a consistent value of the mix temperature.

In addition, a thermographic camera was used to investigate the temperature variation of the mixtures in the truck and during the paving operations. The electromagnetic spectrum encompasses radiation from gamma rays, X-rays, ultraviolet, a thin region of visible light, infrared, terahertz waves, microwaves, and radio waves. All objects emit a certain amount of black body radiation as a function of their temperature. The higher the object's temperature, the more infrared radiation (with wavelengths approximately between 1000 nm and 14,000 nm) is emitted as black body radiation. The thermographic camera can detect this radiation in a way similar to the way in which an ordinary camera detects the visible light.

The contemporary use of multiple devices for temperature monitoring allowed the analysis of different aspects: the infrared and probe thermometers recorded the mix temperature in specific points and gave information of the local behavior, while the thermographic camera showed the temperature of the entire construction site, highlighting the dissimilarities between different areas of the truck bed and the pavement surface.

2.2.4. Indirect Tensile Strength Test

The mechanical resistance of HMA specimens and cores was determined in terms of indirect tensile strength (ITS), according to EN 12697-23 [41]. In the ITS test, cylindrical samples were subjected to a compressive load acting parallel to the vertical diametral plane using a universal hydraulic press. This type of loading produced a relatively uniform tensile stress in perpendicular direction and the sample usually failed by splitting along the loaded plane. The ITS (MPa) was calculated through the following Equation (3).

$$ITS = \frac{2 \times F}{\pi \times L \times D} \tag{3}$$

where F (N) is the peak value of the applied vertical load, L (mm) is the mean thickness of the test sample and D (mm) is the sample diameter. The test was carried out at 25 °C and adopting a constant rate of vertical displacement of 50 mm/min.

2.2.5. Determination of Cracking Tolerance Index

In order to investigate the material resistance to cracking, the Cracking Tolerance index (CT_{index}) was determined according to ASTM D8225-19 [42]. CT_{index} was calculated from the load-vs-displacement curve of ITS tests by means of Equation (4).

$$CT_{index} = \frac{L}{62} \times \frac{l_{75}}{D} \times \frac{G_f}{|m_{75}|} \times 10^6 \quad (4)$$

where l_{75} is the displacement (mm) when the load is 75% of the peak value after the failure, $|m_{75}|$ is the absolute value of the post-peak slope of the load-vs-displacement curve (N/m) when the load is 75% of the peak value after the failure and G_f is the failure energy (J/m^2), i.e., the ratio between the area of the load-vs-displacement curve and the specimen section ($D * L$). It has to be specified that $|m_{75}|$, as suggested by Zhou et al. [43], was determined from Equation (5).

$$|m_{75}| = \left| \frac{F_{85} - F_{65}}{l_{85} - l_{65}} \right| \quad (5)$$

where (l_{85}, F_{85}) and (l_{65}, F_{65}) are the points of the load-vs-displacement curve when the load is 85% and 65% of the peak value after the failure, respectively.

3. Effect of Conditioning Time in the Oven on Lab-Mixed HMA

3.1. Materials and Specimen Preparation

Porphyry gravel (in two fractions, with dimensions between 4 and 8 mm, respectively, and between 8 and 12 mm), coarse RAP (particles with dimensions between 8 and 16 mm), limestone sand and limestone filler were used. With the RAP content fixed, the virgin aggregate proportions were optimized to build a gradation curve for surface layers within the reference envelope [44]. For the RAP, the "white" gradation, i.e., that of the solid particles after binder extraction, was considered. The bitumen content in the RAP was 4.73% by mix weight. Figure 2 shows the gradation curves of the mixtures.

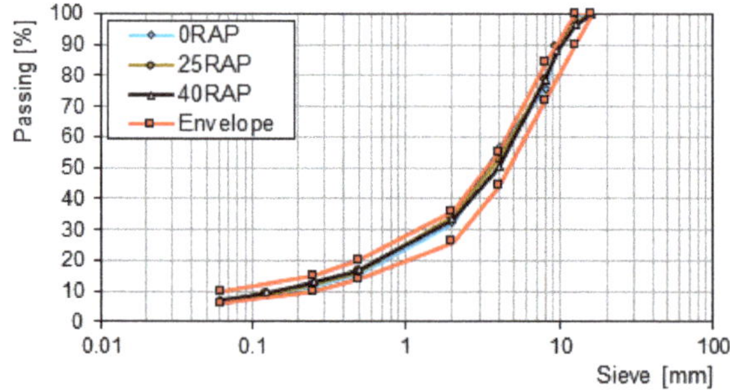

Figure 2. Particle size distribution for the mixes produced in the lab.

The 0RAP and 25RAP-R1 mixtures were produced using Styrene-Butadiene-Styrene (SBS) polymer modified bitumen (PMB) with penetration 50/70 (classified as PMB 45/80-70 according to EN 14023 [45]) and polymer content of 5% by weight (hard modification). The 25RAP-R1 mix included an oil-based rejuvenator, sprayed on the cold RAP before heating, with a dosage of 6% by RAP binder weight. For 25RAP-R2 and 40RAP-R2 mixtures a specific binder, engineered for hot-recycling applications (consisting of a base bitumen with penetration 70/100 modified with SBS and including rejuvenator), was used. In the mix design, a complete blending between RAP and the virgin binder was assumed. All the mixtures had a total bitumen content equal to 6% by mix weight.

According to the EN 12697-35 [46], the laboratory mixing procedure provided the heating at 180 °C of the components (aggregate, RAP and virgin bitumen) and the mixing for 30 s. The loose HMA was conditioned in the oven at 180 °C for different times, from 30 to 180 min. Then, cylindrical specimens with 100 mm diameter and approximately 67 mm height were compacted with 100 revolutions of SGC.

3.2. Results and Discussion from the Lab Investigation

Figure 3 shows the average air voids content (V_m) of the specimens (a) and the trend of the voids content with the conditioning time in the oven (b). All the mixes were able to meet the Italian Specifications (3–6%), except for 40RAP + R2 that had an air voids content lower than the limit. The mixtures contained RAP achieved a greater density than the 0RAP, probably related to the lubricating and fluidifying effect of the rejuvenating techniques. In addition, no significant variation of the air voids content with the conditioning time was observed. This indicated that, even if the bitumen experienced a more severe ageing due to the prolonged conditioning at high temperature, this did not affect the compactability of the mixtures.

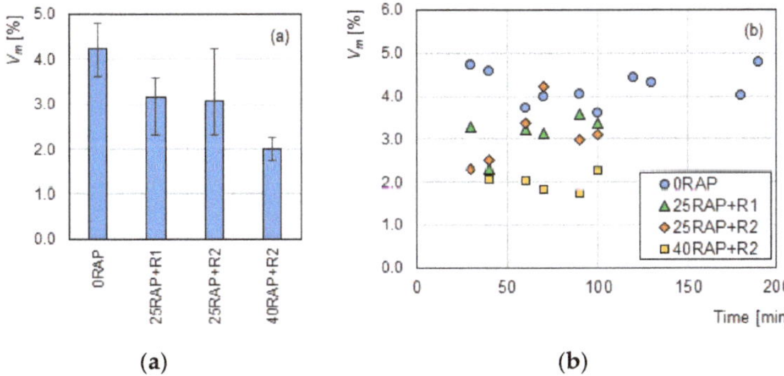

(a) **(b)**

Figure 3. Air voids content V_m of the mixes produced in the laboratory: (**a**) average values; (**b**) V_m as a function of the conditioning time.

Figure 4 shows the ITSM values as a function of conditioning time in the oven, for the different mixtures. The data were interpolated with a linear trend, whose slope (reported in the table within Figure 4) represented a kind of speed of bitumen hardening: the greater the slope, the greater the sensitivity to ageing of the HMA.

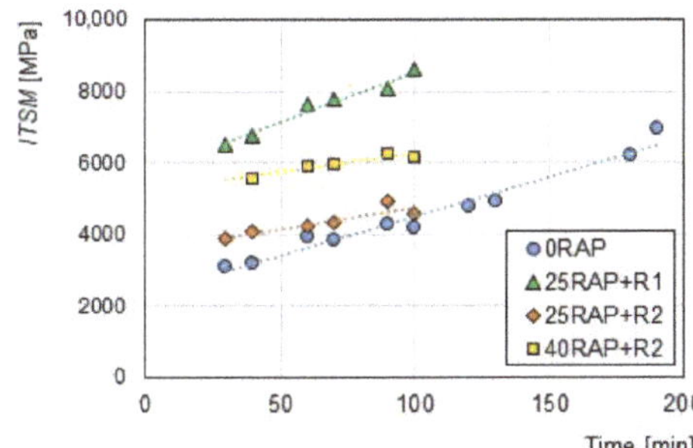

Mix	Slope [MPa/min]	R2
0RAP	22.06	0.964
25RAP + R1	28.74	0.962
25RAP + R2	12.19	0.829
40RAP + R2	10.58	0.918

Figure 4. Indirect tensile stiffness modulus (*ITSM*) of the mixes produced in the laboratory as a function of the conditioning time.

It is very interesting to observe that all the mixtures tested showed a rising trend of stiffness when the conditioning time was increased. In particular, the 0RAP mix showed the lowest ITSM values for short conditioning times (3000 MPa for the conditioning time of 30 min). However, the growth rate for this mix was one of the highest, leading it to reach, after 180 min, an *ITSM* of 7000 MPa (+125%). The 25RAP + R1 mix resulted in the stiffest (*ITSM* approximately between 6500 MPa and 8500 MPa) and the most sensitive to the conditioning time in the oven. Probably, the rejuvenator had a low effectiveness in reducing the bitumen hardness, or it was subjected to evaporation during the conditioning at high temperature. The rejuvenation technique 2 (through a softer PMB including rejuvenator) was able to limit the mix stiffness. In fact, 25RAP + R2 specimens were almost as stiff as 0RAP specimens (*ITSM* approximately between 4000 MPa and 5000 MPa), while the *ITSM* of 40RAP + R2 specimens was slightly higher (approximately between 5500 MPa and 6500 MPa). In addition, the slope of the trends was significantly lower, denoting a lower tendency to ageing. This result is in agreement with other studies [21,33], according to which, the already-aged binder from RAP experiences a less severe ageing than virgin bitumen.

4. Effect of Hauling Time and Truck Type on Plant-Mixed HMA

Once the ageing and stiffening of the loose HMA conditioned in the oven were assessed, field studies were carried out in order to check if the same behavior also occurs on site during the hauling phase. At the same time, the study aimed at verifying whether any temperature segregation happens as a function of truck bed type, when the lay-down site is far from the plant.

4.1. Mix Properties

All the mixtures were produced using the same binder, classified as PMB 45/80-70 according to EN 14023 [45]. The binder content of the mixtures produced in each plant is reported in Table 3, while Figure 5 shows the aggregate distribution compared to the gradation limits for a surface layer defined by the local specifications [44].

Table 3. Binder content of the mixes produced at the plant.

Plant	A	B	C	D
Binder content [% by mix]	6.10	6.45	6.60	5.55

All the gradation curves were collocated near the upper curve of the specifications; this means that every plant produced a mixture slightly finer than expected.

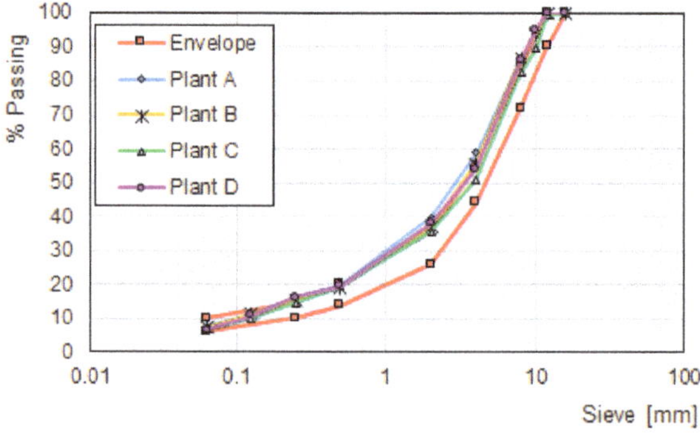

Figure 5. Particle size distribution for the mixes produced in the different HMA plants.

4.2. Result and Discussion from the Site Investigation

4.2.1. Evaluation of Temperatures during Hauling and Paving

The trials in the plants A, B, C and D were carried out between October 2018 and April 2019. In the chosen days, the average air temperature measured during the experimentation was about 15 °C and the weather was sunny.

Table 4 shows, for each system, the hours when the loose mixtures were produced and sampled from the truck bed, and the average temperatures measured on the top corner (C), inside the truck bed (B) and on the loose mixture sampled from the truck body at 50 cm depth (L). The average temperatures measured during the paving phases are shown in Table 5. Moreover, the time elapsed between the production of the mixtures and their lay-down is reported. In the table, the codes "N" and "I", respectively, indicate the normal and the insulated trucks.

Table 4. Temperatures measured during production and loose mix sampling.

Plant	Truck			Production	1st Sampling	2nd Sampling	3rd Sampling	4th Sampling
A	Normal	Hours		9:00	9:12	10:08	11:04	11:27
		T [°C]	C	-	146	121	98	87
			B	-	184	175	171	168
			L	-	165	160	152	146
	Insulated	Hours		9:00	9:40	10:37	11:33	12:27
		T [°C]	C	-	143	124	105	101
			B	-	178	182	170	165
			L	-	165	160	163	154
B	Normal	Hours		8:45	8:52	9:59	10:56	12:30
		T [°C]	C	-	152	112	100	94
			B	-	181	186	184	175
			L	-	150	138	131	110
	Insulated	Hours		8:45	9:21	10:25	11:25	12:30
		T [°C]	C	-	133	106	92	96
			B	-	182	185	180	172
			L	-	165	151	131	111
C	Normal	Hours		9:00	9:15	10:08	11:13	12:18
		T [°C]	C	-	135	101	90	87
			B	-	185	183	178	174
			L	-	147	160	137	117
D	Normal	Hours		8:00	8:15	9:22	10:24	11:18
		T [°C]	C	-	151	109	97	85
			B	-	188	188	186	182
			L	-	169	130	129	109

Table 5. Temperatures measured during paving operations.

	A		B		C	D
	N	I	N	I	N	N
Time elapsed between production and lay-down [min]	210	242	203	230	242	196
Temperature of the loose mix in the paver [°C]	-	-	180	172	121	170
Surface HMA temperature after lay-down (probe) [°C]	-	-	161	140	106	152
Surface HMA temperature after lay-down (infrared) [°C]	-	-	146	181	142	171
Temperature of outside of the truck bed [°C]	-	-	62	155	55	62
Temperature of bottom of the truck bed [°C]	-	-	55	-	75	81

The data in Table 4 show that the cooling of the mix inside the truck bed (B) was very low (10 °C after about 3 h). Moreover, there was no significant difference between the normal and the insulated trucks. Probably, the tarpaulin which protected the top of the normal truck bed allowed the cooling on the material surface to be avoided, similar to the case of the insulated truck. Conversely, the temperatures on the top corner (C) and those measured on the loose mix (L) considerably decreased (up to 66 °C and 60 °C, respectively). For plant A, the insulated truck was proved to reduce the heat loss, particularly in the truck corner. However, for plant B, the lower precision in the measurement did not allow confirmation of this assumption.

From Table 5 it can be noted that the mix produced at plant B showed a comparable temperature in the paver if hauled with a normal or insulated truck. Some temperature differences between the different truck types were observed after the laying. As both the probe and infrared thermometers measured the temperature in localized positions of the road surface, this result was probably related to thermal segregation of the material during hauling. It was difficult to evaluate the influence of the truck type from the temperature determined on the pavement surface. However, the noticeably different temperature outside the truck bed was a sign of the insulated truck's ability to preserve the HMA heat during the 3 h from production to paving.

4.2.2. Thermal Image Analysis

The pictures taken with the infrared camera were of considerable interest. Figures 6 and 7 show the most significant images for the evaluation of HMA cooling and temperature segregation during hauling phase.

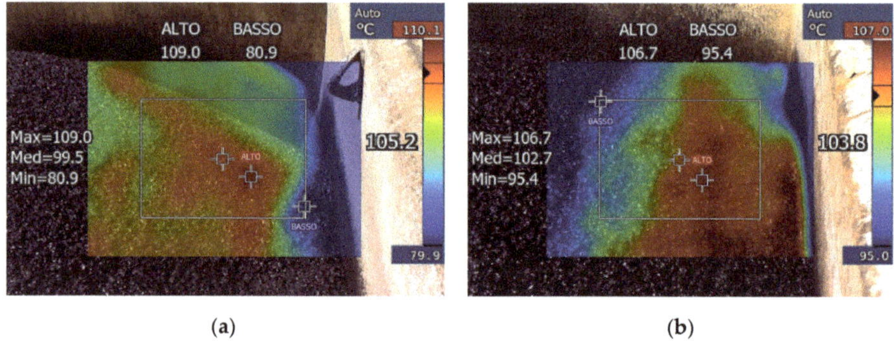

(**a**) (**b**)

Figure 6. Pictures of the bed corner taken with infrared camera 3 h after production: (**a**) Normal truck; (**b**) Insulated truck.

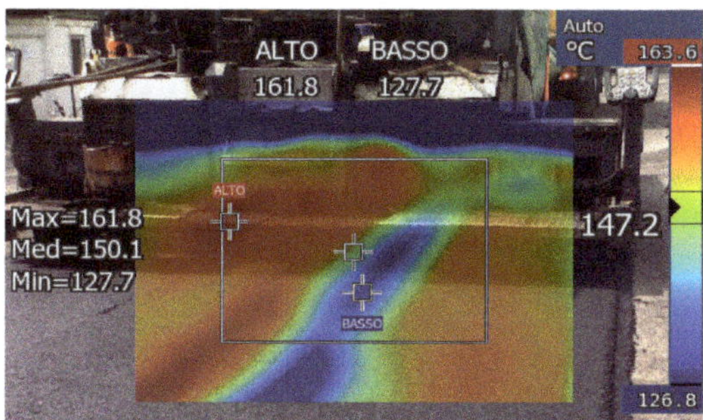

Figure 7. Picture taken with infrared camera showing the paver laying the HMA from plant A hauled with a normal truck.

The strong external cooling compared to the almost zero cooling of the HMA batch determined a huge temperature segregation at the paving site, as evidenced by the measurements with the probe and infrared thermometers and by the photos with the thermal imaging camera. In particular, Figure 6a highlights that 3 h after mix production, near the corner of the normal truck, the loose HMA cooled rather quickly, showing a large temperature difference (approximately 30 °C) between the material in contact with the edge and that in the center of the truck bed. Conversely, in the case of insulated truck (Figure 6b), this temperature segregation was less severe (temperature variation of about 10 °C).

Figure 7 shows that, during the paving, there was a huge difference in the HMA temperature on the pavement surface (higher than 30 °C). This was probably due to the inability of the paving machine to re-mix the loose HMA and disperse the colder parts among the warmer mass. The temperature segregation can negatively affect the HMA compaction, especially in the construction of thin layers (i.e., surface layers) where the colder portions can hardly be heated by the surrounding material, even if much warmer. If the temperature of the colder parts drops below the minimum temperature necessary for a good compaction (typically about 130–140 °C for HMA with polymer modified bitumen), high porosity areas which are more susceptible to rapid degradation (cracking and raveling in particular), can occur.

4.2.3. Laboratory Tests

Figures 8–11 show the average values of V_m, $ITSM$, ITS and CT_{index} measured for the specimens compacted in the plant (immediately after the sampling of the loose asphalt), in the laboratory (after re-heating) and for the cores.

The bar-chart in Figure 8 shows that there was no significant difference in the voids contents of the specimens from the HMA produced in plant A, B and C. In particular, the air voids content of these samples, compacted with SGC in the plant, was about 2–3%, independently from sampling time and truck type. Conversely, the specimens from the HMA produced in plant D showed a higher V_m, approximately 4–5%, but also in this case the influence of sampling time was not very high. This result indicates that the eventual increase in binder viscosity achieved in the truck, due to bitumen ageing or cooling, did not affect the mix compactability, as observed for the HMA produced in the laboratory (Figure 3). The specimens compacted after re-heating showed a higher air voids content (4.2% on average), indicating a certain decrease in mix workability.

The V_m values of the cores showed different trends for the plants A–D: for plants A and B the air voids content of the cores was higher than that of the specimens, for plant C it was comparable, while for plant D it was lower. Moreover, for the HMA produced in plant A there was a significant difference in V_m as a function of the truck type (8% for a normal truck, 5.5% for an insulated truck), while for the mix produced in plant B, the truck type had no influence on V_m (about 4.5%). As the voids content of each core represents the condition of the pavement in the exact place where the cores were taken, the absence of a clear trend and the higher data dispersion probably reflected the temperature segregation and localized cooling observed with the thermal camera.

Figure 9 shows that, for all the plants, the *ITSM* values of the specimens compacted on site were always about 4000 MPa, independently from the truck type and time spent in the truck during the hauling phase (the only exception is represented by the specimens from plant C, whose *ITSM* slightly increased as a function of the sampling time). It is very interesting to note that this result was opposite to what was observed in the mix produced in the laboratory (Figure 4), where ITSM increased as a function of the conditioning time in the oven at 180 °C.

Figure 12 allows a better comparison between the mixes conditioned in the laboratory oven and the mixes kept in the truck bed. The graph shows the normalized *ITSM* as a function of the time between HMA mixing and compaction, where the normalized *ITSM* was calculated as the ratio of the *ITSM* measured at the different conditioning time with the *ITSM* measured at the lowest conditioning time for each mixture. It can be observed that the red lines, which represent the mixes produced at the plant and kept in the truck bed, always have a lower slope than the blue lines, which represent the mixes produced in the laboratory and conditioned in the oven. This clearly indicates a less marked effect of the ageing for the HMA kept in the truck bed. The reason for this is probably in the fact that, in the laboratory, the small amount of HMA in the oven (some kilograms) allowed oxygen to come into contact with most of the loose mix, favoring binder oxidation and loss of volatiles. On site, the HMA was taken out from the truck bed at 50 cm depth from the batch surface, where the temperature remained almost constant and the material was repaired by the one above. In such conditions, the batch surface was exposed to external conditions, but the core was basically isolated and neither oxidation nor loss of volatility could occur, hindering the ageing of the HMA.

Additionally, *ITS* (Figure 10) and CT_{Index} (Figure 11) were approximately constant with sampling time, even if a slight variability between the mixes produced in the different plants was observed. In order to assess whether the groups of data were statistically comparable or different, analysis of variance (ANOVA) was carried out. Table 6 shows the significance values obtained in the comparison between the different data (V_m, *ITSM*, *ITS* or CT_{Index}) for each examined variable (type of truck, sampling time and HMA manufacturing plant): for values lower than 5% (in bold characters) the null hypothesis is rejected, meaning that the groups of data are statistically different. The significance values obtained through the ANOVA test showed that all the measured properties (V_m, *ITSM*, *ITS* or CT_{Index}) were not dependent from the truck type, i.e., the voids, stiffness, strength and cracking tolerance were comparable for the specimens from the normal and insulated truck bed. The properties V_m, *ITSM* and *ITS* were also independent from the sampling time (significance values higher than 0.05), indicating that there was not any correlation between the different quantities and the time between HMA mixing and compaction. This is also valid for CT_{Index}, even if the significance value was lower than 0.05, because the statistically different data groups did not set a monotonic (increasing or decreasing) trend. Finally, the very low significance values obtained for the data from the different plants indicate that the volumetric and mechanical properties of the HMA specimens varied according to the plant where they had been mixed.

From Figures 8–11 it can be noted that the cores had lower *ITSM* and *ITS* and higher CT_{index} with respect to the specimens from the same plant, compacted on site, probably related to different air voids content. Instead, for the HMA compacted in the laboratory after re-heating at 170 °C, the stiffness modulus considerably increased (growth greater than 100%). At the same time, *ITS* noticeably increased (up to 2.1 MPa) and CT_{Index} fell (lower than 30). This result confirms that a severe ageing happened during the HMA re-heating in the oven.

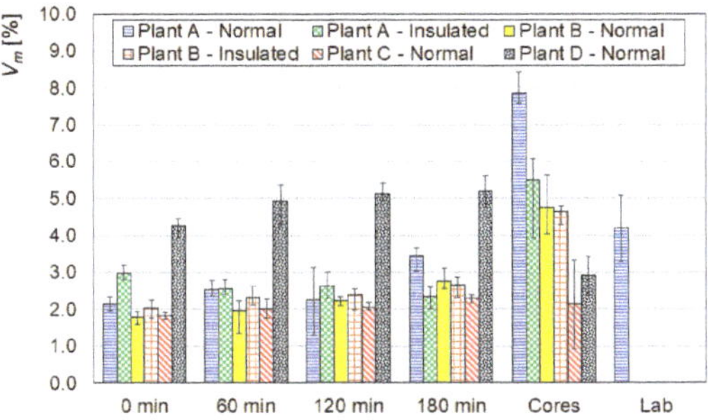

Figure 8. V_m of the mixes produced at the plants: field- and lab-compacted specimens and cores.

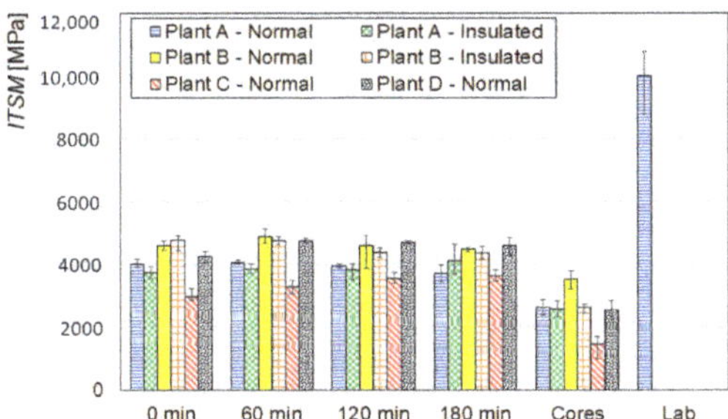

Figure 9. *ITSM* of the mixes produced at the plants: field- and lab-compacted specimens and cores.

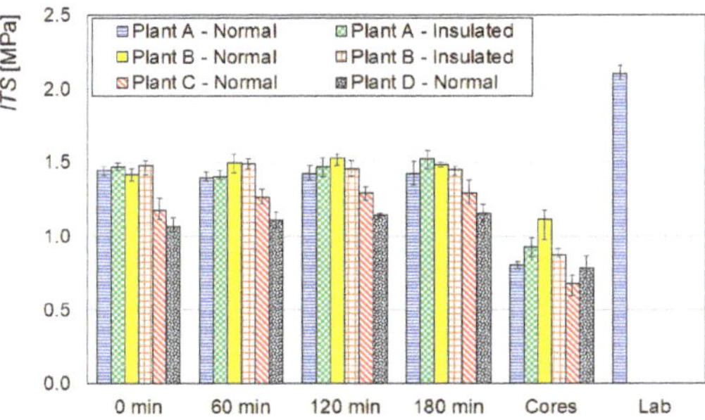

Figure 10. *ITS* of the mixes produced at the plants: field- and lab-compacted specimens and cores.

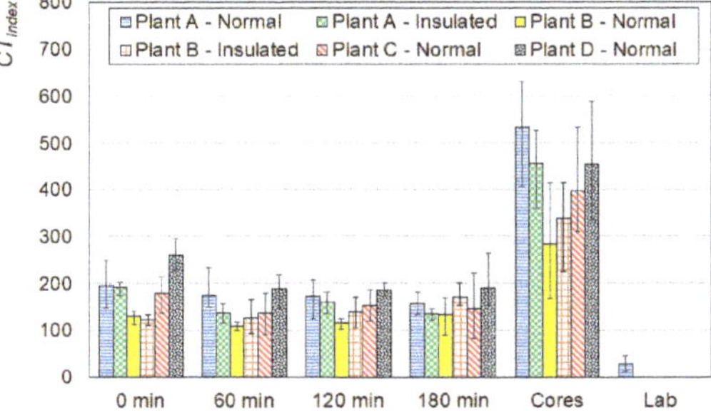

Figure 11. Cracking Tolerance index (CT_{Index}) of the mixes produced at the plants: field- and lab-compacted specimens and cores.

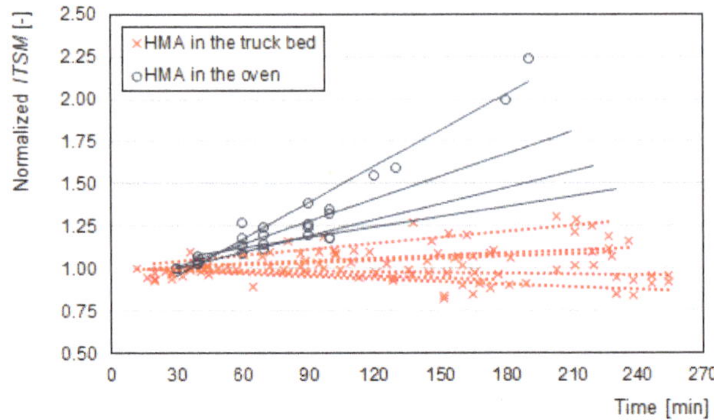

Figure 12. Normalized *ITSM* of the mixes conditioned in the laboratory oven or in the truck bed as a function of the time between HMA mixing and compaction.

Table 6. Analysis of variance (ANOVA) of the V_m, *ITSM*, *ITS* and CT_{Index} data: significance values.

	Variable		
Property	**Truck Type**	**Sampling Time**	**HMA Manufacturing Plant**
V_m	0.438	0.235	7.0×10^{-25}
ITSM	0.552	0.615	9.7×10^{-22}
ITS	0.266	0.655	6.9×10^{-27}
CT_{Index}	0.967	0.032	5.0×10^{-7}

5. Conclusions

The present research aimed at determining the influence of the conditioning time, when the loose HMA is kept in the oven at a high temperature, on the mix properties. Moreover, the ageing/cooling that happens on site during HMA hauling, as a function of time and type of truck, was investigated. Temperatures were monitored using a thermal camera and different probes, and gyratory compactor specimens were produced after sampling some HMA from the trucks every 1 h for a total of 3 h. Air voids content, *ITSM*, *ITS* and CT_{Index} were measured on the specimens compacted on site, on the cores taken from the pavement and on the loose HMA, re-heated and compacted in the laboratory.

In light of the results obtained, the following conclusions can be drawn:

- the compactability of the HMA was not influenced by the time during which the material is conditioned in the oven before compaction. Conversely, all the mixtures tested showed a rising trend of stiffness when the conditioning time was increased. The growing rate was higher for the mix with no RAP (22.1 MPa/min) with respect to those including RAP and a special bitumen engineered for hot recycling applications (12.2 and 10.6 MPa/min respectively for the mix with 25% RAP and 40% RAP);
- the temperature monitoring proved that the mix inside the truck bed did not significantly cool after 3 h of hauling (temperature decrease between 6 °C and 16 °C), while the temperatures on the top corner of the truck and those measured on the sampled loose mix considerably decreased (approximately 60 °C cooling);
- the thermal camera images highlighted a large temperature segregation at the paving site, with discrepancies higher than 30 °C both on the truck bed (normal truck) and on the road surface after paving. However, in the case of the insulated truck the severity of this phenomenon was reduced;

- HMA compactability, stiffness, strength and cracking tolerance did not change when hauling time increased. Inversely from the laboratory oven (where most of the HMA volume was exposed to air), the core of the HMA in the truck bed was basically isolated by the surface material and neither oxidation nor loss of volatility could occur, hindering the ageing of the HMA;
- the cores showed lower *ITSM* (−39% on average) and *ITS* (−37% on average) and higher CT_{index} (about 2.5 times on average) with respect to the specimens compacted on site, probably due to a higher air voids content;
- when the loose HMA was re-heated and compacted in the laboratory, *ITSM* and *ITS* (+150% and +50%, respectively) considerably increased, whereas CT_{Index} decreased (−84%), denoting that a severe ageing happened through this specimen preparation procedure.

The research demonstrated that HMA handling in the laboratory, during quality assurance and quality controls, is extremely important, since keeping the material in the oven for a prolonged time or re-heating it may significantly affect the test results. However, when the HMA is in the truck bed it does not experience as much severe ageing, even if it stays at a high temperature for a long time. Differently, the plant trial showed that on site cooling and temperature segregation represent a higher risk for HMA than ageing. Therefore, the use of insulated trucks (to avoid temperature segregation) and remixing material transfer vehicles is recommended in the case of long hauling distances.

Author Contributions: Conceptualization, M.B.; methodology, M.B. and E.B.; software, E.B. and E.P.; validation, M.B., E.B. and E.P.; formal analysis, E.B. and E.P.; investigation, E.B. and E.P.; resources, M.B. and V.M.; data curation, E.B. and E.P.; writing—original draft preparation, E.P.; writing—review and editing, E.B.; visualization, E.P.; supervision, M.B.; project administration, V.M. All authors have read and agreed to the published version of the manuscript.

Funding: This research received no external funding.

Conflicts of Interest: The authors declare no conflict of interest.

References

1. Kök, B.V.; Yilmaz, M.; Alatas, T. Evaluation of the mechanical properties of field and laboratory Compacted hot-mix asphalt. *J. Mater. Civ. Eng.* **2014**, *26*. [CrossRef]
2. Morovatdar, A.; Ashtiani, R.S.; Licon, C. Development of a mechanistic framework to predict pavement service life using axle load spectra from Texas overload corridors. In Proceedings of the International Conference on Transportation and Development 2020, Seattle, WA, USA, 26–29 May 2020.
3. Lu, X.; Isacsson, U. Effect of ageing on bitumen chemistry and rheology. *Constr. Build. Mater.* **2002**, *16*, 15–22. [CrossRef]
4. Miró, R.; Martínez, A.H.; Moreno-Navarro, F.; Rubio-Gámez, C. Effect of ageing and temperature on the fatigue behaviour of bitumens. *Mater. Des.* **2015**, *86*, 129–137. [CrossRef]
5. Hung, A.M.; Fini, E.H. Absorption spectroscopy to determine the extent and mechanisms of aging in bitumen and asphaltenes. *Fuel* **2019**, *242*, 408–415. [CrossRef]
6. Le Guern, M.; Chailleux, E.; Farcas, F.; Dreessen, S.; Mabille, I. Physico-chemical analysis of five hard bitumens: Identification of chemical species and molecular organization before and after artificial aging. *Fuel* **2010**, *89*, 3330–3339. [CrossRef]
7. Mousavi, M.; Pahlavan, F.; Oldham, D.; Hosseinnezhad, S.; Fini, E.H. Multiscale investigation of oxidative aging in biomodified asphalt binder. *J. Phys. Chem. C* **2016**, *120*, 17224–17233. [CrossRef]
8. Mirwald, J.; Werkovits, S.; Camargo, I.; Maschauer, D.; Hofko, B.; Grothe, H. Understanding bitumen ageing by investigation of its polarity fractions. *Constr. Build. Mater.* **2020**, *250*, 118809. [CrossRef]
9. Siddiqui, M.N.; Ali, M.F. Studies on the aging behavior of the Arabian asphalts. *Fuel* **1999**, *78*, 1005–1015. [CrossRef]
10. Lemarchand, C.A.; Schrøder, T.B.; Dyre, J.C.; Hansen, J.S. Cooee bitumen: Chemical aging. *J. Chem. Phys.* **2013**, *139*. [CrossRef]

11. Pahlavan, F.; Samieadel, A.; Deng, S.F.E. Exploiting synergistic effects of intermolecular interactions to synthesize hybrid rejuvenators to revitalize aged asphalt. *ACS Sustain. Chem. Eng.* **2019**, *7*, 15514–15525. [CrossRef]
12. Lesueur, D.; Gerard, J.-F.; Claudy, P.; Letoffe, J.-M.; Planche, J.-P.; Martin, D. Structure related model to describe asphalt linear viscoelasticity. *J. Rheol.* **1996**, *40*, 813–836. [CrossRef]
13. Ongel, A.; Hugener, M. Impact of rejuvenators on aging properties of bitumen. *Constr. Build. Mater.* **2015**, *94*, 467–474. [CrossRef]
14. Mazzoni, G.; Bocci, E.; Canestrari, F. Influence of rejuvenators on bitumen ageing in hot recycled asphalt mixtures. *J. Traffic Transp. Eng. Engl. Ed.* **2018**, *5*, 157–168. [CrossRef]
15. Nakhaei, M.; Ziari, H.; Korayem, A.H.; Hajiloo, M. Aging evaluation of amorphous carbon-modified asphalt binders using rheological and chemical approach. *J. Mater. Civ. Eng.* **2020**, *32*. [CrossRef]
16. Hofko, B.; Falchetto, A.C.; Grenfell, J.; Huber, L.; Lu, X.; Porot, L.; You, Z.; Falchetto, A.C. Effect of short-term ageing temperature on bitumen properties. *Road Mater. Pavement Des.* **2017**, *18*, 108–117. [CrossRef]
17. Dessouky, S.; Reyes, C.; Ilias, M.; Contreras, D.; Papagiannakis, A.T. Effect of pre-heating duration and temperature conditioning on the rheological properties of bitumen. *Constr. Build. Mater.* **2011**, *25*, 2785–2792. [CrossRef]
18. Lolly, R.; Zeiada, W.; Souliman, M.; Kaloush, K. Effects of short-term aging on asphalt binders and hot mix asphalt at elevated temperatures and extended Aging Time. *MATEC Web Conf.* **2017**, 07010. [CrossRef]
19. Lemke, Z.; Sadek, H.; Swiertz, D.; Reichelt, S.; Bahia, H.U. Effects of reheating procedure and oven type on performance testing results of asphalt mixtures. *Transp. Res. Rec.* **2018**, *2672*, 124–133. [CrossRef]
20. Kidd, A.; Stephenson, G.; White, G. Implications of reheating of asphalt mixes on performance testing. In Proceedings of the 18th AAPA International Flexible Pavements Conference 2019, Sydney, New South Wales, Australia, 18–21 August 2019.
21. Daniel, J. How mixture, fabrication, and plant production parameters affect mixture properties. *Transp. Res. Circ.* **2018**, *E-C234*, 1–20.
22. Roberts, F.L.; Kandhal, P.S.; Brown, E.R.; Lee, D.Y.; Kennedy, T.W. *Hot Mix Asphalt Materials, Mixture Design, and Construction*, 3rd ed.; National Asphalt Paving Association Education Foundation: Lanham, MD, USA, 2016.
23. Muhammad, M.; Syuhada, A.; Huzni, S.; Fuadi, Z. Study on Heat loss through dump truck wall insulated by sengon wood. *J. Adv. Res. Fluid Mech. Therm. Sci.* **2019**, *58*, 126–134.
24. Cho, Y.K.; Bode, T.; Song, J.; Jeong, J.-H. Thermography-driven distress prediction from hot mix asphalt road paving construction. *ASCE J. Constr. Eng. Manag.* **2012**, *138*, 206–214. [CrossRef]
25. Mahoney, J.P.; Muench, S.T.; Pierce, L.M.; Read, S.A.; Jakob, H.; Moore, R. Construction-related temperature differentials in asphalt concrete pavement: Identification and assessment. *Transp. Res. Rec.* **2000**, *1712*, 93–100. [CrossRef]
26. Amirkhanian, S.N.; Putman, B.J. *Laboratory and Field Investigation of Temperature Differential in HMA Mixtures Using an Infrared Camera*; Clemson University Department of Civil Engineering: Clemson, SC, USA, 2006.
27. Brock, J.D.; Jakob, H. *Temperature Segregation/Temperature Differential Damage*; Technical Paper T-134; Astec Industries: Chattanooga, TN, USA, 2009.
28. Kim, M.; Phaltane, P.; Mohammad, L.N.; Elseifi, M. Temperature segregation and its impact on the quality and performance of asphalt pavements. *Front. Struct. Civ. Eng.* **2017**, *12*, 536–547. [CrossRef]
29. Delgadillo, R.; Bahia, H.U. Effects of temperature and pressure on hot mixed asphalt compaction: Field and laboratory study. *J. Mater. Civ. Eng.* **2008**, *20*, 440–448. [CrossRef]
30. Hayat, A.; Hussain, A.; Afridi, H.F. Determination of in-field temperature variations in fresh HMA and corresponding compaction temperatures. *Constr. Build. Mater.* **2019**, *216*, 84–92. [CrossRef]
31. Khan, R.; Grenfell, J.; Collop, A.; Airey, G.; Gregory, H. Moisture damage in asphalt mixtures using the modified SATS test and image analysis. *Constr. Build. Mater.* **2013**, *43*, 165–173. [CrossRef]
32. Plati, C.; Georgiou, P.; Loizos, A. Use of infrared thermography for assessing HMA paving and compaction. *Transp. Res. Part C Emerg. Technol.* **2014**, *46*, 192–208. [CrossRef]
33. Bocci, E.; Mazzoni, G.; Canestrari, F. Ageing of rejuvenated bitumen in hot recycled bituminous mixtures: Influence of bitumen origin and additive type. *Road Mater. Pavement Des.* **2019**, *20*, 127–148. [CrossRef]
34. *Test Methods for Hot Mix Asphalt. Part 31: Specimen Preparation by Gyratory Compactor*; EN 12697-31; European Committee for Standardization (CEN): Brussels, Belgium, 2019.

35. *Test Methods for Hot Mix Asphalt. Part 8: Determination of Void Characteristics of Bituminous Specimens*; EN 12697-8; European Committee for Standardization (CEN): Brussels, Belgium, 2019.
36. *Test Methods for Hot Mix Asphalt. Part 5: Determination of the Maximum Density*; EN 12697-5; European Committee for Standardization (CEN): Brussels, Belgium, 2019.
37. *Test Methods for hot Mix Asphalt. Part 6: Determination of the Bulk Density of Bituminous Specimens*; EN 12697-6; European Committee for Standardization (CEN): Brussels, Belgium, 2020.
38. Cerni, G.; Bocci, E.; Cardone, F.; Corradini, A. Correlation between asphalt mixture stiffness determined through static and dynamic indirect tensile tests. *Arab. J. Sci. Eng.* **2017**, *42*. [CrossRef]
39. *Test Methods for Hot Mix Asphalt. Part 26: Stiffness*; EN 12697-26; European Committee for Standardization (CEN): Brussels, Belgium, 2018.
40. Graziani, A.; Bocci, E.; Canestrari, F. Bulk and shear characterization of bituminous mixtures in the linear viscoelastic domain. *Mech. Time Depend. Mater.* **2014**, *18*, 527–554. [CrossRef]
41. *Test Methods for Hot Mix Asphalt. Part 23: Determination of the Indirect Tensile Strength of Bituminous Specimens*; EN 12697-23; European Committee for Standardization (CEN): Brussels, Belgium, 2017.
42. *Standard Test Method for Determination of Cracking Tolerance Index of Asphalt Mixture Using the Indirect Tensile Cracking Test at Intermediate Temperature*; ASTM D8225-19; ASTM International: West Conshohocken, PA, USA, 2019. [CrossRef]
43. Zhou, F.; Im, S.; Sun, L.; Scullion, T. Development of an IDEAL cracking test for asphalt mix design and QC/QA. *Road Mater. Pavement Des.* **2017**, *18*, 405–427. [CrossRef]
44. *Direttive Tecniche per Pavimentazioni Bituminose*; Amministrazione Provincia Bolzano: Bolzano, Italy, 2016. (In Italian)
45. *Framework Specification for Polymer Modified Bitumens*; EN 14023; European Committee for Standardization (CEN): Brussels, Belgium, 2010.
46. *Test Methods for Hot Mix Asphalt. Part 35: Laboratory Mixing*; EN 12697-35; European Committee for Standardization (CEN): Brussels, Belgium, 2016.

Publisher's Note: MDPI stays neutral with regard to jurisdictional claims in published maps and institutional affiliations.

Article

Analysis of Contact Stresses and Rolling Resistance of Truck-Bus Tyres under Different Working Conditions

Minrui Guo [1,2,*], Xiangwen Li [1], Maoping Ran [1], Xinglin Zhou [1] and Yuan Yan [1]

[1] College of Automotive and Transportation Engineering, Wuhan University of Science and Technology, Wuhan 430065, China; catwin86@wust.edu.cn (X.L.); ranmaoping@wust.edu.cn (M.R.); zhouxinglin@wust.edu.cn (X.Z.); sahara1990@163.com (Y.Y.)

[2] College of Mechanical and Energy Engineering, Huanghuai University, Zhumadian 463000, China

* Correspondence: 20131293@huanghuai.edu.cn

Received: 13 October 2020; Accepted: 15 December 2020; Published: 18 December 2020

Abstract: In this work, to analyse the changing characteristics of contact stresses in the tyre–pavement interface and the functional relationship between rolling resistance and the working conditions of truck-bus tyres, a three-dimensional tyre–pavement model is established and used to predict the distribution of contact stresses and rolling resistance under different working conditions of the tyre, comprising various tyre loads, inflation pressures, and velocities. Results show that the magnitude relationship between transverse and longitudinal contact stresses is related to rolling conditions, and overload and low tyre pressure are important contributors to the wear of the tyre shoulder. In addition, the proposed exponential equation presents a method that can be used to forecast rolling resistance related to the working conditions of the truck-bus tyre, and a similar method can be used to predict the rolling resistances of other types of tyres.

Keywords: contact stresses; rolling resistance; braking; free rolling; load; inflation pressure; speed

---

1. Introduction

Different simulation studies [1,2] and experimental results [3,4] have shown that contact stresses between tyre and pavement not only have a great influence on vehicle handling and stability but also significantly affect the prediction of pavement responses and performance [5–9]. Regardless of the type of tyre or pavement, the contact stresses or forces in the tyre–pavement interface can be resolved into three orthogonal directions, namely, longitudinal (X), lateral (Y), and vertical (Z) contact stress or force distributions [10], as shown in Figure 1. The directions of the coordinate systems adopted by different countries and professions may be different. In this paper, the orientation of these contact stresses is defined according to the Society of Automotive Engineers' (SAE) coordinate system [11], or in accordance with the right-hand rule. In addition, the tyre rolling resistance (RR) is responsible for 20% of the fuel consumption of truck tyres [12], and RR predictions generate better estimations of fuel consumption and greenhouse gas emissions [13].

Previous studies have demonstrated that the vertical contact stress of the tyre–road interface has a non-uniform distribution, and horizontal contact stress are developed [14]. Some studies have regarded vertical contact stress as uniformly distributed [15–17], some have only considered non-uniformly distributed vertical stress and ignored tangential stress [18–20], and some have obtained the distribution of the three orthogonal contact stresses under both static and vehicle manoeuvring conditions [21–23]. However, the change trends of contact stresses during rolling conditions have not been studied and depend on many factors, including the tyre structure (geometry, tread pattern, rubber constitutive model and reinforcement), pavement surface conditions, tyre working conditions (tyre load, inflation pressure, and speed) and rolling state (accelerating, free rolling, braking). Cho [24] predicted the RR of a patterned tyre (205/60R15) from a passenger car by utilising static tyre contact analysis. Aldhufairi [25] investigated the RR of a radial passenger-car tyre (225/55R17) based on material viscoelasticity. Rafei [26] implemented a simulation of the RR of a passenger-car tyre (185/65R14) using the finite element method. Ghost [12] analysed the RRs of truck-bus tyres with a carbon black tread as well as a silica tread under rated conditions. However, variation of the RR of truck-bus tyres under different working conditions and the functional relationship between RR and tyre working conditions have not been studied.

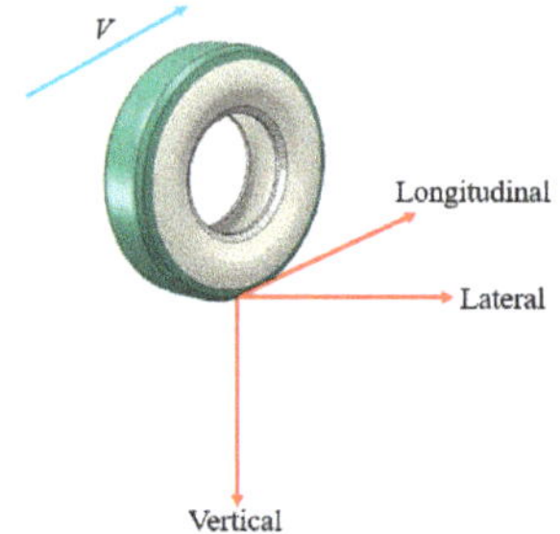

Figure 1. Three orthogonal directions of contact stresses.

Some scholars have used experimental equipment to measure contact stresses and RR. For example, Anghelache et al. [27] used 30 strain gauged sensing elements to measure the tri-axial stress distribution of a 205/55R16 tyre of a passenger car with a 240 kPa inflation pressure at a speed of 0.54 km/h. Anghelache et al. [28] used the same measuring device to measure the stress distribution of a 11R22.5 truck tyre with a 780 kPa inflation pressure at a speed of 3.2 km/h. De Beer and Fisher [29] applied 63 strain measurement channels to capture the tri-axial tyre-road interaction, and the working conditions of the 12R22.5 truck tyre were 20 and 25 kN vertical loads with a 520 kPa inflation pressure at a free rolling speed of 1.08 km/h. Chen [30] applied a pressure-sensitive film to determine the contact region and stress distribution between a 215/75R15 tyre of a light vehicle and asphalt pavement under static conditions. Ghost [12] investigated the RR values of three tyres in drum-type RR testing equipment, and a good correlation between the simulated and measured RR values was observed. However, the cost of experimental equipment is high, the installation is complex and time-consuming, and substantial amounts of manpower and material resources are required. In addition, many factors affect the accuracy and reliability of the experimental test results, including sensor aspects (the number, direction, and arrangement), tyre aspects (the type, structure, materials, working conditions, and rolling state), and the road aspect. During the testing process, due to the different combinations of measured variables and because the actual driving conditions of a tyre are difficult to achieve, it is very challenging to take into account all the working conditions of the tyre and achieve different driving conditions.

Based on preceding analysis, this study primarily develops a theoretical methodology for the prediction of the contact stresses and RR of the truck-bus radial tyre (275/70R22.5) under rolling conditions and explores the change rule of tri-axial contact stresses and the functional relationship between RR and tyre working conditions. The findings provide valuable information about the distribution of contact stresses, the location of tyre wear, and the functional expression of tyre RR, especially in terms of tyre and pavement contact mechanics and fuel-economic tyre production/design.

2. Establishment and Verification of Tyre–Pavement Finite Element Model

Two-dimensional (2D) and three-dimensional (3D) tyre models can be used to solve tyre dynamics issues. Simplified 2D models have been used in tyre dynamics analysis to forecast the characteristics and handling stability of a tyre [31]. For example, the classical point contact model can simplify the tyre into a spring-damping system to analyse the vertical forces acting on the tyre [32]. Although 2D tyre models can solve many problems in tyre mechanics and have some successful applications, these simplified models are not suitable for the prediction and calculation of the contact stresses between the tyre and pavement. They also cannot be used to analyse the complex composition of each part of the tyre in detail or the nonlinear characteristics of tyre materials [22]. In this study, a complete 3D tyre model that considers the actual dimensions and nonlinear characteristics of the tyre is established with the assistance of ABAQUS software.

2.1. The Constitutive Model of Rubber Material

Rubber material has a hyperelastic property that exhibits an obvious nonlinear response, and the rubber constitutive model is expressed in terms of the strain potential energy. The general expression of this model is as follows:

$$U = \sum_{i+j=1}^{N} C_{ij}(I_1-3)^i(I_2-3)^j + \sum_{i=1}^{N} \frac{1}{D_i}(J-1)^{2i} \tag{1}$$

where U is strain energy, C_{ij} is the constant of material properties, N is the polynomial order, I_1 and I_2 are the distortion measurement of materials, D_i is the material parameter that introduces the compression feature, and J is the bulk modulus.

When $N = 1$, Equation (1) is rewritten as:

$$U = C_{10}(I_1-3) + C_{01}(I_2-3) + \frac{1}{D_1}(J-1)^2 \tag{2}$$

Equation (2) is the Mooney–Rivlin material model. When $C_{01} = 0$, the Mooney–Rivlin material model is transformed into the Neo-Hookean model, namely:

$$U = C_{10}(I_1-3) + \frac{1}{D_1}(J-1)^2 \tag{3}$$

The Neo-Hookean material model can predict both the moderate and small strain of rubber material and is characterised by good reliability and stability. Therefore, Equation (3) is applied to analyse the hyperelasticity characteristics of rubber material. In addition, rubber material also has viscoelastic properties, which are usually expressed by Prony series parameters.

2.2. Rolling Resistance of the Tyre

The RR is equal to the energy loss per revolution of rolling divided by the distance of each revolution [33]. The energy loss of each element volume of tyre material in each cycle is described by Equation (4).

$$\Delta W = \pi \cdot E' \cdot \varepsilon_0^2 \cdot \tan\delta \tag{4}$$

$$\tan\delta = \frac{E''}{E'} \tag{5}$$

The rolling energy loss of the tyre in each cycle is then as follows:

$$E_{Loss} = \sum_i \Delta W_i \cdot v_i \tag{6}$$

Finally, the calculation expression of the RR is as follows:

$$RR = \frac{E_{Loss}}{2\pi r} = \frac{\sum_i \Delta W_i \cdot v_i}{2\pi r} \tag{7}$$

where ΔW represents the energy loss of each element, E'' and E' represent storage and loss moluli, ε_0 represents strain amplitude, $\tan\delta$ represents the loss factor, v_i represents the volume of element i of tyre material, E_{Loss} represents total energy loss, and r represents the tyre rolling radius.

2.3. Development of a Complete 3D Tyre Model

Figure 2 presents the tyre–pavement contact model and the components of a complex radial tyre (275/70R22.5), namely a truck-bus radial tyre containing four longitudinal grooves, and each tyre component is distinguished by a different colour. The outer diameter is 958 mm, the cross-sectional width is 276 mm, and steel-belt-1 and steel-belt-2 have orientation angles of 20°and −20°, respectively. The length of the mesh is 16 mm, and the width is 9 mm in the contact area. Steel- belt-1 and steel-belt-2 are located approximately 15 and 17 mm away from the outer surface of the tread, respectively. To optimise computational efficiency and ensure better convergence, the tread of the tyre in contact with the road surface is selected as a dense mesh, and the other non-contact locations of the tyre are selected as a sparse mesh. The material parameters of the tyre composition are based on the authors' previous study [34].

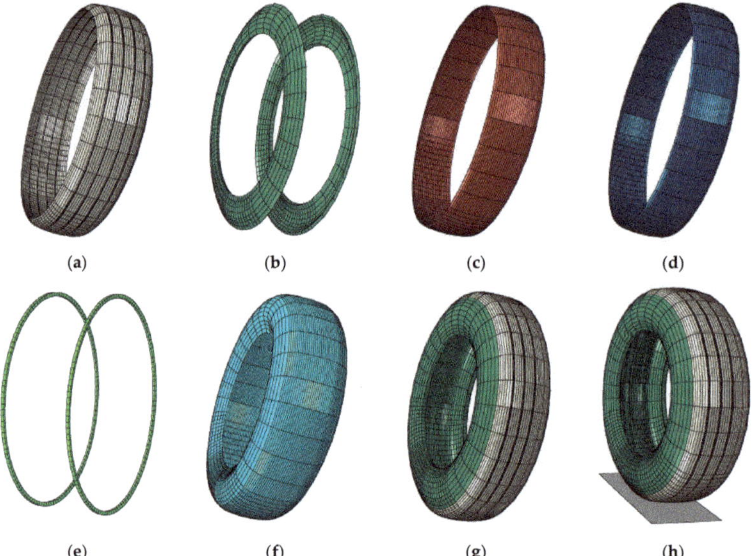

Figure 2. Meshes of tyre components. (**a**) Tread with four longitudinal grooves; (**b**) sidewall; (**c**) steel-belt-1; (**d**) steel-belt-2; (**e**) bead; (**f**) carcass; (**g**) full tyre; (**h**) the contact model of tyre–pavement.

2.4. Modelling Tyre-Pavement Contact

Complex, non-linear contact problems must be considered in the study of tyre–pavement contact. There are many factors that prevent successful modelling of tyre–pavement contact; for example, some challenges that arise when modelling with a two-solid mode include the non-linear contact characteristics of the tyre–pavement, material characteristics of the tyre, fast moving conditions, great deformation of the tyre, and the nonlinear frictional relationship between the friction coefficient and velocity [35]. It is clear that dealing with the tyre–pavement problem with a two-solid mode is not sufficient. In this study, when the tyre was in contact with the road surface, the degree of the deflection of the tyre was much larger than that of the pavement. It can be presumed that the pavement is a non-deformable rigid surface to achieve better astringency and stability, and the accuracy and reliability of contact stresses analysis can still be guaranteed. In previous research, the pavement has been considered as a rigid surface that can successfully capture contact stresses [22,36–38]. The 3D model of tyre–pavement contact is shown in Figure 2h.

2.5. Validation of the 3D Contact Model

The tyre–pavement interaction model can be validated by the tyre load–displacement relationship in a static condition. Figure 3 presents the calculation comparisons of tyre deflections at a tyre load of 25 kN and different inflation pressures. When the tyre load was constant, the lower the tyre inflation pressure, the greater the displacement. The tyre loaded measurement system is presented in Figure 4a and was used to measure the deformation data of tyres with different loads. Figure 4b shows comparisons between calculated and measured tyre displacements at different tyre loads and an inflation pressure of 530 kPa. When the inflation pressure was constant, the larger the tyre load, the larger the displacement. Figure 4c shows comparisons at different tyre loads and an inflation pressure of 730 kPa. Figure 4 reveals that good consistency was achieved between the calculated and tested displacements at various tyre loads and inflation pressures.

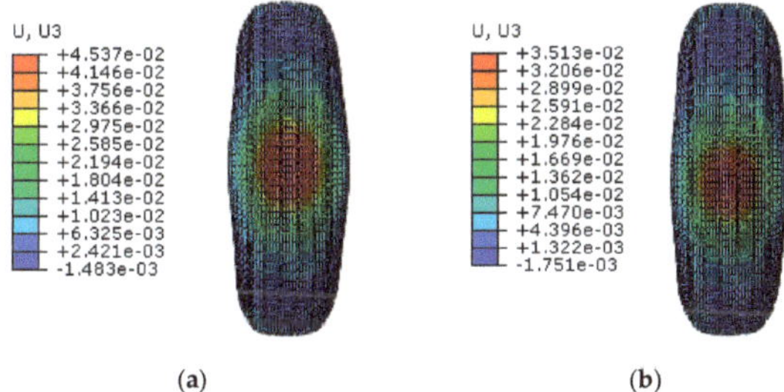

Figure 3. Comparisons of deflections at a tyre load of 25 kN and inflation pressures of (**a**) 530 kPa and (**b**) 730 kPa.

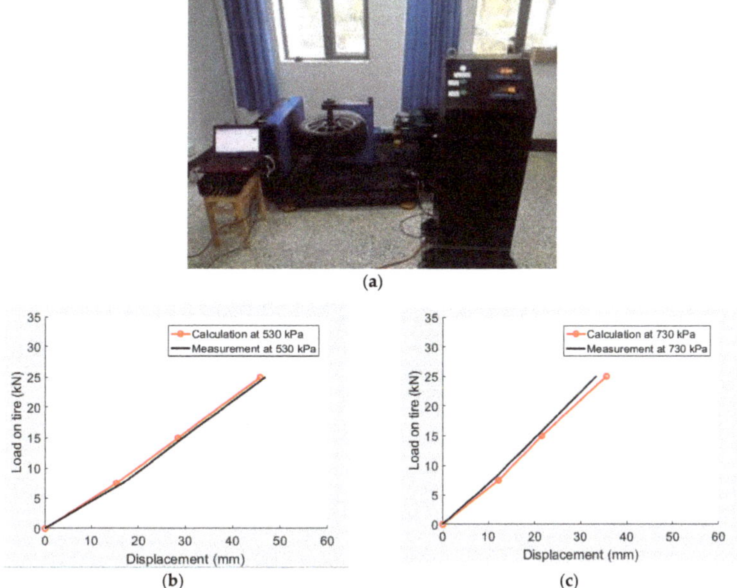

Figure 4. (**a**) Tyre static loaded test. (**b**) Comparisons between calculated and measured tyre displacements at different tyre loads and inflation pressures of 530 kPa and (**c**) 730 kPa.

3. Effects of Tyre Working Conditions on Contact Stresses

3.1. Simulation Condition Design of Tyre–Pavement Contact

For a two-wheel group, an axle load of 100 kN is the Chinese standard in the design of asphalt and town pavement [39]; thus, in the simulation, the vertical load (*F*) of each tyre was 25 kN in accordance with the standard axle load, and the vertical overloads of 30% and 60% were 32.5 and 40 kN, respectively. The rated inflation pressure (*P*) for a 275/70R22.5 tyre was 730 kPa [40], and the inflation pressures of 530 and 930 kPa respectively correspond to low and high inflation pressures. In addition, the friction coefficient between the tyre and road surface was 0.6 [34].

3.2. Realisation of the Steady Rolling State

Analysis of contact stresses requires not only detailed information about tyre materials and components but also knowledge of non-linear characteristics and dynamic analysis. When the tyre is in the breaking state, the direction of braking torque (T) is opposite to the orientation of tyre rolling, as shown in Figure 5. When the tyre is in a free-rolling state, there is no braking torque. Figure 6 shows the longitudinal reaction force at different angular velocities for a linear velocity of V = 15 km/h. At the moment, the angular velocity of the free-rolling tyre corresponds to the angular velocity at zero torque. When the tyre is in the breaking state, the braking torque is transmitted to the tyre, and the angular velocity is less than that in the free-rolling condition, as shown in Figure 6. When the longitudinal reaction force has a positive value, it means that the tyre is in the braking state. When the longitudinal reaction force is basically constant, it means that the tyre is under full-braking conditions. Because the braking and acceleration forces are similar, this study focuses on comparing the change characteristics of the contact stresses and contact area with basic working conditions under free-rolling and full-braking conditions. The basic working condition was a tyre load of 25 kN, an inflation pressure of 730 kPa, and a speed of 15 km/h. Based on the described research method and results, the functional relationship between RR and tyre working conditions are explored.

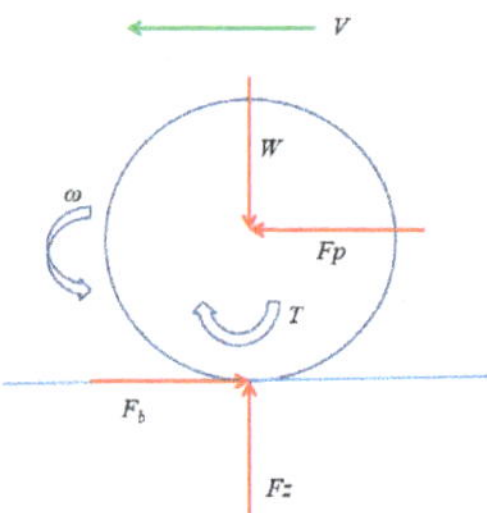

Figure 5. Tyre force diagram during braking.

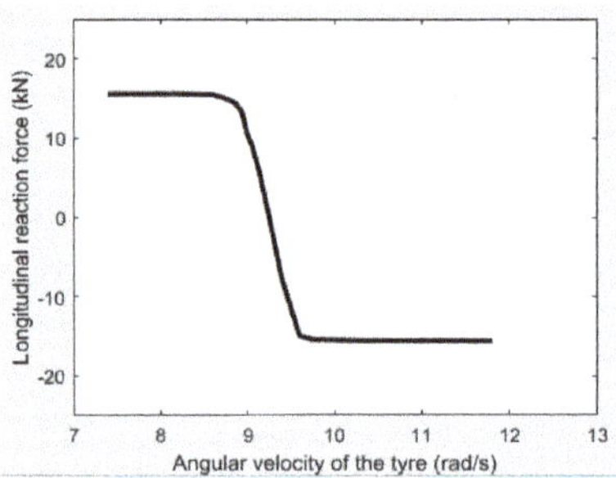

Figure 6. Longitudinal force at different angular velocities for a linear velocity of V = 15 km/h.

3.3. *Effect of Tyre Load on the Change Characteristics of Contact Stresses*

The inflation pressure of 730 kPa and a tyre speed of 15 km/h were considered to be invariant, while the tyre load was varied to examine the effect of tyre load on tyre–pavement contact stresses. Figure 7 shows the vertical contact stresses (CPRESS) under free rolling and full braking with tyre loads of 25 and 40 kN. With the gradual increase of the tyre load under the condition of steady rolling, the shape of the contact region was found to transition from approximately circular to rectangular, and the contact area also gradually increased. Regardless of whether the tyre was in the free-rolling or full-braking condition, the vertical stress at the tread of the tyre was found to change only slightly. However, the vertical stress was found to increase at the tyre shoulder. When the tyre load was small, the maximum value of vertical contact stress was near the centre of the tread, whereas when the load became large, the maximum value of vertical contact stress moved toward the tyre shoulder. Under different load conditions, the vertical stress was found to be greater under the full-braking condition than under the free-rolling condition. When the tyre load changed from 25 to 40 kN, the vertical stress of the free-rolling tyre increased from 1145 to 1425 kPa, an increase of 24.5%, while the vertical stress during full braking increased from 1243 to 1642 kPa, an increase of 32.1%. It is evident that once the overload is excessive during full braking, the maximum value of vertical stress occurring at the tyre shoulder will increase the damage to the tyre.

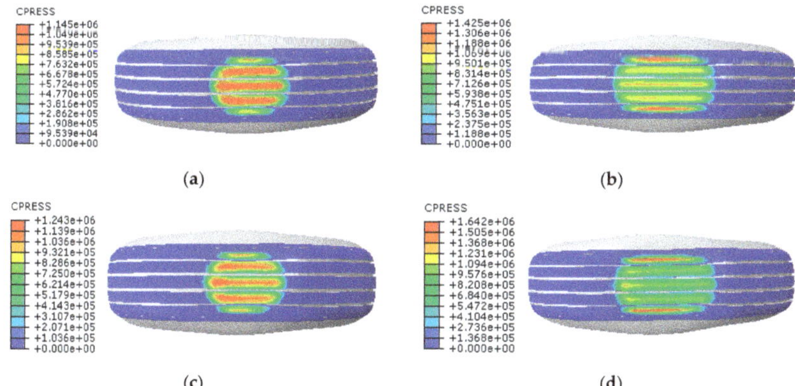

Figure 7. Vertical contact stresses under the free-rolling condition with tyre loads of (**a**) 25 kN and (**b**) 40 kN; vertical contact stresses under the full-braking condition with tyre loads of (**c**) 25 kN and (**d**) 40 kN.

Figure 8 shows the longitudinal contact stresses (CSHEAR1) in the free-rolling and full-braking conditions with tyre loads of 25 and 40 kN. The shapes of the contact region under free-rolling and full-braking conditions were very different. When the longitudinal contact stress was calculated under the full-braking condition, the contact area shape and stress distribution were similar to those when the vertical stress was calculated. However, when the longitudinal contact stress was calculated under the free-rolling condition, the shape of the contact region and stress distribution varied greatly from when the vertical stress was calculated. According to Figure 8a,b, the longitudinal contact stress had positive and negative values during free rolling, and the positive and negative values were relatively close; this is mainly because the torque applied to the tyre was 0. It can be seen from Figure 8c,d that the longitudinal contact stress during full braking was positive. With the increase of load, the longitudinal contact stress gradually increased, and the position at which the maximum value of longitudinal stress occurred also moved from near the centre of the tread to the tyre shoulder, which was similar to the trend of vertical stress. As the tyre load increased, the difference between longitudinal stress during full braking and that during free rolling became increasingly larger. This illustrates that overload under in the braking condition not only does great harm to the tyre but can also accelerate the emergence of ruts and cracks in the pavement.

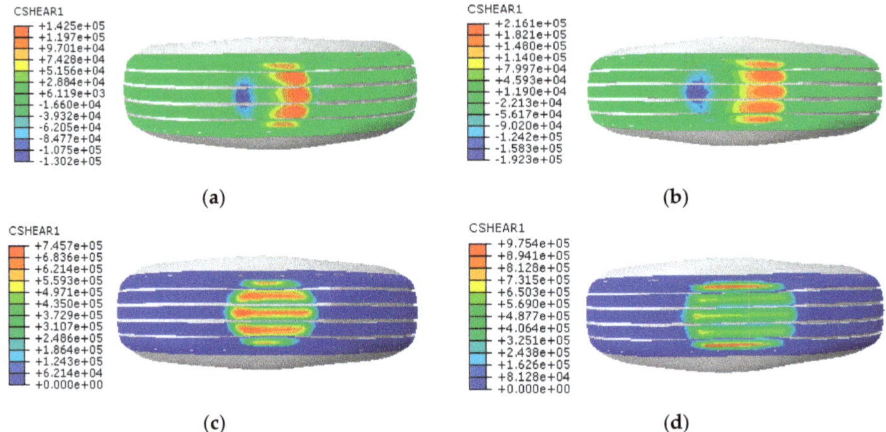

Figure 8. Longitudinal contact stresses under the free-rolling condition with tyre loads of (**a**) 25 kN and (**b**) 40 kN; longitudinal contact stresses under the full-braking condition with tyre loads of (**c**) 25 kN and (**d**) 40 kN.

Figure 9 shows the lateral contact stresses (CSHEAR2) under free-rolling and full braking conditions with tyre loads of 25 and 40 kN. The shape of the contact region for calculating transverse stress was different from those for calculating vertical and longitudinal contact stresses under the two working conditions. Transverse contact stress had positive and negative values that were close to each other under the two working conditions. This is mainly because during the longitudinal movement process of the tyre, the tyre did not undergo sideslip, and the transverse force of the tyre was close to the equilibrium state. When the tyre load was 25 kN, the lateral contact stress during free rolling was greater than that during full braking. When the load continued to increase, the lateral contact stresses during free rolling and full braking were closer, and the difference between them became smaller.

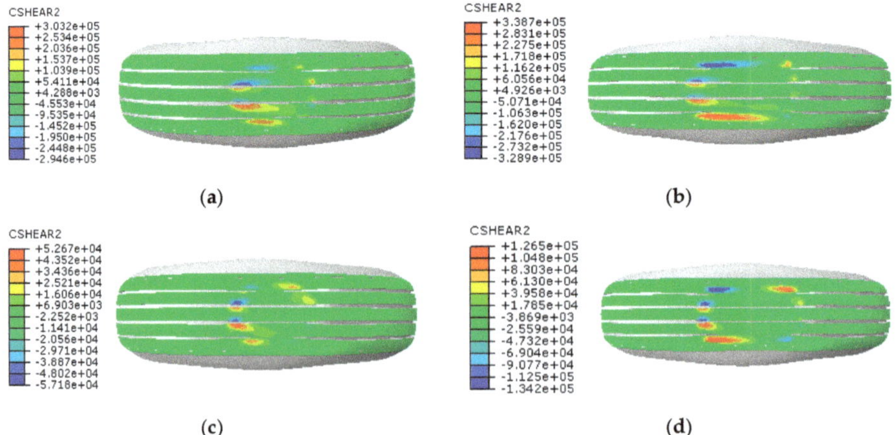

Figure 9. Lateral contact stresses under the free-rolling condition with tyre loads of (**a**) 25 kN and (**b**) 40 kN; lateral contact stresses under the full-braking condition with tyre loads of (**c**) 25 kN and (**d**) 40 kN.

Taking the tyre load of 32.5 kN as an example, the spatial distribution of contact stresses under the free-rolling condition is shown in Figure 10. As presented in Figure 10b, longitudinal contact stress under the free-rolling condition was symmetrically distributed in a longitudinal direction, whereas, as presented in Figure 10c, the lateral contact stress exhibited an almost antisymmetric distribution in a longitudinal direction. The longitudinal contact stress was negative at the front end of the contact area and positive at the back end, and the distribution area of the negative value of the contact area was smaller than that of the positive value.

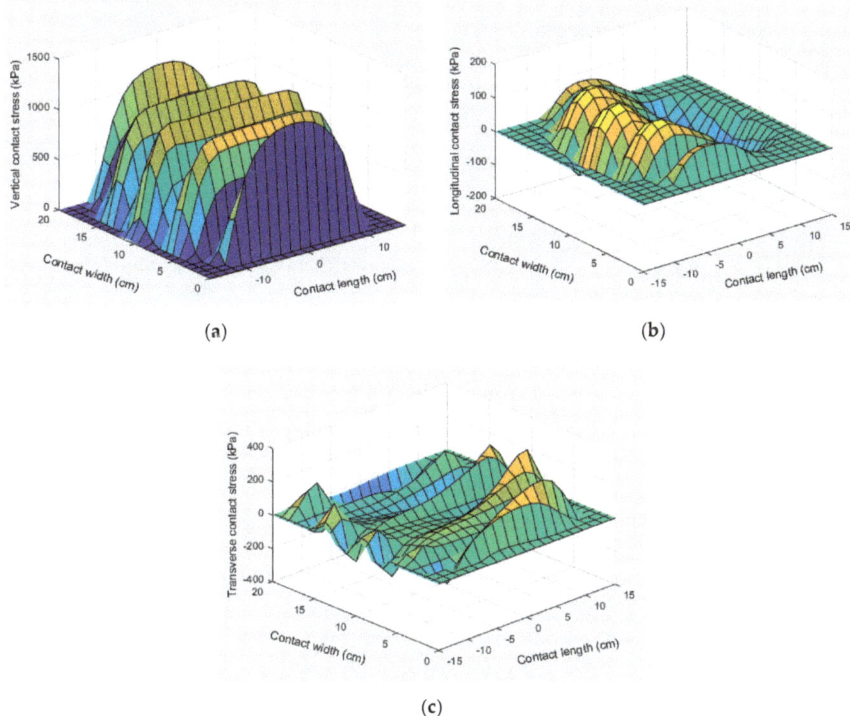

Figure 10. Spatial distribution characteristics of (**a**) vertical, (**b**) longitudinal, and (**c**) transverse contact stresses under the free-rolling condition with the load of 32.5 kN.

3.4. Effect of Inflation Pressure on the Change Characteristics of Contact Stresses

The tyre load of 25 kN and the speed of 15 km/h were considered to be invariant, while the inflation pressure was varied to examine the effect of inflation pressure on tyre–pavement contact stresses. Table 1 presents maximum contact stresses under different inflation pressures under free-rolling and full-braking conditions.

Table 1. Comparison of maximum contact stresses under different inflation pressures.

Inflation Pressure (kPa)	Maximum Contact Stress during Free Rolling (kPa)			Maximum Contact Stress during Full Braking (kPa)		
	Vertical	Longitudinal	Lateral	Vertical	Longitudinal	Lateral
530	931	194	240	1052	631	40
730	1145	142	303	1243	745	52
930	1386	86	381	1495	896	66

As the tyre inflation pressure was increased, the contact area decreased, and the vertical stress increased. Regardless of whether the tyres were in the free-rolling or full-braking condition, the vertical stress was positive, indicating that the direction of stress was downward. It also indicates that the tyre tread was compressed and therefore bore compressive stress. To clearly observe the change trend, Figure 11 exhibits maximum contact stresses under different inflation pressures in the form of a colour histogram.

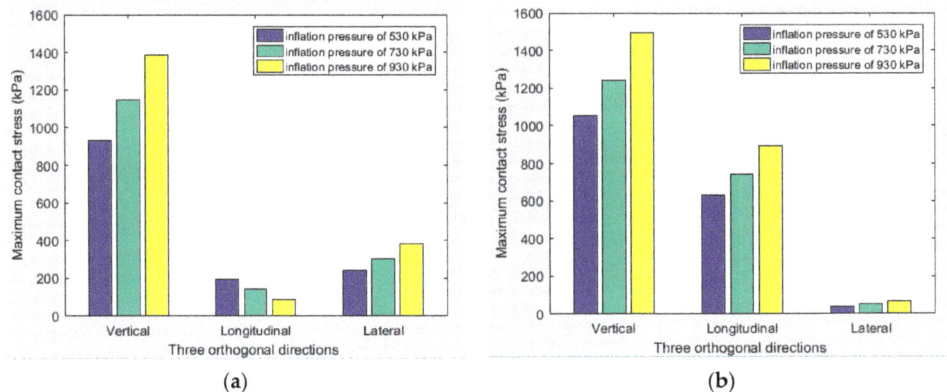

Figure 11. Maximum contact stresses under different inflation pressures during (**a**) free rolling and (**b**) full braking.

When the inflation pressure was 530 kPa, the maximum vertical stress was located in the tyre shoulder. As the inflation pressure increased from 730 to 930 kPa, the maximum vertical stress under the free-rolling condition increased from 1145 to 1386 kPa, an increase of 21.0%. The maximum vertical stress under the full-braking condition increased from 1243 to 1495 kPa, an increase of 20.3%, and the position of the maximum values moved from the tyre shoulder to the crown. Under different inflation pressures, the contact area during the free-rolling condition was similar for different tyre loads. When the inflation pressure was high, the magnitude of the positive value of longitudinal stress was slightly less than that of the negative value. With the increase of inflation pressure, longitudinal stress decreased, and lateral stress increased. Under the full-braking condition, the lateral stress also increased with the increase in tyre pressure, but it was not obvious.

Taking the tyre pressure of 930 kPa as an example, the contact stress value of the node in the contact area under the full-braking condition was extracted, and the spatial distribution characteristics of the complex contact stresses under a high tyre pressure are exhibited in Figure 12. The vertical stress under a high tyre pressure was greater than that under the basic working condition, and the spatial distribution was large in the middle and small on both sides. Longitudinal contact stresses under the full-braking condition were symmetrically distributed in a longitudinal direction, while the lateral contact stress presented an almost antisymmetric distribution in a longitudinal direction. In addition, the front end of the vertical and longitudinal stresses was greater than the back end, which was determined by the motion state of the tyre. Overall, the lateral contact stress was less than 100 kPa and therefore could almost be ignored.

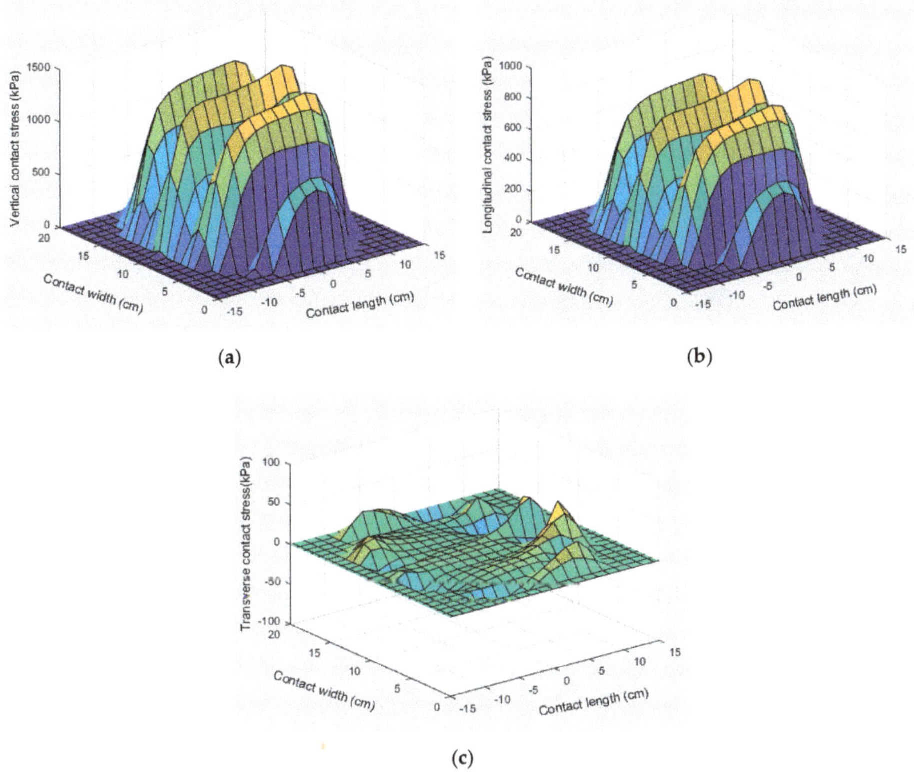

Figure 12. Spatial distribution characteristics of (**a**) vertical, (**b**) longitudinal, and (**c**) transverse contact stresses under the full braking condition with inflation pressure of 930 kPa.

3.5. Effect of the Tyre Speed on the Change Characteristics of Contact Stresses

The tyre load of 25 kN and the inflation pressure of 730 kPa were considered to be invariant, while the speed was varied to examine the effect of speed on tyre–pavement contact stresses. Table 2 presents the maximum contact stresses under different speeds during free rolling and full braking. To clearly observe the change characteristics, Figure 13 shows the maximum contact stresses under different inflation pressures, in the form of a colour histogram. As the speed increased, the vertical and longitudinal stresses gradually decreased under the free-rolling condition, but the reduction was very small. The lateral stress remained essentially unchanged. Under the full-braking condition, vertical stress increased slightly. The longitudinal stress remained basically unchanged, and the transverse stress gradually increased, but the value of transverse stress was negligible compared with the value of vertical stress. Compared with the tyre load and inflation pressure, the impact of tyre speed on the maximum contact stresses is relatively weak.

Table 2. Comparison of maximum contact stresses under different speeds.

Speed (km/h)	Maximum Contact Stress During Free Rolling (kPa)			Maximum Contact Stress During Full Braking (kPa)		
	Vertical	Longitudinal	Lateral	Vertical	Longitudinal	Lateral
15	1145	142	303	1243	745	52
45	1120	116	308	1254	752	82
90	1117	87	315	1256	753	175

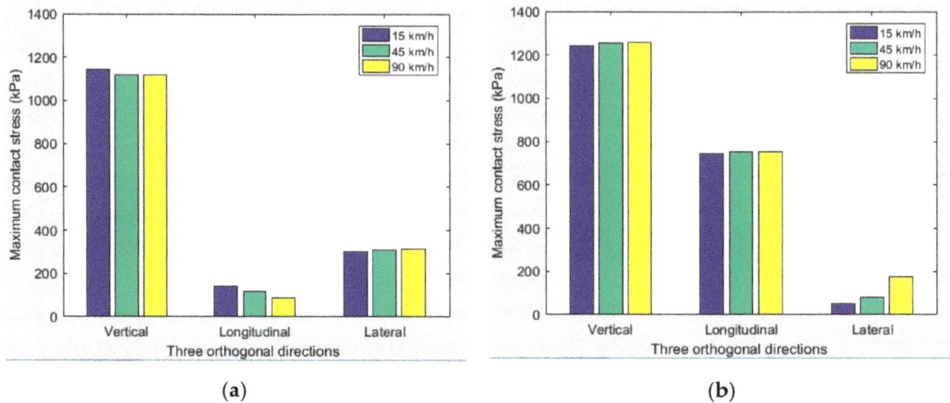

Figure 13. Maximum contact stresses under different speeds during (**a**) free rolling and (**b**) full braking.

4. Effects of Tyre Working Conditions on Rolling Resistance

Due to the viscoelasticity of tyre rubber material, the processes of compression and rebound are not the same in the rolling process, and therefore the sum of external forces on the tyre i.e., the RR, is not 0 [41]. Figure 14 presents changes in RR with the tyre load, inflation pressure, and speed. Figure 14a shows the change in RR with the tyre load when the inflation pressure (P) was 730 kPa, and the speed (V) was 45 km/h. With the increase in tyre load, tyre deformation increased significantly, and the overall RR presented an upward trend, which would cause the vehicle to consume more fuel. Figure 14b shows the change in RR with the pressure when the tyre load (F) was 25 kN and the speed (V) was 45 km/h. With the increase in pressure, tyre stiffness increased, and deformation decreased under the same load; thus, the overall RR presented a downward trend, which would result in the vehicle consuming less fuel. Figure 14c shows the change in RR with the tyre load and speed when the pressure (P) was 730 kPa, and it was found that the increased tyre load led to increased RR. Moreover, as the speed increased, the loss factor of tyre rolling and deformation decreased [42]. However, the increased value of RR was greater than the decreased value, so the overall RR still increased.

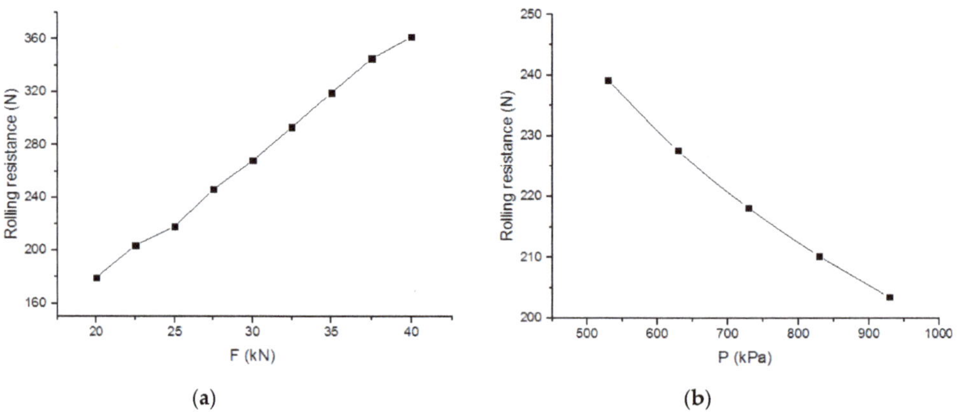

Figure 14. *Cont.*

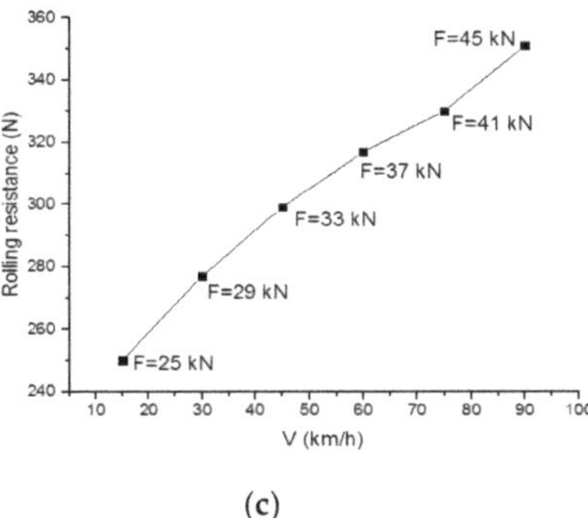

(c)

Figure 14. Rolling resistance (RR) changes with (**a**) tyre load, (**b**) inflation pressure, and (**c**) tyre load and speed.

The exponential Equation (8) is used to fit the functional relationship between RR and tyre load, inflation pressure, and speed parameters, and variable combinations were considered, as shown in Figure 14. The exponential equation is as follows:

$$RR = (a + b \cdot V + c \cdot V^2) \cdot F^m \cdot P^n \quad (8)$$

where a, b, and c are velocity correlation coefficients, m is the tyre load exponent, and n is the inflation pressure exponent. According to these sets of variable combinations, the following relationship is obtained by fitting the functional relationship:

$$RR = (0.13153 - 0.00211 \cdot V + 0.00001 \cdot V^2) \cdot F^{1.13258} \cdot p^{-0.28753} \quad (9)$$

The international standard unit was adopted for the unit of each parameter in the process of fitting. The parameters in Equation (9) are different for different types of tyres, but Equation (9) describes a method that can calculate the RR related to tyre working conditions. The tyre studied in this article is one of the most commonly used truck-bus tyres in China. Therefore, Equation (9) can be used to forecast the RR of the tyre studied in this article in any combination of working conditions, and a similar method can be used to forecast the RRs of other types of tyres.

5. Conclusions

The tyre-pavement contact model developed in this work shows the potential to predict contact stresses and rolling resistance under different rolling conditions. Based on the preceding analysis, the following conclusions can be made and provide valuable suggestions and viewpoints for tyre–pavement contact mechanics and fuel-economic tyre production/design.

(1) The maximum value of the transverse contact stress is greater than the longitudinal contact stress under the free-rolling condition. However, the situation under the full-braking condition is the opposite.
(2) Longitudinal contact stresses under the free-rolling and full-braking conditions were symmetrically distributed in a longitudinal direction, while the lateral contact stress presented an almost antisymmetric distribution in a longitudinal direction. Under the full-braking condition, longitudinal stress is the main component of the horizontal contact stresses.
(3) The tyre load and inflation pressure have significant impacts on contact stresses. Overload and low tyre pressure are important contributors to the wear of the tyre shoulder. Properly increasing the inflation pressure can effectively relieve damage to the tyre shoulder caused by overloading.

Additionally, compared with the tyre load and inflation pressure, the impact of tyre speed on contact stress is relatively weak.

(4) The proposed exponential equation describes a method that can forecast the RR related to the working conditions of truck-bus tyres, and a similar method can be used to predict the RRs of other types of tyres.

It is evident from the full analysis presented in this study that the change rules of contact stresses and RR are significantly affected by tyre working conditions under different rolling conditions. Conventional methods focus on the RR of patterned tyres of passenger cars or on the adoption of special materials to reduce the RR of truck tyres. Our new approach considers the different working conditions of the tyre by investigating changes in contact stresses (forces) to pinpoint the functional relationship between the RR and working conditions of the truck-bus tyre. The proposed exponential equation method can be usefully applied to predict the RR, and better estimations of fuel consumption and greenhouse gas emissions can then be made.

Author Contributions: Conceptualization, M.G. and X.L.; methodology, M.G.; software, M.G.; validation, M.G. and X.L.; formal analysis, M.G. and X.L.; investigation, M.G. and X.L.; resources, M.G. and X.L.; data curation, M.G. and X.L.; writing—original draft preparation, M.G. and X.L.; writing—review and editing, M.R. and Y.Y.; visualization, M.R. and Y.Y.; supervision, X.Z.; project administration, M.G. and X.L.; All authors have read and agreed to the published version of the manuscript.

Funding: This research was funded by the National Major Scientific Research Instrument Development Project of China (No. 51827812), National Natural Science Foundation of China (Nos. 51778509 and 51578430).

Conflicts of Interest: The authors declare that there is no conflict of interest.

References

1. Guan, Y.J.; Zhao, G.C.; Cheng, G. FEA and testing studies on static camber performance of the radial tire. *J. Reinf. Plast. Comp.* **2007**, *26*, 1921–1936.
2. Guan, Y.J.; Zhao, G.C.; Cheng, G. 3-Dimensional non-linear FEM modeling and analysis of steady-rolling of radial tires. *J. Reinf. Plast. Comp.* **2011**, *30*, 229–240. [CrossRef]
3. Park, S.; Yoo, W.; Cho, J.; Kang, B. Pressure-sensing pad test and computer simulation for the pressure distribution on the contact patch of a tyre. *Proc. Inst. Mech. Eng. Part D J. Automob. Eng.* **2007**, *221*, 25–31. [CrossRef]
4. Wang, W.; Yan, S.; Zhao, S. Experimental verification and finite element modeling of radial truck tire under static loading. *J. Reinf. Plast. Comp.* **2013**, *32*, 490–498. [CrossRef]
5. De Beer, M.; Fisher, C. Evaluation of non-uniform tire contact stresses on thin asphalt pavements. In Proceedings of the 9th International Conference on Asphalt Pavements, Copenhagen, Denmark, 17–22 August 2002.
6. Elseifi, M.A.; Al-Qadi, I.L.; Yoo, P.J.; Janajreh, I. Quantification of pavement damage caused by dual and wide-base tires. *Transp. Res. Rec.* **2005**, *1940*, 125–135. [CrossRef]
7. Al-Qadi, I.L.; Yoo, P.J. Effect of surface tangential contact stress on flexible pavement response. In Proceedings of the Journal of Association of Asphalt Paving Technologists: From the Proceedings of the Technical Sessions, San Antonio, TX, USA, 11–14 March 2007.
8. Al-Qadi, I.L.; Wang, H.; Yoo, P.J.; Dessouky, S.H. Dynamic analysis and in-situvalidation of perpetual pavement response to vehicular loading. *Transp. Res. Rec.* **2008**, *2087*, 29–39. [CrossRef]
9. Wang, H.; Al-Qadi, I.L. Evaluation of surface-related pavement damage due to tire braking. *Road. Mater. Pavement.* **2010**, *11*, 101–122. [CrossRef]
10. Wang, G.; Roque, R. Three-dimensional finite element modeling of static tire–pavement interaction. *Transp. Res. Rec.* **2010**, *2155*, 158–169. [CrossRef]
11. Wang, G.; Roque, R. Impact of wide-based tires on the near-surface pavement stress states based on three-dimensional tire-pavement interaction model. *Road. Mater. Pavement.* **2011**, *12*, 639–662. [CrossRef]
12. Ghosh, S.; Sengupta, R.A.; Kaliske, M. Prediction of rolling resistance for truck bus radial tires with nanocomposite based tread compounds using finite element simulation. *Rubber. Chem. Technol.* **2014**, *87*, 276–290. [CrossRef]

13. Hernandez, J.A.; Al-Qadi, I.L. Semicoupled Modeling of Interaction between Deformable Tires and Pavements. *J. Transp. Eng. A Syst.* **2017**, *143*, 1–9. [CrossRef]
14. De Beer, M.; Fisher, C.; Jooste, F.J. Determination of pneumatic tire pavement interface contact stresses under moving loads and some effects on pavements with thin asphalt surfacing layers. In Proceedings of the 8th International Conference on Asphalt Pavements, Seattle, WA, USA, 10–14 August 1997; pp. 179–227.
15. Assogba, O.C.; Tan, Y.; Sun, Z.; Lushinga, N.; Zheng, B. Effect of vehicle speed and overload on dynamic response of semi-rigid base asphalt pavement. *Road. Mater. Pavement.* **2019**, 1–31. [CrossRef]
16. Si, C.; Chen, E.; You, Z.; Zhang, R.; Qiao, P.; Feng, Y. Dynamic response of temperature-seepage-stress coupling in asphalt pavement. *Constr. Build. Mater.* **2019**, *211*, 824–836. [CrossRef]
17. Alireza, S. Numerical comparison of flexible pavement dynamic response under different axles. *Int. J. Pavement. Eng.* **2016**, *17*, 377–387.
18. Tarefder, R.A.; Ahmed, M.U.; Islam, M.R. Impact of cross-anisotropy on embedded sensor stress–strain and pavement damage. *Eur. J. Environ. Civ. Eng.* **2014**, *18*, 845–861. [CrossRef]
19. Ahmed, M.U.; Rahman, A.; Islam, M.R.; Tarefder, R.A. Combined effect of asphalt concrete cross-anisotropy and temperature variation on pavement stress–strain under dynamic loading. *Constr. Build. Mater.* **2015**, *93*, 685–694. [CrossRef]
20. Tarefder, R.A.; Ahmed, M.U.; Rahman, A. Effects of cross-anisotropy and stress-dependency of pavement layers on pavement responses under dynamic truck loading. *J. Rock. Mech. Geotech. Eng.* **2016**, *8*, 366–377. [CrossRef]
21. Wang, H.; Al-Qadi, I.L.; Stanciulescu, I. Simulation of tyre–pavement interaction for predicting contact stresses at static and various rolling conditions. *Int. J. Pavement Eng.* **2012**, *13*, 310–321. [CrossRef]
22. Wang, H.; Al-Qadi, I.L.; Stanciulescu, I. Effect of surface friction on tire-pavement contact stress during vehicle maneuvering. *J. Eng. Mech.* **2014**, *140*, 1–8. [CrossRef]
23. Zhou, H.; Wang, G.; Ding, Y. Effect of Friction Model and Tire Maneuvering on Tire-Pavement Contact Stress. *Adv. Mater. Sci. Eng.* **2015**, *2015*, 1–12. [CrossRef]
24. Cho, J.R.; Lee, H.W.; Jeong, W.B.; Jeong, K.M.; Kim, K.W. Finite element estimation of hysteretic loss and rolling resistance of 3-D patterned tire. *Int. J. Mech. Mater. Des.* **2013**, *9*, 355–366. [CrossRef]
25. Aldhufairi, H.S.; Essa, K. Tire rolling-resistance computation based on material viscoelasticity representation. *Adv. Automot. Eng.* **2019**, *2*, 167–183.
26. Rafei, M.; Ghoreishy, M.H.R.; Naderi, G. Computer simulation of tire rolling resistance using finite element method: Effect of linear and nonlinear viscoelastic models. *Proc. Inst. Mech. Eng. Part D J. Automob. Eng.* **2019**, *233*, 2746–2760. [CrossRef]
27. Anghelache, G.; Moisescu, R.; Sorohan, S.; Buretea, D. Measuring system for investigation of tri-axial stress distribution across the tyre–road contact patch. *Measurement* **2011**, *44*, 559–568. [CrossRef]
28. Anghelache, G.; Moisescu, R. Measurement of stress distributions in truck tyre contact patch in real rolling conditions. *Vehicle. Syst. Dyn.* **2012**, *50*, 1747–1760. [CrossRef]
29. De Beer, M.; Fisher, C. Stress-In-Motion (SIM) system for capturing tri-axial tyre–road interaction in the contact patch. *Measurement* **2013**, *46*, 2155–2173. [CrossRef]
30. Chen, B.; Zhang, X.; Yu, J. Impact of contact stress distribution on skid resistance of asphalt pavements. *Constr. Build. Mater.* **2017**, *133*, 330–339. [CrossRef]
31. Knothe, K.; Wille, R.; Zastrau, B.W. Advanced contact mechanics-Road and rail. *Veh. Syst. Dyn.* **2001**, *35*, 361–407.
32. Hurtford, S. Improving the Quality of Terrain Measurement. Master's Thesis, Virginia Polytechnic Institute and State University, Blacksburg, VA, USA, 2009.
33. Cho, J.R.; Lee, H.W.; Jeong, W.B.; Jeong, K.M.; Kim, K.W. Numerical estimation of rolling resistance and temperature distribution of 3-D periodic patterned tire. *Int. J. Solids. Struct.* **2013**, *50*, 86–96.
34. Guo, M.; Zhou, X. Tire-Pavement Contact Stress Characteristics and Critical Slip Ratio at Multiple Working Conditions. *Adv. Mater. Sci. Eng.* **2019**, *2019*, 1–11. [CrossRef]
35. Laursen, A.; Stanciulescu, I. An algorithm for incorporation of frictional sliding conditions within a steady state rolling framework. *Commun. Numer. Methods Eng.* **2006**, *22*, 301–318. [CrossRef]
36. Zhang, X. Nonlinear Finite Element Modeling and Incremental Analysis of a Composite Truck Tire Structure. Ph.D. Dissertation, Concordia University, Montreal, QC, Canada, 2001.
37. Meng, L. Truck Tire/Pavement Interaction Analysis by the Finite Element Method. Ph.D. Dissertation, Michigan State University, East Lansing, MI, USA, 2002.

38. Ghoreishy, M.H.R.; Malekzadeh, M.; Rahimi, H. A parametric study on the steady state rolling behaviour of a steel-belted radial tyre. *Iran. Polym. J.* **2007**, *16*, 539–548.
39. China Communications Construction Group Road and Bridge Technology Co., Ltd. *Specifications for Design of Highway Asphalt Pavement: JTG D50—2017[S]*; People's Transportation Press: Beijing, China, 2017.
40. Lv, J.; Mao, J.; Feng, Y.; Xiang, Z. *Size Designation, Dimensions, Inflation Pressure and Load Capacity for Truck Tyres: GB/T 2977—2016 [S]*; China National Standardization Administration: Beijing, China, 2016.
41. Ye, J. *Finite Element Simulation and Validation of All-Steel Radial Truck Tire Rolling Resistance*; Tsinghua University: Beijing, China, 2007.
42. Schuring, D.J. *The Rolling Loss of Pneumatic Tires*; The Firestone Tire and Rubber Company, Central Research Laboratories: Akron, OH, USA, 1980.

Article

Permeability and Strength of Pervious Concrete According to Aggregate Size and Blocking Material

Vu Viet Hung [1], Soo-Yeon Seo [2,*], Hyun-Woo Kim [2] and Gun-Cheol Lee [2]

[1] Civil Engineering Department, University of Transport and Communications, Ho Chi Minh 700000, Vietnam; hungvv_ph@utc.edu.vn
[2] School of Architecture, Korea National University of Transportation, Chungju 27389, Korea; khw0029@ut.ac.kr (H.-W.K.); gclee@ut.ac.kr (G.-C.L.)
* Correspondence: syseo@ut.ac.kr

Abstract: The purpose of this study is to identify the differences in porosity and permeability coefficients when the mixing ratio of aggregates is different and to present the mixing ratio satisfying the strength requirement of compressive specified in a specification of Korea. Three mix ratios were suggested by considering various aggregate sizes and three cylinders were made for each ratio. The porosities of those cylinders were evaluated through the compression and water permeability test, measuring the weight of specimens in underwater and analysis of the pictured Computed Tomography (CT) image. Experiments have shown that it is best to mix 50% for 5–10 mm aggregates, 45% for 2–5 mm aggregates, and 5% for sand in terms of strength and permeability. In addition, as the proportion of fine aggregates increased, the porosity and permeability decreased. Moreover, the effectiveness of maintenance method was also examined in this study.

Keywords: porosity; permeability coefficients; mixing ratio; aggregate size; compressive strength; computed tomography (CT) image

Citation: Hung, V.V.; Seo, S.-Y.; Kim, H.-W.; Lee, G.-C. Permeability and Strength of Pervious Concrete According to Aggregate Size and Blocking Material. *Sustainability* **2021**, *13*, 426. https://doi.org/10.3390/su13010426

Received: 30 November 2020
Accepted: 30 December 2020
Published: 5 January 2021

1. Introduction

Recently, the area of buildings and roads has increased due to the rapid industrial development, which has reduced the permeable area of surface runoff during heavy rains. As a result, the flow of rainwater that has evaporated or flowed into the groundwater in the past has now changed to flow into the stream through the drainage system or accumulated on the surface of the road. Therefore, the flooding damage of cities is increasing, the groundwater level is lowered due to the decrease of the surface penetration of rainwater, and the environmental problems such as heat island phenomenon in urban areas are generated. As such, the amount of damage caused by heavy rains in the last five years (2013–2017) in Korea has been about 439 billion won ($0.37 billion), which is 52% of the total damage of 8,486 billion won ($7.07 billion) in five years related to the disaster [1]. One of the solutions to this problem is to increase the permeable pavement area of urban ground and restore the reduced drainage performance.

Pervious concrete, known as no-fines concrete or permeable concrete, is an environmentally friendly paving material, which has been well recognized as one of the key elements of low-impact sustainable development [2]. In general, it consists of cement, water, uniform/single-sized coarse aggregate, and little or no fine aggregate, resulting in a large, open pore structure. As a result, pervious concrete shows better permeability than conventional impermeable pavements due to the porosity between aggregates [3]. Generally, the permeability of pervious concrete is due to its macropore structure. However, as the use of pavement concrete becomes longer, the pores of pervious concrete are easily clogged by various small particles [4–7]. The permeability of pervious concrete helps to reduce heat island phenomena, traffic noise, and provides better condition for recharging the underground water source. In addition, it has the advantage of ensuring the driver's safety by preventing water splash and reducing/eliminating the water film phenomenon on driving surfaces [8]. However, when the sediments, such as leaves, soil, and dust, penetrate the pores, it can be easily noticed that the pores may be blocked and the water permeability decreases. Therefore, continuous or regular maintenance of the pervious concrete is necessary to maintain the permeability performance [9,10].

One of the methods to solve the pore blockage in pervious concrete is the optimum design using the appropriate size aggregate [11–13]. Accumulation of the fine particles in the pores of the pervious pavement causes blockages and deposits, which are mainly related to the size of the clogging particles [4,10,11,14]. Although a series of studies [4,7,14,15] on the particle size have been conducted, there is little literature associated with the combined influence of particle size and pore size on clogging because it is difficult to accurately measure the pore size of pervious concrete. In order to solve this problem, recently, research has been conducted to determine the pore blockage of pervious concrete using the CT scanning method [16–19].

The purpose of this study is to compare and analyze the differences between porosity and permeability coefficient by varying the aggregate mixing ratio of pervious concrete, and to present a mixing ratio that meets the criterion of compressive strength and secures permeability as well. Moreover, in order to investigate the effect of blocking materials to the permeability, experiments on the maintenance of pervious concrete and recovery of permeation performance are carried out by conducting experiments on the pore blocking phenomenon. In addition, pores were identified using the CT scanning method and compared with experimental results.

2. Experiment

2.1. Mixture of Test Specimen

The pervious concrete was designed with a water-to-cement ratio (w/c) of 0.3, the Ordinary Portland cement was used as a binder and a poly-carboxylate high-range water-reducing admixture (AD) was applied in this research. The mix ratios of concrete considered in this study for finding the optimum pores and compressive strength using impact crush type aggregates are shown in Table 1. Three mix ratios were suggested by considering various aggregate sizes: 100% aggregate (D5–10) (Mix 1); 80% aggregate (D5–10) + 15% aggregate (D2–5) + 5% sand (Mix 2); and 50% aggregate (D5–10) + 45% aggregate (D2–5) + 5% sand (Mix 3). When manufacturing test specimens, the component materials were mixed in order of aggregate (D5–10), aggregate (D2–5), sand, AD, cement, and water. Three cylinder specimens with ϕ 100 mm × 200 mm were manufactured for each mixing ratio, and their permeability and compressive strength were measured. For Mix 1, one more cylinder was made for CT scanning and three block specimens with 200 mm × 200 mm × 150 mm for blockage testing. The specimen preparation procedures and their shapes are shown in Figure 1.

Table 1. Mixing ratio of pervious concrete.

Mix Case	Unit Weight (kg/m^3)				
	Cement	Water	Aggregate (D5–10)	Aggregate (D2–5)	Sand
Mix 1	358	107	1595	-	-
Mix 2	431	129	1222	229	76
Mix 3	431	129	764	687	76

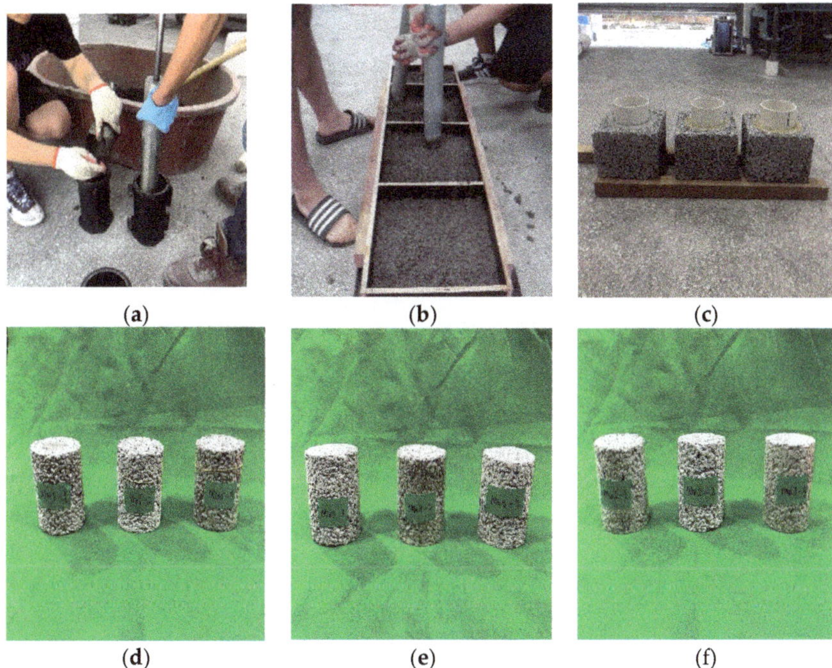

Figure 1. Specimen preparation procedure and their shapes. (**a**) cylinder fabrication; (**b**) block specimen fabrication; (**c**) block specimens; (**d**) mix 1 specimens; (**e**) mix 2 specimens; (**f**) mix 3 specimens.

Immediately after the end of compaction and surface finishing process, the concrete cylinders were covered with a plastic layer to prevent moisture loss and cured in the room under a temperature of 20–25 °C and a relative humidity of 50–60% within 24 hours. Then, after removing the mold, the cylinders were cured in the water environment of 20–25 °C during 14 days and in air under experimental room conditions during the rest period as mentioned above.

2.2. Test for Measuring Void Content of Hardened Pervious Concrete

The void content for each cylinder specimen was measured at 21 days of age by using the volume displacement method: the underwater mass and the dry mass of concrete were measured in accordance with ASTM C1754/C1754M-12 [20] as shown in Figure 2, and the void ratio was calculated by using Equation (1):

$$\mathrm{VCR} = \left(1 - \left[\frac{K \times (A - B)}{\rho_w \times D^2 \times L}\right]\right) \times 100 \quad (1)$$

where VCR is Void content ratio (%), K is 1,273,240 ((mm^3·kg)/(m^3·g)), A and B are dry mass and underwater mass of specimen (g), respectively, ρ_w is density of water (kg/m^3), D and L are average diameter and length of specimen (mm), respectively.

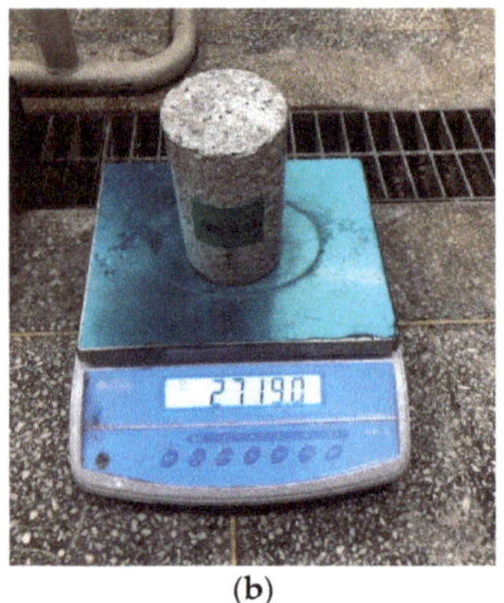

(**a**) (**b**)

Figure 2. Measurement of mass in underwater and dry conditions. (**a**) in underwater; (**b**) in dry condition.

2.3. Evaluation of Void Content of Hardened Concrete by CT Scan

It is worth noting that detailed capture and measurement of internal component, such as porosity, is often vital for quality control, failure analysis, and material research. The measurement method of void ratio using underwater mass has the disadvantage of not being able to identify the void ratio isolated inside the test specimen. Therefore, CT scans were conducted to evaluate pores more accurately. Figure 3a shows the specimens for the CT scan after cutting the cylinder specimen to a height of 50 mm, making it into three specimens at the age of 21 days. Figure 3b,c represent the equipment for CT scan and CT imaging process, respectively. Figure 4 is a representation of the CT-taped image, with the center portion of the specimen cut to 50 mm designated as 0 mm and the position above +5 mm, +10 mm, +15 mm, +20 mm, +25 mm, −5 mm, −10 mm, −15 mm, −20 mm, and −25 mm below. To obtain the void content of each image, the pixel of the area where the void portion was deleted was calculated through Photoshop program and converted to the entire void. For the evaluation of CT scan results, a void content test was performed on the CT scan specimen at the age of 21 days by using the ASTM C1754/C1754M-12 [20] method, and the results were compared with CT results.

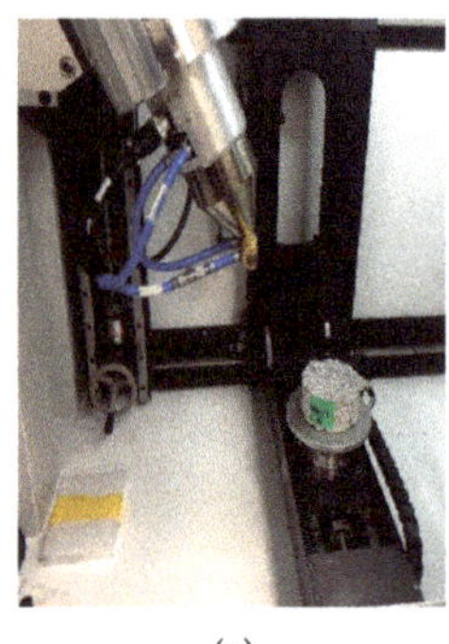

(**a**) (**b**) (**c**)

Figure 3. Truncated specimens and CT imaging. (**a**) truncated specimen; (**b**) CT imaging equipment; (**c**) CT imaging process.

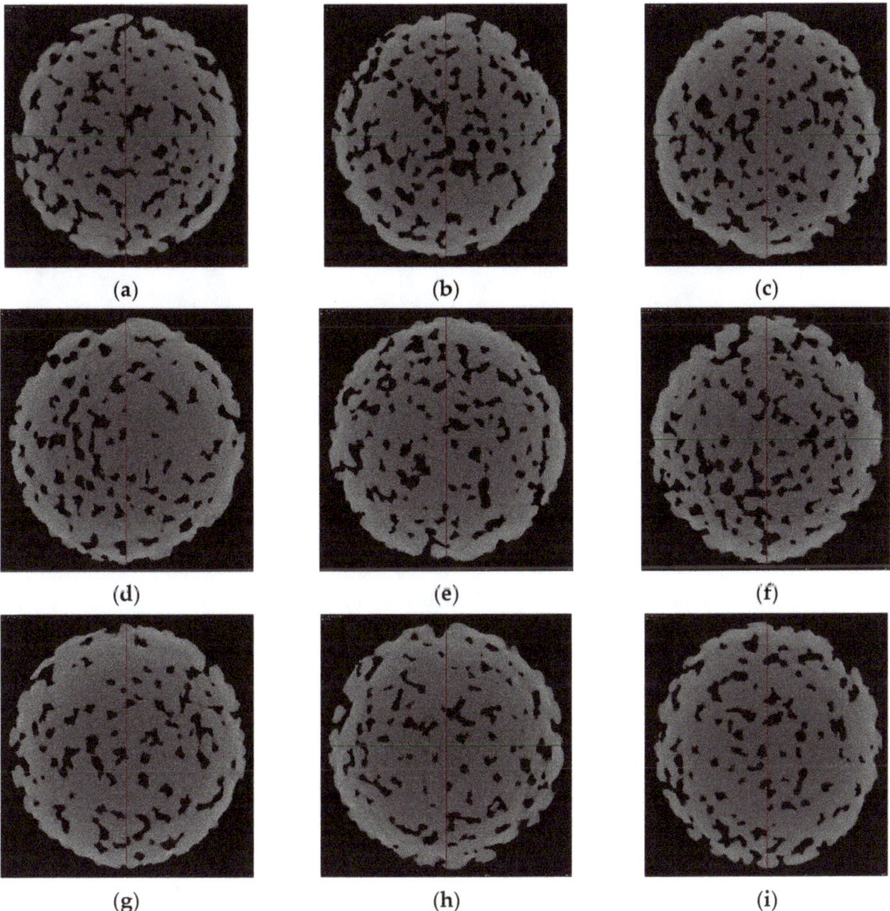

Figure 4. Images with removed void from CT imaging. (**a**) Center; (**b**) +5 mm; (**c**) −5 mm; (**d**) + 10 mm; (**e**) –10 mm; (**f**) +15 mm; (**g**) −15 mm; (**h**) +20 mm; (**i**) −20 mm.

2.4. Water Permeability Test

The water permeability test is an experiment to obtain the permeability coefficient of pervious concrete. In accordance with the ASTM C1701/1701M-17a [21] infiltration test for pervious concrete, the sides of the specimens were wrapped with tape and rubber packing to block the water flowing out of the specimens, and the specimens were fixed on the pipe top in Figure 5a. The experiment was carried out while keeping the water of 1.2 L constant at a height of 10–15 mm at the age of 28 days. In this process, the total time of 1.2 L water passing through the specimen was measured, and the permeability coefficient was calculated by substituting this in Equation (2).

$$\mathrm{I} = \frac{KM}{D^2 \times t} \tag{2}$$

where I is the infiltration ratio (mm/h), M is the mass of infiltrated water (kg), D is the inside diameter of infiltration ring (mm), t is time required for measured amount of water to infiltrate the concrete (s), K is 4,583,666,000 ((mm^3·s)/(kg·h)).

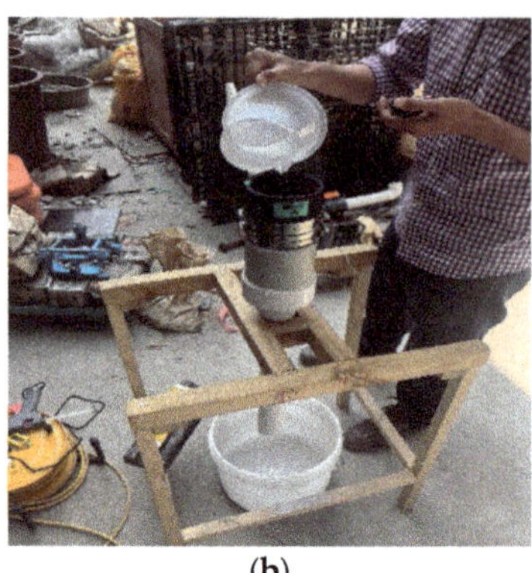

(a) (b)

Figure 5. Equipment and progress for the permeability test. (**a**) device for test; (**b**) test process.

2.5. Compressive Strength Test

In order to know the compressive strength of the pervious concrete for each mix case, the cylinder specimens of ∅100 mm × 200 mm size were cured for 28 days and then tested according to the KS F 2405-2005 [22] test method for cylindrical concrete specimens. In the compression test, the surface of the specimen was polished to be the same load condition for each test specimen, and the test was carried out by padding rubber un-bonded caps on the top and bottom. Figure 6 shows the rubber pad used in the test and the compressive strength test.

(a) (b)

Figure 6. Compressive strength test. (**a**) rubber pad; (**b**) test setup.

2.6. Void Clogging Test

In pervious concrete, "pore blockages" occur in which pores are blocked by deposits such as leaves, sand, and dust on the road. In this study, to verify the effectiveness of the maintenance method of pervious concrete, permeability tests were carried out by accumulating 8.3 g/L of contaminant substance of 0.15 mm or less four times assuming the sandy soils at the age of 56 days as shown in (a) and (b) of Figure 7. After that, high pressure water and vacuum cleaning were performed as shown in (c) and (d) of Figure 7 to recover the permeability. This process was repeated three times to examine the permeability coefficient of the pervious concrete block.

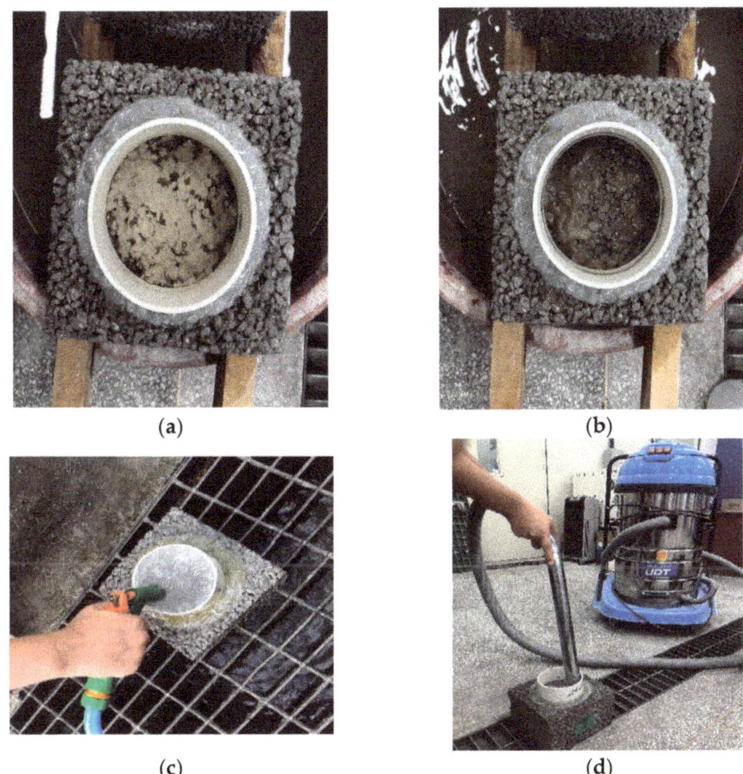

Figure 7. Void clogging test. (**a**) application of the blocking material; (**b**) after water penetration; (**c**) high pressure watering; (**d**) vacuum cleaning.

3. Test Results

3.1. Porosity of Hardened Pervious Concrete

Table 2 shows the information and calculated results for each specimen measured to calculate the porosity. The average void content for Mixes 1, 2, and 3 was measured at approximately 22.2%, 13.7%, and 11.6%, respectively. As a result of measuring the porosity of specimens by mass in water, it was observed that the porosity decreased as the proportion of fine aggregates increased. The reason for this is that, when the amount of fine aggregate increases, the surface area in which the aggregates contact each other increases. Figure 8 shows the porosity graph for each position obtained by analyzing CT images of Mix 1 case. At locations +25 mm and −25 mm, the breaking of concrete was serious once cutting so the results were not included. As a result of analyzing, the porosity after photographing specimens, the porosity ranged from 17.62% to 21.94% and the average was 19.57%. Before the CT measurement as mentioned previously, the void content test for the specimen was also performed by [20]. The porosity obtained from the test was 17.1%, which is 87% of the CT scan result. The reason for this difference is that isolated pores inside the specimen, which could not be measured in the underwater mass, can be identified by CT scan.

Table 2. Measurement results of porosity.

Mix Case		ρ_ω (kg/m^3)	A (g)	B (g)	Porosity Ratio (%)
Mix 1	1	999.26 (14.2 °C)	2555	1600	22.2
	2		2493	1556	22.3
	3		2515	1564	22.0
	Avg.		2521	1573	22.2
Mix 2	1	997.44 (23.4 °C)	2589	1557	13.3
	2		2683	1614	13.5
	3		2609	1571	14.2
	Avg.		2627	1581	13.7
Mix 3	1	997.51 (23.1 °C)	2716	1662	11.6
	2		2724	1670	11.4
	3		2724	1668	11.7
	Avg.		2721	1667	11.6

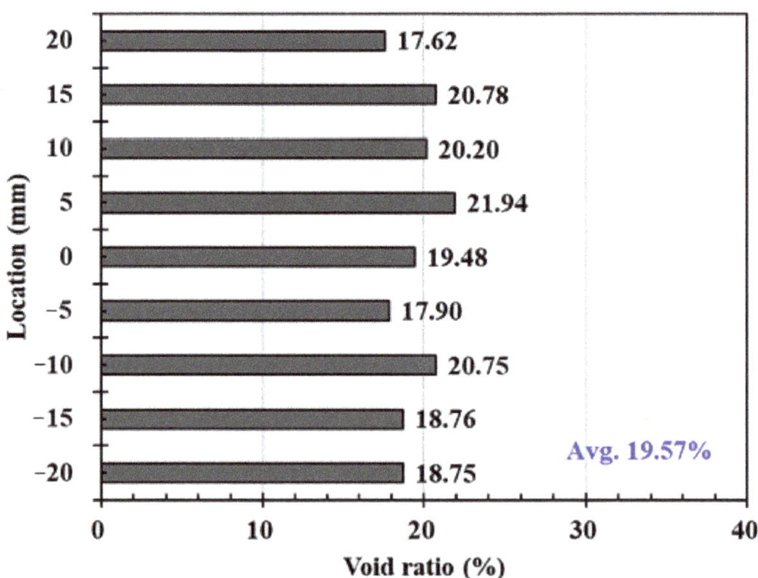

Figure 8. Void ratio at each truncated section from CT image.

3.2. Water Permeability Coefficient of Pervious Concrete

Table 3 shows the permeability coefficients obtained by substituting the results obtained from the water permeability tests of three specimens for each mixing ratio into Equation (2). The diameter of infiltration pipe and weight of infiltrated water for all specimens were 100 mm and 1.2 kg, respectively.

Table 3. Permeability coefficients calculated by Equation (2) using water permeability test results.

No. of Specimen	No. of Tests	Mix 1		Mix 2		Mix 3	
		Time (s)	Permeability Coefficient (mm/s)	Time (s)	Permeability Coefficient (mm/s)	Time (s)	Permeability Coefficient (mm/s)
1	1	94.5	1.62	183.7	0.83	183.7	0.15
	2	88.8	1.72	172	0.89	172	0.18
2	1	36	4.13	225.4	0.68	225.4	0.10
	2	37	4.23	219.4	0.70	219.4	0.09
3	1	63.3	2.41	147.5	1.04	147.5	0.22
	2	61.4	2.19	155.5	0.98	155.5	0.23
Average permeability coefficient		2.72 mm/s		0.85 mm/s		0.16 mm/s	

The average permeability coefficient for each mix case is 2.72 mm/s for Mix 1, 0.85 mm/s for Mix 2, and 0.16 mm/s for Mix 3, respectively. It can be seen that all the mix ratios satisfy the permeability coefficient of 0.1 mm/s or more according to Guide Specifications of Korea Land & Housing Corporation [23]. However, despite the same porosity, there was a difference in the permeability coefficient. This is because the permeability coefficient is affected when the voids in the specimens are not connected smoothly or are isolated by themselves. In other words, the infiltration rate of pervious concrete depends on pore connectivity rather than porosity [24]. In addition, as the proportion of fine aggregates increased, the porosity and permeability decreased. It can be explained that, when the blended aggregates of different sizes were used in pervious concrete mixture, the initial porosity between aggregates after compaction is smaller than that of single size due to the particle packing effect. As a result, the total void content for pervious concrete with an appropriate aggregate mix is to be lower. Therefore, the water permeability coefficient decreases.

3.3. Compressive Strength

Table 4 shows the results of 28-day compressive strength of specimens. The average compressive strengths of Mix 1, Mix 2, and Mix 3 were 11.0 MPa, 13.3 MPa, and 18.4 MPa, respectively. Among them, the average compressive strength of the specimens of Mix 3 satisfied 18 MPa that can be applied to sidewalks, bicycle roads, and other civil facilities without considering the traffic loads [23]. It can be seen that, when the porosity increases, the compressive strength of pervious concrete decreases. In this study, different aggregate mixing ratios significantly influence the void content and thus affect the strength of hardened pervious concrete. By substituting smaller and appropriate-sized aggregates in blended aggregate system, it is believed that the void content becomes lower due to further particle contacting effect, resulting in higher compressive strengths. Similar result was also observed and well-explained in another research [25].

Table 4. Compressive strength results of specimens.

Mix Case	No.	Porosity Ratio (%)	Density (kg/m^3)	Load (N)	Compressive Strength (MPa)
Mix 1	1	22.2	2078	81,200	10.3
	2	22.3	2065	84,000	10.7
	3	22.0	2060	93,200	11.9
	Average	22.2	2068	86,133	11.0
Mix 2	1	13.3	2170	102,400	13.0
	2	13.5	2167	114,200	14.5
	3	14.2	2152	97,200	12.4
	Average	13.7	2163	104,600	13.3
Mix 3	1	11.6	2272	163,000	20.8
	2	11.4	2284	138,000	17.6
	3	11.7	2271	132,800	16.9
	Average	11.6	2276	144,600	18.4

3.4. *Void Clogging Test Results*

Table 5 and Figure 9 represent the void clogging test results. The average permeability coefficient, which was 1.72 mm/s before the test without blocking materials, decreased to 0.77 mm/s after one cycle, and recovered 1.11 mm/s after the internal cleaning; the coefficient is reduced by 55% compared to before the test in the first cycle. After completing the second and third cycles, the permeability coefficients were 0.65 and 0.68, respectively. These values represent 38% and 40%, respectively, as ratios of the permeability coefficient drawn from the permeability test without blocking materials. The experimental results show that the vacuum cleaning and high pressure watering methods are effective in the recovery of permeability. Similar results were also observed in [9,10]. However, further experiments will be needed to restore the exact limit life, maintenance cycle, and permeability of pervious concrete.

Table 5. Variation of permeability coefficient corresponding to accumulation of blocking material.

Test Cycle		No. of Test				
		1	2	3	4	5
First cycle	Permeability coefficient (mm/s)	1.72 *	1.42	1.21	0.93	0.77
	Ratio (%)	100	83	70	54	45
Second cycle	Permeability coefficient (mm/s)	1.11 +	0.93	0.83	0.73	0.65
	Ratio (%)	65	54	48	42	38
Third cycle	Permeability coefficient (mm/s)	0.98 +	0.81	0.78	0.72	0.68
	Ratio (%)	57	47	45	42	40

* Before adding blocking material; + After cleaning the blocking materials of previous cycle.

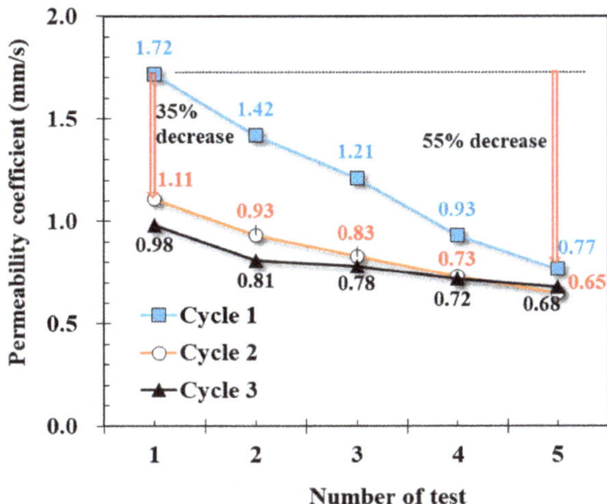

Figure 9. Permeability corresponding to accumulation of blocking material.

4. Conclusions

An experimental investigation is conducted to examine the influence of blended aggregate and blocking materials on the mechanical and permeability characteristics of pervious concrete in this study. The void ratio, density, compressive strength, water permeability coefficient, and the pore clogging phenomenon of pervious concrete are investigated. This contribution will support the introduction and popularization of permeable concrete technology to sustainable development, minimizing the adverse effects of heavy rainfall and urban heat island. Based on observed data in this study, the following conclusions have been drawn:

1. The mixing ratio of pervious concrete Mix 3 (kg/m^3), which contains 431 kg of cement, 129 kg of water, 764 kg of D5–10 mm aggregate (50%), 687 kg of D2–5 mm aggregate (45%), and 76 kg sand (5%), was considered as the most appropriate ratio that can satisfy the permeability coefficient of 0.1 mm/s, the porosity of 8%, and the strength of 18 MPa.
2. Comparing the porosity through the experiment of mass underwater and the CT image analysis, it can be seen that the porosity by the CT image is to be 115% higher. This is because it is difficult to measure the voids isolated inside in the case of the test of underwater mass, but it is possible to measure all the voids inside in the case of CT imaging.
3. From the permeability test results, it was shown that the permeability coefficient is about 2.72 mm/s for Mix 1 with an average porosity of 22.2%, 0.85 mm/s for Mix 2 with an average porosity of 13.7%, and 0.16 mm/s for Mix 3 with an average porosity of 11.6%. From the above results, it can be concluded that, by blending smaller and appropriate-sized aggregates into concrete mix, the porosity of hardened concrete decreases resulting in the reduction of the water permeability coefficient of pervious concrete. As a result of the compressive strength test, it was found that the compressive strength increased as the porosity decreased. This is because the specific surface and adhesion area between aggregates and cement paste increase as the aggregate size gets smaller, which is beneficial to the compressive strength of pervious concrete.
4. Analyzing the results of laboratory research conducted within this study on the pore clogging experiment presented in Section 3.4 and in Table 5, it can be concluded that the water infiltration rate of pervious concrete was restored at a constant rate when the permeation performance was reduced and then cleaned by vacuum cleaning

and high pressure spraying. However, it was not possible to restore the original permeability performance. Further experimentation and research are needed.

Author Contributions: V.V.H.: Conceptualization, Methodology, Investigation, Writing—original draft, Writing—review and editing. S.-Y.S.: Conceptualization, Methodology, Investigation, Writing—original draft, Writing—review and editing. H.-W.K.: Investigation, review and editing. G.-C.L.: Investigation, Writing. All authors have read and agreed to the published version of the manuscript.

Funding: This research was supported by Basic Science Research Program through the National Research Foundation of Korea (NRF) funded by the Ministry of Education (No. 2018R1A4A1025953) and 2018 NRF Postdoctoral Fellowship Program for Foreign Researchers from the National Research Foundation of Korea.

Informed Consent Statement: Not applicable.

Data Availability Statement: The data presented in this study are available on request from the corresponding author. The data are not publicly available due to the continuation of research and other manuscripts that are in process.

Conflicts of Interest: The authors declare no conflict of interest.

References

1. National Disaster Safety Portal. Natural Disaster Statistics. Available online: http://www.safekorea.go.kr (accessed on 13 July 2018).
2. Yang, J.; Jiang, G. Experimental study on properties of pervious concrete pavement materials. *Cem. Concr. Res.* **2003**, *33*, 381–386. [CrossRef]
3. Chindaprasirt, P.; Hatanaka, S.; Chareerat, T. Cement Paste Characteristics and Porous Concrete Properties. *Constr. Build. Mater.* **2008**, *22*, 894–901. [CrossRef]
4. Lin, W.; Park, D.-G.; Ryu, S.W.; Lee, B.-T.; Cho, Y.-H. Development of Permeability Test Method for Porous Concrete Block Pavement Materials Considering Clogging. *Constr. Build. Mater.* **2016**, *118*, 20–26. [CrossRef]
5. Zhang, N. *Experimental Investigation on Clogging Mechanism of Pervious Concrete*; Shandong University: Jinan, China, 2014.
6. Haselbach, L.M. Potential for Clay Clogging of Pervious Concrete under Extreme Conditions. *J. Hydrol. Eng.* **2010**, *15*, 67–69. [CrossRef]
7. Yuan, J.; Chen, X.; Liu, S.; Li, S.; Shen, N. Effect of Water Head, Gradation of Clogging Agent, and Horizontal Flow Velocity on the Clogging Characteristics of Pervious Concrete. *J. Mater. Civ. Eng.* **2018**, *30*, 04018215. [CrossRef]
8. Koo, Y. A Study on Runoff Analysis of Urban Watershed by Hydrologic Infiltration Experiment of Permeable Pavement. *J. Korean Soc. Civ. Eng.* **2013**, *33*, 559–571. [CrossRef]
9. Hein, M.F.; Dougherty, M.; Hobbs, T. Cleaning Methods for Pervious Concrete Pavements. *Int. J. Constr. Educ. Res.* **2013**, *9*, 102–116. [CrossRef]
10. Kia, A.; Wong, H.S.; Cheeseman, C.R. Clogging in Permeable Concrete: A review. *J. Environ. Manag.* **2017**, *193*, 221–233. [CrossRef] [PubMed]
11. Haselbach, L.M.; Valavala, S.; Montes, F. Permeability Predictions for Sand-clogged Portland Cement Pervious Concrete Pavement Systems. *J. Environ. Manag.* **2006**, *81*, 42–49. [CrossRef]
12. Kiran, V.K.; Anand, K.B. Study on Identically Voided Pervious Concrete Made with Different Sized Aggregates. *IOP Conf. Ser. Mater. Sci. Eng.* **2018**, *310*, 012064. [CrossRef]
13. Haselbach, L.M.; Dutra, V.P.; Schwetz, P.F.; da Silva Filho, L.C.P. A Pervious Concrete Mix Design Based on Clogging Performance in Rio Grande do Sul. In Proceedings of the 3rd International Conference on Best Practices for Concrete Pavements, Rodovia, Brazil, 28–30 October 2015.
14. Walsh, S.P.; Rowe, A.; Guo, Q. Laboratory Scale Study to Quantify the Effect of Sediment Accumulation on the Hydraulic Conductivity of Pervious Concrete. *J. Irrig. Drain. Eng.* **2014**, *140*, 04014014. [CrossRef]
15. Kia, A.; Wong, H.S.; Cheeseman, C.R. Defining Clogging Potential for Permeable Concrete. *J. Environ. Manag.* **2018**, *220*, 44–53. [CrossRef] [PubMed]
16. Zhou, H.; Li, H.; Abdelhady, A.; Liang, X.; Wang, H.; Yang, B. Experimental Investigation on the Effect of Pore Characteristics on Clogging Risk of Pervious Concrete Based on CT Scanning. *Constr. Build. Mater.* **2019**, *212*, 130–139. [CrossRef]
17. Yu, F.; Sun, D.; Hu, M.; Wang, J. Study on the Pores Characteristics and Permeability Simulation of Pervious Concrete Based on 2D/3D CT Images. *Constr. Build. Mater.* **2019**, *200*, 687–702. [CrossRef]
18. Sumanasooriya, M.S.; Bentz, D.P.; Neithalath, N. Planar Image-Based Reconstruction of Pervious Concrete Pore Structure and Permeability Prediction. *ACI Mater. J.* **2010**, *107*, 413–421.
19. Deo, O.; Neithalath, N. Compressive Behavior of Pervious Concretes and a Quantification of the Influence of Random Pore Structure Features. *Mater. Sci. Eng. A* **2015**, *528*, 402–412. [CrossRef]

20. *Standard Test Method for Density and Void Content of Hardened Pervious Concrete*; ASTM C1754; ASTM International: West Conshohocken, PA, USA, 2012.
21. *Standard Test Method for Infiltration Rate of in Place Pervious Concrete*; ASTM C1701; ASTM International: West Conshohocken, PA, USA, 2017.
22. *Method of Test for Compressive Strength of Concrete*; KS F 2405; Korean Standards Association: Seoul, Korea, 2005.
23. Korea Land & Housing Corporation. *34023 Concrete Pavement with Permeable Cement in Guide Specifications*; Korea Land & Housing Corporation: Jinju-si, Korea, 2012.
24. Sonebi, M.; Bassuoni, M.; Yahia, A. Pervious Concrete: Mix Design, Properties and Applications. *RILEM Tech. Lett.* **2016**, *1*, 109–115. [CrossRef]
25. Fu, T.C.; Yeih, W.; Chang, J.J.; Huang, R. The Influence of Aggregate Size and Binder Material on the Properties of Pervious Concrete. *Adv. Mater. Sci. Eng.* **2014**, *2014*, 963971. [CrossRef]

MDPI

Article

Modernisation of Regional Roads Evaluated Using Ex-Post CBA

Petr Halámek [1,*], Radka Matuszková [2,*] and Michal Radimský [2]

[1] Faculty of Economics and Administration, Masaryk University, 602 00 Brno, Czech Republic
[2] Faculty of Civil Engineering, Brno University of Technology, 602 00 Brno, Czech Republic; radimsky.m@vut.cz
* Correspondence: halamek@econ.muni.cz (P.H.); matuszkova.r@fce.vutbr.cz (R.M.); Tel.: +420-602-513-254 (P.H.); +420-724-573-969 (R.M.)

Abstract: The aim of this evaluation is to verify the telling value of the Cost and Benefits Analysis (CBA) of regional roads modernisation based on an ex-post evaluation of the investments and their impacts on the incidence of traffic accidents. A set of 144 projects were the subject of evaluation. The analysis of the actual investment costs confirmed the assumption that the majority of projects were planned with a sufficient provision. When compared with the costs foreseen for the entire set of projects, the total reduction of actual costs spent was over 11%. The investigation of project impacts on traffic accidents was based on an analysis done prior to and after construction by using the Czech Police database. The measurement results show only minimum changes in the incidence of traffic accidents in the scenario prior to and after project completion. This however strongly contradicts the project goals declared, because the projects were anticipating almost zero accidents with a fatality and a 50% reduction of accidents with health consequences. However, a slight increase in road fatalities and in light and serious injuries was measured. These facts have a significant impact on the Net Present Value (NPV) and the weighted profitability index for the entire set of projects dropped from 16.7% to −2.8%. The key recommendation is to eliminate the impact on traffic accidents in the case of project evaluations processed ex-ante for projects focused only on a reconstruction or modernisation of existing roads.

Keywords: ex-post CBA; road modernisation; incidence of traffic accidents; decision-making process

Citation: Halámek, P.; Matuszková, R.; Radimský, M. Modernisation of Regional Roads Evaluated Using Ex-Post CBA. *Sustainability* **2021**, *13*, 1849. https://doi.org/10.3390/su13041849

Academic Editor: Edoardo Bocci
Received: 31 December 2020
Accepted: 4 February 2021
Published: 8 February 2021

Publisher's Note: MDPI stays neutral with regard to jurisdictional claims in published maps and institutional affiliations.

1. Introduction

Ex-post evaluation of projects in the public sector does not currently enjoy much popularity. The money was spent and any possible discrepancy between the project impacts declared and quantified and the reality can have a negative impact on the institutions (or individuals) responsible for project preparation and implementation. When making decisions about whether a project, supported by a CBA, shall be implemented or not, there is actually no argument against the verification of costs and benefits even through ex-post CBA. Boardman, Mallery and Vining state that the primary benefit of an ex-post CBA is the possibility of defining the actual socio-economic value of the project evaluated and the benefits of major experience for the evaluation of similar projects [1].

Odeck and Kjerkreit mention a systematic approach to ex-post project evaluation in the UK (referred to as POPE—Post Opening Project Evaluation) or New Zealand [2]. Ex-post evaluation based on a sample of large projects has been also adopted by the European Commission [3]. A systematic approach to ex-post project evaluation is presently lacking in the Czech Republic. The assessment of project impacts is almost always based on the monitoring of compliance with project indicators (programmes) for projects financed from European Structural and Investment Funds (particularly the European Regional Development Fund (ERDF) and the Cohesion Fund (CF)) in the sustainability phase (usually five years since formal project completion).

CBA ex ante evaluations are however often an integral part of the decision-making process on whether to implement a project or not. Integrating the CBA into the decision-making process for the spending of public or private resources only makes sense if its results are highly reliable. This evaluation shall assess whether (and how) CBA results change and whether this fact should affect the decision-making process regarding relevant projects.

The analysis rests upon a relatively homogeneous set of regional road modernisation projects in the Czech Republic. All ex-ante evaluations were prepared in the uniform programme environment eCBA 1.0 [4]. The investment volume, residual value, operating cash flow gap and anticipated socio-economic impacts including their valuation are among key inputs for the evaluation.

In ex-ante evaluations investments are usually entered based on the civil design and itemised budget and/or building volume, construction budget and other related expenditure. Ex-post verification of investment volumes is usually quite easy, because sufficient data is available for projects financed from operational programmes. The costs actually spent are usually part of the final report of any project. On the contrary, it is problematic to determine the residual value already for ex-ante project applications. The linear loss of value method and/or estimation of operating cash flow since evaluation completion until the end of the service life are usually adopted [5]. Ex-post analyses after several years of operation can improve the precision in determining the residual value only to a very limited degree. The ability to verify the operating cash flow usually depends on the type of the project. If a new civic amenities project is to be measured (e.g., construction of a new school building, sports stadium, etc.), the ex-post identification of operating income and expenses is usually easy, because the investor monitors the income and expenses in detail. It is however problematic to identify operating expenses in such projects as road infrastructure modernisation and this requires a detailed identification of the costs spent on the respective road sections.

Time savings, lower costs for vehicle owners and fewer external negative factors in transport are among the key inputs for the socio-economic analysis of road infrastructure modernisation projects (particularly less noise, lower emissions and higher safety). The ex-post verification of most such inputs is very problematic. Time measurements prior to construction and other road traffic data, allowing for speed measurements prior to modernisation, are not easy to obtain. Speed can have a significant impact on the incidence of traffic accidents, because the higher the speed the faster the driver must react to a risk stimulus. The central traffic intensity and flows monitoring system based on "floating car data" is only now being introduced in the Czech Republic. The use of other systems (e.g., data from navigation service providers or mobile phone operators could not be obtained due to the higher number of projects investigated, start of construction of the first projects dating back to 2008 and 2009 and construction only on regional roads) cannot be achieved. Similarly, accurate measurements of emissions from transport and noise are not available. Given the specifics of the pool of projects evaluated (modernisation of regional road infrastructure during the period 2008–2015) in the Czech Republic, the only impact that can be at least partly qualified is the change in the incidence of traffic accidents.

These facts define the objective of this article. The main aim of this evaluation is to verify the impacts of the road infrastructure modernisation projects on the incidence of traffic accidents. And subsequently together with the ex-post verification of the investment cost to analyse the impact on CBA results. The findings should be helpful to decision-making process concerning the regional road maintenance and modernisation.

The importance of ex-post evaluations of investments has been repeatedly stressed in the applicable literature. Boardman says that without an ex-post evaluation the benefits of the CBA for the decision-making process cannot be evaluated either [1]. The ability to verify whether the benefits declared actually materialised answers many important questions regarding the telling value of the CBA. He uses the example of the Coquihalla Highway to demonstrate the size of the deviation in the resulting evaluation from expected development and highlights assumptions that can significantly affect preliminary evaluation results.

Anguera followed up on Boardman's study with an ex-post economic assessment of the Channel Tunnel connecting France with the UK and says that the British economy would be doing better without the tunnel than with the tunnel [6]. He says that the total cost invested in the tunnel is higher than the benefits arising from the tunnel's operation. He identifies lower transport demand, which does not correspond to the rather optimistic assumptions when making the decision about the infrastructure project, as one of the main causes. Other ex-post evaluations stated in the literature include the Stockholm metro analysis where—contrary to the previous example—it is little surprise that significant socio-economic benefits were demonstrated, including urbanisation impacts [7].

The significance of the ex-post evaluation for the decision-making process is also confirmed by Odeck and Krejkreit [2]. They perform an ex-post analysis of a set of 27 road infrastructure projects in Norway. He then benchmarks the results of ex-post analyses done after five years since the opening date against the ex-ante results used for decision-making. The comparison shows an undervaluation of the NPV results by an average of more than 50% in absolute terms and 0.14% on the profitability index level. i.e., the Net Present Value constituting the investment unit. Traffic intensity and the traffic intensity growth index were among the undervaluated inputs. The positive impact of growing traffic volumes on CBA indicators was partly compensated by higher investment costs, but this effect was low enough not to significantly affect the undervaluation of the preliminary NPV.

An ex-post evaluation of a set of 10 projects financed from ERDF and CF resources was carried out by Jong, Vignetti and Pancotti [8]. The aim of this evaluation was to verify the investment costs and to estimate the demand and expected benefits. The largest deviations were identified mostly in monetised benefits. One half of the projects had to be marked as underachieving, but none of the projects dropped below the failure level. The study brought an interesting proposal: to categorise projects based on the ex-ante and ex-post comparison in six classes bearing the name of an object in space—from clear stars to black holes.

Besides the impacts of CBA, Welde also introduces other factors that affect the decision-making process in project selection [9]. Besides inaccuracies and deviations from inputs, he also mentions the limited capabilities of the CBA in covering key impacts (monetised or non-monetised) that must be taken into account in decision-making. The effectiveness in reaching the goals declared can be more important than the economic efficiency of the projects evaluated. What he sees as a separate item in ex-post evaluations is also the cost performance comparing investment cost estimations with actual, real-world compliance. Long-term project impacts are taken into account as strategic success, composed of the Relevance criterion (project impact on economic development and national trends in the incidence of traffic accidents), the Other Impacts criterion (e.g., insufficiently monetised negative traffic impacts on the environment) and the Sustainability criterion (robust infrastructure with respect to future transport volumes). Welde and Meunier & Welde also adopt this method [10,11].

Nicolaisen and Driscoll mention the absence of a uniform approach to ex-post evaluations of transportation projects [12]. According to these findings, a missing standard methodology for retrospective evaluations is the major issue making it hard to draw mutual comparisons of results and having negative impacts on feedback for future evaluations.

2. Materials and Methods

To ensure a correct evaluation, the aspects leading to correct quantification and accounting for a change in the incidence of traffic accidents in the CBA must be correctly defined as well. This includes the definition of a reference scenario and estimated increments/changes resulting from project implementation (or planned project), consideration of time impacts and valuation (quantification) of impacts on the incidence of traffic accidents based on shadow prices.

It is also important to note that negative CBA results (both ex-post and ex-ante) do not automatically mean that the projects implemented are bad projects and/or that their costs are not compensated by possible socio-economic benefits. This analysis does not aim to question the decision-making process as such. According to Sudiana the possible loss of trust of the public in the decision-making process in the public sector can invalidate this process for the future [13]. Therefore, the analysis aims to improve the telling value of the costs and benefits.

The crucial data for the evaluation is the list of the projects collected in the eCBA 1.0 system and the database of the accidents in the Czech Republic [4,14]. The accident database makes it possible to verify the change in accidents on modernized sections of regional roads and to determine the mean percentage error. Subsequently, using the standard methodology for project evaluation, it is possible to calculate the impact on the CBA results.

2.1. *Identification of Reference Scenario and Impact Value Estimation*

The reference scenario (zero variant—no project implemented) was defined in accordance with the considerations of Florio et al. and the European Commission as "business as usual" [5,15]. The change in the number of traffic accidents was observed by comparing the situation prior to and after project opening. The statistics of Czech Police, summarised in the Uniform Traffic Vector Map, were used to determine the number of accidents [14]. Due to data availability, 1 January 2007 was set as the starting point for the measurements. The monitoring period prior to project implementation is also limited from the other side by the date directly preceding the commencement of physical construction (i.e., usually the site handover day) for each individual project. The incidence of traffic accidents during construction is not followed. The incidence of traffic accidents after project implementation, i.e., since the day directly following the day of construction completion (site handover day) until the present (namely data processing as of 31 December 2019) is determined in a similar manner. Figures 1 and 2 show the incidence of traffic accidents of a selected project.

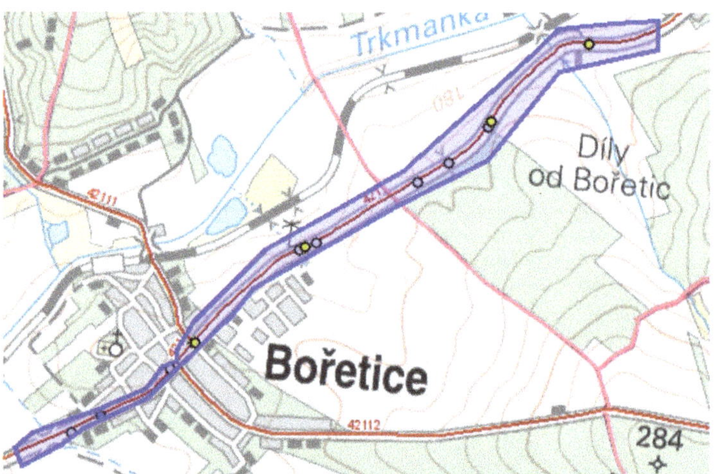

Figure 1. Example of the incidence of traffic accidents prior to construction (II/421 Bořetice through road—Kobylí, 1 January 2007 until 28 February 2012). Inputs from Czech Republic Police map and data [14], the authors' own method.

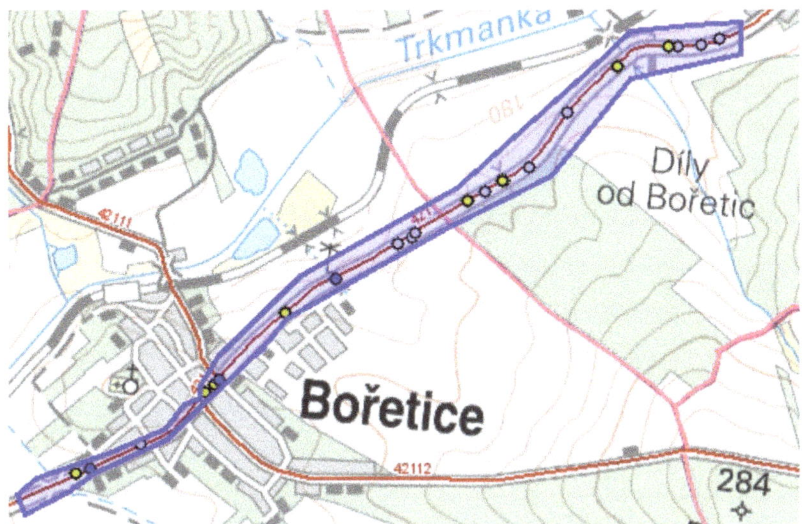

Figure 2. Example of the incidence of traffic accidents after construction (II/421 Bořetice through road—Kobylí, 1 September 2015 until 31 December 2019). Inputs from Czech Republic Police map and data [14], the authors' own method.

The inputs allow—besides the incidence of traffic accidents—the identification of fatalities (within 24 h), seriously injured persons (within 24 h) and lightly injured persons (within 24 h). Given the usual issues with statistics tracking the incidence of traffic accidents and the potential impact of external and internal factors, for example Lord and Mannering say that this method is highly simplified [16]. But when considering the goal followed (i.e., to identify the deviation in NPV estimation), it appears to be acceptable. For the sake of complexity, the national trend in the development of road accidents and their impacts was identified (Figure 3).

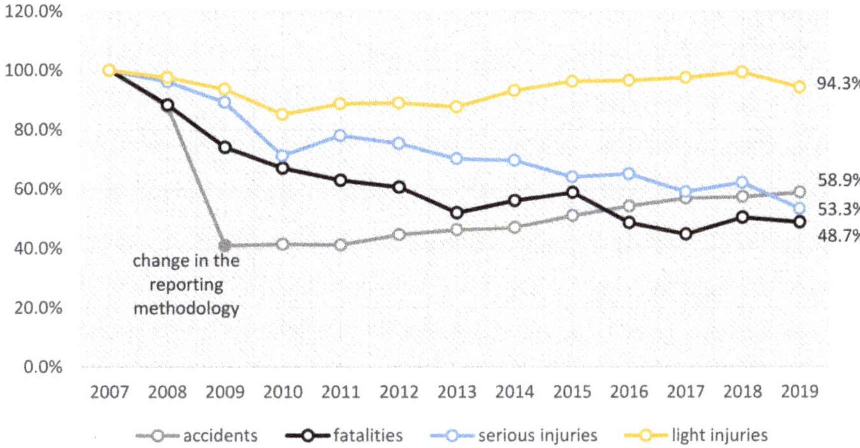

Figure 3. National trend of traffic accidents (2007 = 100%). Input data from Czech Republic Police [17], authors' own method.

According to the statistics of the Czech Republic the general trend in road fatalities, seriously and lightly injured persons has been positive, i.e., declining, since 2007 [17].

According to the Czech Statistical Office ČSÚ the following are the main factors causing the positive traffic accident trends in the Czech Republic: gradual car fleet improvements, more safety features in vehicles, development of transport infrastructure, traffic education and awareness campaigns in media, demographic changes (significant demographic groups become more responsible) and introduction of the penalty-points system [18].

The national component goes in the same direction as the projects evaluated (less injured and fewer fatalities). An adjustment for the national trend is theoretically possible, e.g., based on the shift-share analysis [19,20], but highly problematic due to a change in the traffic-accident reporting methodology. Also Figure 3 shows a significant drop in traffic accidence between 2008 and 2009, caused by the change in legislation and reporting since 1 January 2009. This drop is therefore probably only formal. Also with regard to these conclusions (minimum impact of the projects investigated on the incidence of traffic accidents), the shift-share analysis is not applied on the data measured.

2.2. Inaccuracy Measurement

To measure the inaccuracy of the estimation, the percentage error (*PE*) is used, expressing the percentage deviation of the estimation from the reality measured as of the ex-post analysis date. The mean percentage error (*MPE*) and mean absolute percentage error (*MAPE*) are also defined on the entire pool of projects. Indicators were defined according to Odeck and Kjerkerit [2].

$$PE_{relevant\ indicator} = \frac{(relevant\ indicator_{ex-post} - relevant\ indicator_{ex-ante})}{relevant\ indicator_{ex-ante}} \times 100 \quad (1)$$

$$MPE = 1/n \sum_{i=1}^{n} PE_i \quad (2)$$

$$MAPE = 1/n \sum_{i=1}^{n} |PE_i| \quad (3)$$

To determine the deviation of the variables followed (number of accidents and change in investment costs), their impacts on *NPV* changes must be identified. The variables followed track the change in investment costs and incidence of traffic accidents (fewer fatalities, seriously and slightly injured persons and lower material damage). The resulting impact on *NPV* changes can be defined as follows.

$$\Delta NPV = NPV_{ex-ante} - \Delta NPV(I) + \Delta NPV(accidents) \quad (4)$$

The data from the set of projects however contain the investment value prior to construction and after completion. To duly calculate $\Delta NPV(I)$, its spreading in time (in years) must be defined as well. The taxable supply date on invoices issued during construction is the main criterion of investment classification in time. As these data are not available for the set of projects, the investment was equally spread in time during construction depending on the number of days of construction during the different calendar years. The change in investment (i.e., difference between ex-post and ex-ante) in a year (*t*) is therefore calculated according to the formula shown below.

$$\Delta I_t = \frac{\Delta I(total)}{p\ (costruction\ days\ total)} \times q\ (construction\ days\ in\ the\ year\ t) \quad (5)$$

If we know the investment during the different years, we can calculate, i.e., measure the change in the *NPV* (ex-ante vs. ex-post) in a standard manner.

$$NPV(\Delta I) = \sum_{t=0}^{s} I_t \times 1/(1+r)^t \quad (6)$$

To determine the impact of the incidence deviation on the change of the *NPV* in the project, shadow prices must be defined, making it possible to determine their socio-economic value in money. The ex-ante evaluation was based on unit prices on the 2008 price level and no changes in prices were assumed for the 2007–2015 period. All evaluations were counting with real prices valid at the time of the evaluation (2008–2013) and prepared by valuating impacts on the incidence on the 2008 price level. Given that there were only marginal changes in prices during this period, the method appears to be acceptable. Identical valuation as in the case of the ex-ante evaluation was used for the ex-post evaluation. For a comparison, the current method used in the Czech Republic for infrastructure project evaluations was applied for the change in incidence [21]. The valuation available in the current methodology was converted (for the sake of comparison with the ex-ante methodology) to the 2008 price level (Table 1).

Table 1. Valuation of impacts on the incidence of traffic accidents Data [4,21,22], authors' own method.

	Ex-Ante Method	Current Method	Current Method
price level	2008	2017	2008
material damage	48,500	344,900	282,973
fatalities	9,662,427	20,790,000	17,057,114
serious injury	3,243,737	5,033,600	4,129,807
light injury	364,577	649,800	533,127

The evaluation parameters primarily involving the discount date (r), evaluation period and conversions of price levels are preserved in line with the ex-ante evaluation. The discount rate is determined in real terms as 5.5% p.a. and the valuation period is 25 years. No testing of changes to the discount rate was applied [14]. Conversion factors were not used for the ex-ante and ex-post evaluations.

All projects with a positive NPV (i.e., when the net present value of benefits exceeds the net present value of costs) are suitable for implementation and, if a choice shall be made, such a combination of projects shall be chosen which maximises net benefits [22]. Since the CBA is not perfect (impact valuation, non-quantifiable impacts), this principle and potential externalities (e.g., political will) are not always reflected in practice. The projects were supported from the resources of the European Structural and Investment Funds (ESIF) under the Regional Operational Programme South-East. The CBA results were taken into account in the evaluation criteria in the section dealing with project quality in terms of benefits and adequacy in the criterion "adequacy of the investment in terms of the outputs achieved, the socio-economic outputs and the standard situation". The criterion was defined as a point criterion (possible gain: 0 to 12 points out of 100) and negative NPV led to a zero point gain in this criterion [23]. The influence of the ex-post change in NPV was therefore only investigated based on whether NPV drops below zero tolerance where the non-quantifiable benefits associated with implementation must be proven [24]. The actual change in points with regards to the limited number of authorised applicants in this field of support (only two regional road administration units) had only a marginal impact on project selection.

2.3. *Set of Projects Evaluated*

A set of 144 projects implemented between 2008 and 2015 within the regions of South Moravia and Vysočina in the Czech Republic with the support of resources from the Regional Operational Programme South-East was processed. All projects apply to category II and III roads and primarily involved the reconstruction and modernisation of the road surface, increase in load-bearing capacity, restoration of horizontal and vertical traffic signs, road shoulder adjustments, drainage, bridge and turn-off lane upgrades. Adjacent pedestrian pavements, bus stops, safety islands, parking and long-term parking areas and other local road connections were also usually built as part of these project activities. The direction and elevation of the road changed only in very few projects. Roads were relocated only in exceptional cases (only four projects). In some projects (mostly in the Vysočina region) the road was widened, usually to category S9.5. The total length of the sections modernised was 564.6 km and the total investment was CZK 8.1 billion (actual costs spent). Given the marginal representation of new roads (relocated roads) in the set of projects, no conclusions can be made regarding the impact of new construction on a change in the incidence of traffic accidents.

Higher safety and smoother traffic, time savings, less noise and dust and lower emissions, higher travel convenience and lower operating costs (lower fuel consumption and slower wear) are among the most frequent project goals. A project list can be found in Appendix A.

3. Results

When considering the arguments mentioned in this analysis, the ex-post evaluation aims at a change of investment costs, project impacts on the incidence of traffic accidents and their impacts on evaluation results.

3.1. Ex-Post Evaluation of Investment Costs

The main source of data for the ex-post identification of investment costs are final reports on project implementation and sustainability control reports and/or summary of data from investors in the system of the control body of the programme [25]. Ex-ante values come from eCBA 1.0 [4]. The project units were evaluated—due to system deployment only during 2008—only in the final phase of construction. The minimum 0.0% deviation is also from this period when investment costs were entered in the economic evaluation only after signing the contract with the contractor and the financial limit of the contract was met without any changes.

The results (Table 2) indicate a relatively large drop in investment costs during tendering and construction. The average size of the project decreased from CZK 63.1 million to CZK 56.1 million after implementation, which corresponds to a decrease of approximately 11.2%. The average value decreased from CZK 54.5 million to CZK 45.9 million, the maximum size of the project increased slightly from CZK 199.7 million to CZK 201.4 million.

Table 2. Investment cost deviation (amounts in million CZK).

	Ex-Ante	Ex-Post	MPE(I)	MAPE(I)
average	63.08	56.13	−11.19%	17.40%
median	54.46	45.84	−2.25%	8.55%
min	5.83	3.88	−60.97%	0.00%
max	199.66	201.38	46.10%	60.97%
reference deviation	43.13	42.73	22.91%	18.62%
projects in total	144	144	144	144
cost increase			53	
cost decrease			91	

3.2. Ex-Post Evaluation of the Incidence of Traffic Accidents

Increased safety was one of the major decision-making factors. The ex-ante evaluation quantified a positive impact on safety in 123 projects (86%) and fewer injuries were quantified in 4 projects (63%). No negative impact on safety was mentioned for any of the projects evaluated. A reduction in fatalities in traffic accidents of 5.9 persons p.a., a reduction in the number of seriously injured of 15.0 persons p.a., a reduction in the number of light injuries of 100.2 persons p.a. and a reduction in the number of traffic accidents total of 300.7 were assumed for the entire pool of projects. The ex-ante estimates were verified based on Czech Republic Police data [14] and measurements done prior to project commencement and after project completion. The average incidence monitoring period prior to project commencement was 1795 days (i.e., 4.9 years) and the average incidence monitoring period after project completion was 2573 days (i.e., 7.0 years). The average monitoring time therefore exceeded the time (3 years) recommended by Ambros for determining the long-term mean value [26]. Separate analyses were done for the reduction of fatalities, serious injuries, light injuries and incidence of traffic accidents as such. The results are shown in Table 3.

Table 3. Summary values prior to and after project and ex-ante target values (all figures are p.a.).

	Traced Prior to Project (Number)	Traced after Project Completion (Number)	Change (Number)	Change (%)	Ex-Ante Goal (Number)	Deviation (Number)	PE
accidents total	658.6	590.0	−68.7	−10.4%	358.0	−300.7	64.8%
light injuries	235.3	277.1	41.8	17.8%	135.1	−100.2	105.2%
serious injuries	30.7	30.6	−0.1	−0.4%	15.7	−15.0	94.6%
fatalities	6.6	6.9	0.3	4.8%	0.7	−5.9	921.6%

The analyses show no significant impact on the incidence of traffic accidents. The average annual number of accidents total dropped by 10.4%, but when taking the change in the reporting method in 2009 into account (see above), the change is absolutely marginal. The number of fatalities and heavy injuries remained unchanged. The slight growth in fatalities by 4.8% and decrease in serious injuries by 0.4% (comparison of all annual values) lies within statistical errors. The total number of light injuries saw a more distinct growth by 17.8%. It is obvious based on the summary data that ex-ante target values regarding lower incidence of traffic accidents as a result of project implementation could not be reached. The CBAs assumed almost complete elimination of fatalities, but this did not happen—their number has not changed after project completion. The deviation in the original estimation therefore reaches 900%. The target value in serious and light injuries was about one half of the original volume. Even this goal was missed, because the number of serious injuries did not change and now there are even more light injuries than before. The deviation in estimation was therefore 94.6% for heavy injuries and 105.2% for light injuries. The reduction in the number of traffic accidents did not materialise, even with a big help in the form of the change in accident reporting methodology. The status quo or even a slight increase of the existing numbers shall be assumed without the methodical impacts.

A breakdown of the comparison of ex-ante estimates and ex-post measurements for light injuries, serious injuries an fatalities is shown in Figures 4–6. There is a graphic representation of ex-ante estimates and ex-post measurements (actual values recorded). Axis (x) shows the anticipated decrease in the number of injuries/fatalities after project completion. The estimation of the reduction of injuries for all projects was negative, i.e., all projects sit to the left from axis (y). The actual changes, as measured (i.e., difference between injuries prior to project commencement and after project completion), often go to negative values, i.e., the number of injuries/fatalities go up rather than down after project completion. These projects are shown in the figures below the line representing positive impact on incidence of traffic accidents, i.e., below axis (x). The portion of projects where injuries grew at least little when compared with the original situation exceeds 55%, 32% for serious injuries and 20% for fatalities.

The correlation between ex-ante estimate and ex-post measurement is shown in the so-called "matching line". Projects lying close to this line show no or minimum deviations, meaning that the original estimate was accurate. Projects lying below this line show how overstated the expected positive impact on the incidence was (75% of the projects evaluated). In projects lying above this line the expected positive impact on light injuries was understated (20% of the projects evaluated). The evaluation for serious injuries (Figure 5) and fatalities (Figure 6) is represented in the same manner.

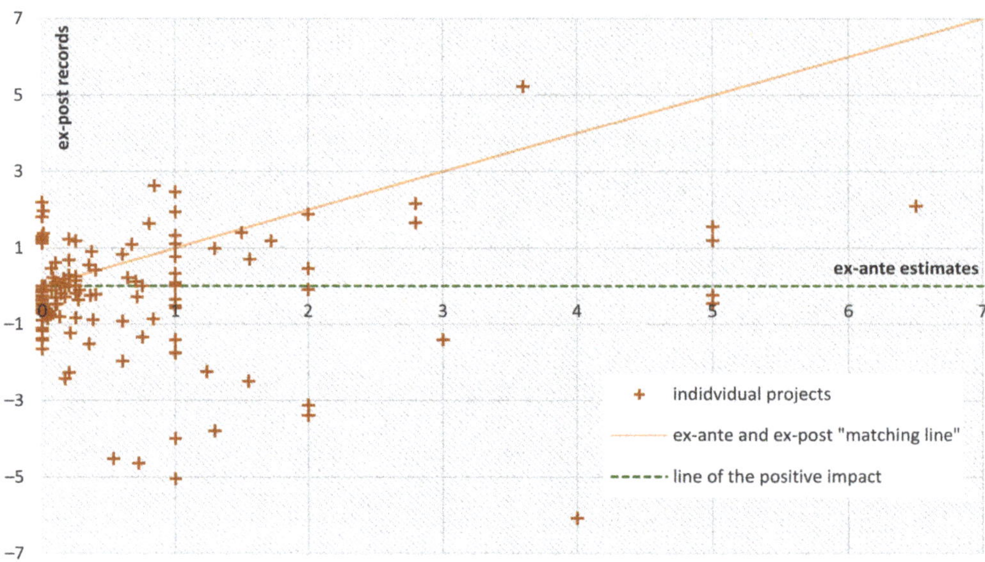

Figure 4. Comparison of ex-ante and ex-post values representing the reduction in the number of light injuries (p.a. values).

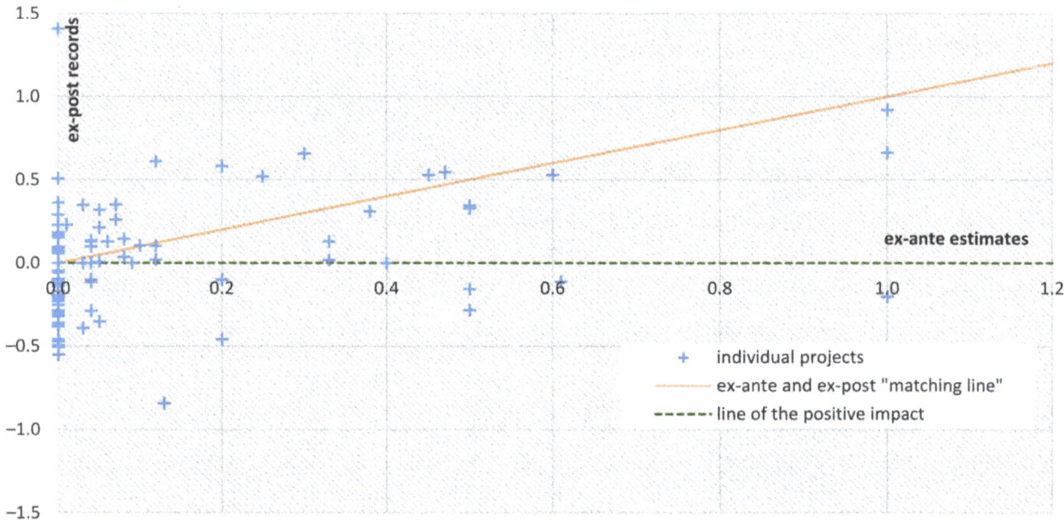

Figure 5. Comparison of ex-ante and ex-post values representing the reduction in the number of serious injuries (p.a. values).

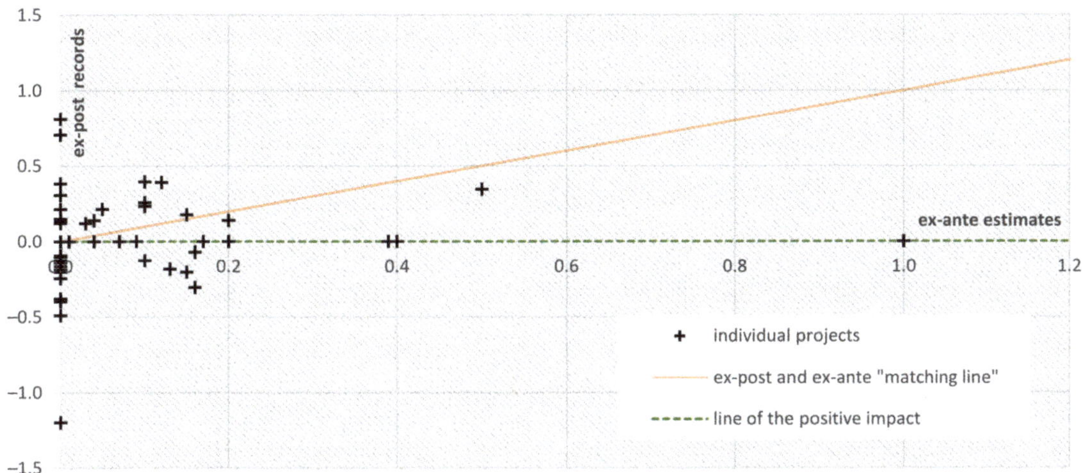

Figure 6. Comparison of ex-ante and ex-post values representing the reduction in the number of deadly injuries (p.a. values).

3.3. Impact on NPV and Decision-Making about Project Implementation

The positive NPV as a criterion was only a part of the decision-making process—it was not a pre-condition for project delivery. The total net present value (ex-ante) of all projects was CZK 1081.2 million and the profitability index (weighted average according to investment volume) reached 16.7% (Table 4). One half of the projects (50.0%) achieved positive NPV. The other half reached negative NPV values. As required for this methodology, the investment volume and impact on incidence of traffic accidents according to reality were corrected.

Table 4. NPV error.

	Ex-Ante (Million CZK)	Ex-Post (Million CZK)	MPE(NPV)	MAPE(NPV)
average	7.51	−1.70	3.8%	267.2%
median	−0.30	−5.67	−3.1%	58.8%
min	−145.17	−237.97	−5521.2%	0.2%
max	333.44	303.20	5261.7%	5521.2%
reference deviation	65.06	74.32	763.7%	715.5%
number of projects	144	144	–	–
total	1081.2	−245.0	–	–

The ex-post correction of the NPV calculation resulted in significantly worse results of all projects. The NPV total for all projects dropped below zero to CZK −256 million. The summary profitability index (weighted average for the entire set) dropped below zero to −2.8%. The average positive NPV in the amount of CZK 7.5 million dropped to −1.7 million. The mean percentage error (MEP) of the NPV is 3.8%. Very high values of minimum and maximum changes were mostly observed in projects with an ex-ante NPV close to zero (i.e., even a small change in NPV can lead to huge changes in the percentage indicator).

The comparison of ex-ante and ex-post NPV shows that the revision of the investment volume and of the negative impact on the incidence of accidents according to reality (and/or ex-post data) had no significant impact on 52 projects (i.e., 36%)—their NPV remained positive. The NPV revision had a critical impact only on 20 projects (14%) where

NPV dropped below zero. These projects were partly compensated by 15 projects (10%) where negative NPV increased to positive NPV and their positive impact on the society could be proven only ex-post. Most projects remained in negative values (40%).

The correlation between the ex-ante and ex-post evaluation is represented in the Figure 7 where the profitability index was applied (social rate of return per one investment unit). Projects meeting the conditions of a positive evaluation in the design phase and after construction are shown in the top-right segment (to the right of axis y and at the same time above axis x). Projects where their positive impact on the society could only be proven ex-post lie in the bottom-left segment (to the left of axis y and at the same time above axis x). For projects below axis x, no positive CBA results were proven through quantification (valuation).

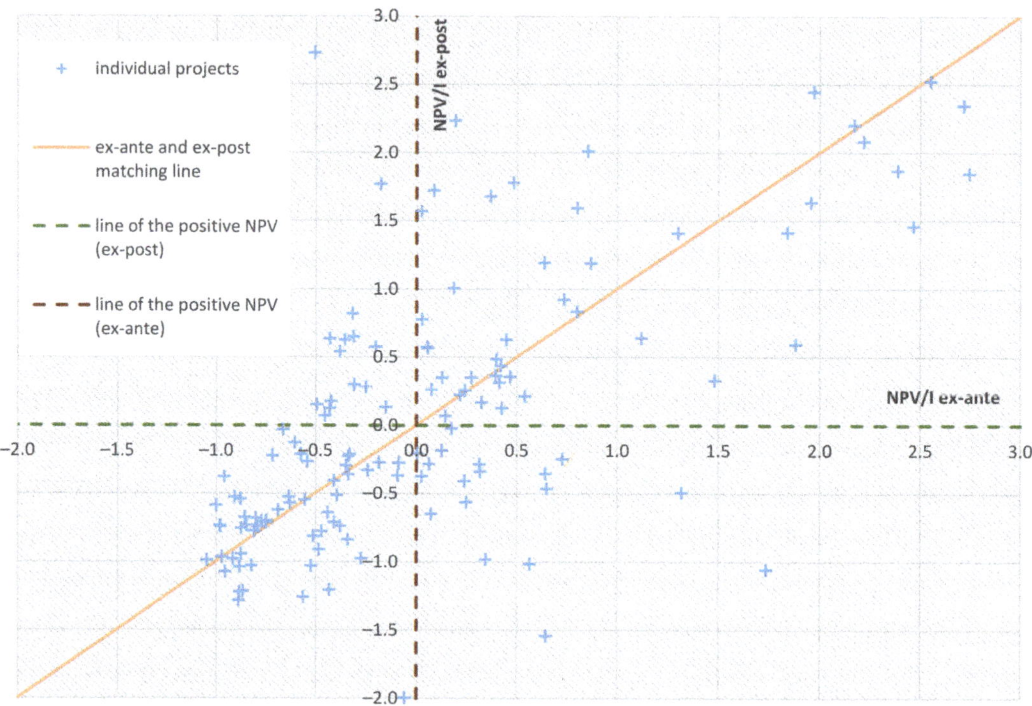

Figure 7. Comparison of ex-ante and ex-post profitability index (NPV/I).

The change in NPV (and/or the profitability index) would therefore have an impact on project evaluation. Contrary to the NPV/I shift towards the bottom, the points rating grew. The average point gain per one project increased from 3.3 to 3.8 points (Table 5). This fact was due to the way the point system is set up: projects with negative NPV/I results gain zero points (except for a few ratings assigned manually). In the 0–100% range the evaluation is linear and projects with NPV/I over 100% are already evaluated equally, with the same full rating.

Table 5. Change in NPV results and the points rating.

Compliance with Criterion NPV > 0	Number of Projects	Share (%)	Ex-Ante Points Average	Ex-Post Points Average
prior to = yes, after = yes	52	36.1%	6.6	8.8
prior to = yes, after = no	20	13.9%	4.9	0.0
prior to = no, after = yes	15	10.4%	0.0	5.7
prior to = no, after = no	57	39.6%	0.5	0.0
total	144	100.0%	3.3	3.8

4. Discussion

The obtained results clearly demonstrate the importance of ex-post CBA for the decision-making process in accordance with the experience described in the literature [1–3,6–12]. The results of the evaluation indicate a reduction in investment costs, a slight negative impact on the incidence of traffic accidents, especially in the number of light and severe injuries and fatalities. The most significant impact on the NPV values is presented by the unfulfilled ex-ante expectations of the reduction of the injuries and fatalities in the traffic accidents.

The average decrease in investment costs by 11.2% contradicts the results published earlier. In Odeck and Kjerkreit an average cost increase of 9.3% on a set of 27 transportation projects in Norway was given, past investigations in Norway showed an average increase of 7.9% [2,27]. Odeck further introduces a meta-analysis of 21 studies investigating the change in investment costs and none of them comes to a negative change (i.e., no investment cost saving) [28]. Flyvbjerg investigated the correlation between the cost increase depending on the size and time of construction and the type of investor [29]. This dependence was confirmed only for construction duration: every year that passed between the moment of decision to build the project and the actual construction increased the investment by 4.64%. On the contrary, Subramani sees the cause of the investment cost increase primarily in project management [30]. The likely cause of the drop in investment costs is corresponding control by the investor and control bodies and/or the impact of the economic cycle (a considerable portion of the projects were tendered at a time when supply was exceeding demand on the building market between 2010 and 2012), as indicated by the negative trend in construction work prices [21,31].

Huge ex-ante expectations of reduction in the number injuries and fatalities remained unfulfilled. The reason can be found in the sense of the modernization. After modernization, the road has a homogenized width dimensions, in many cases the road has been widened. The new pavement is designed with corresponding centripetal cross fall and skid resistance. As part of the modernization, the sight distance is improved, e.g., by reducing road vegetation. The new traffic signs and road marking have a higher retroreflection and are complemented by delineator posts. These facts create a feeling of greater comfort and safety in the drivers, which leads to smoother traffic, but also an increase in average speed. It is the higher speed that brings a higher risk of an accident and it probably compensates the construction benefits. Another factor that could have affected the incidence of traffic accidents may be the increase of traffic volume due to the induction effect, when a newer, higher quality route of the road takes vehicles from the surrounding network. Unfortunately, we are not able to verify this for the evaluated sample of regional roads, because the ongoing data of traffic volume are available in the Czech Republic only for the superior network (highways).

However, we can say that if the incidence of traffic accidents remained the same after project completion but traffic is now more smooth, the project would have a positive impact. The fact that speed/smooth traffic can have an impact on the number of accidents prior to and after project completion can be seen in the different values of the mean incidence values p.a. where the numbers are broken down into rural areas (undeveloped land) and urban areas (developed land). The number of accidents decreased in urban areas by 1.74 p.a., while it grew by 0.6 in out-of-town areas.

These findings can be supported by the evaluation of correlation between relative number of The Figure 8 shows the positive effect of modernized crossroads on the number of people with injuries. A parameter of the number of crossroads per 1 km was created for each modernized road. It is obvious that in sections with a larger number of crossroads there is a reduction in the number of injuries and fatalities.

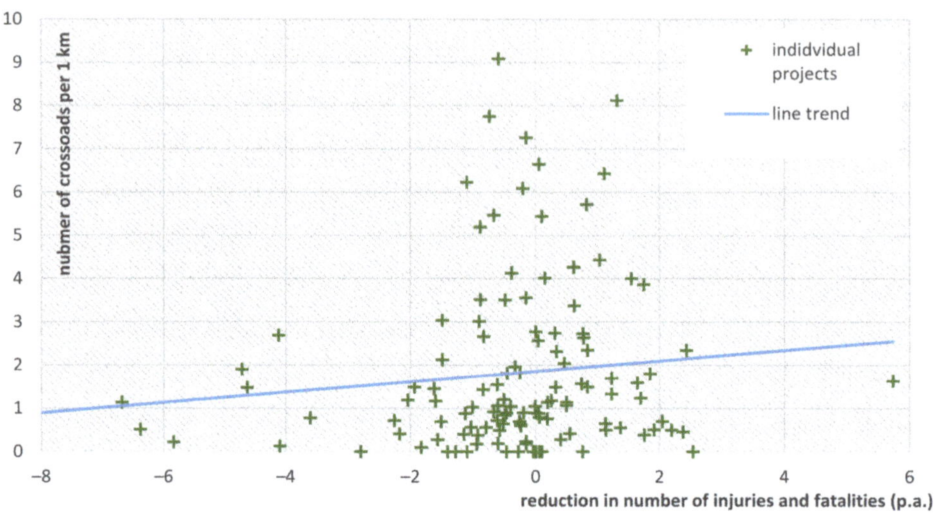

Figure 8. Reduction in number of injuries and fatalities in correlation to relative number of crossroads.

In accordance with the study results, positive impact on the incidence of traffic accidents should not be used as a general argument for the implementation of regional road modernization projects. The consideration of the anticipated accident reduction in the assessment (CBA) should be conditional on a safety audit of a specific section, including a specific assessment of the possible effects on incidence of traffic accidents and taking into account the real conditions at the site. In general, it can be assumed that a positive effect on incidence of traffic accidents can be expected mainly in connection with the elimination of the spot defects (unsatisfactory crossroads etc.).

A similar study conducted on a set of 27 road projects in Norway shows an opposite trend in the incidence of traffic accidents [2]. This fact is probably due to the structure of projects. While the evaluated set only includes projects for the modernization of regional roads, the study in Norway included mainly large road projects such as sub-sea tunnels, bypasses and trunk roads. This question may be further investigated. Further investigation should focus on the changes in the incidence of traffic accidents in connection with the implementation of the modernization of the national roads and highways. Alternatively, it would be relevant to make a comparison with the construction of new roads.

5. Conclusions

The ex-post CBA was prepared for a set of 144 regional road modernisation projects in the Czech Republic. The main objective of the evaluation was to verify the real impact of the implemented projects in comparison to its ex-ante goals. The verification was aimed at the investment costs and the incidence of traffic accidents. The ex-post data was used to rectify the ex-ante CBA results. The purpose of the study is to improve the decision-making process concerning the regional roads maintenance and modernisation.

The ex-post verification of the incidence of traffic accidents was performed with use of the Czech Republic Police database and was based on an analysis prior to and after project completion. Ex-post values of injuries and fatalities were measured slightly higher prior to project completion, although the aim of the project was to increase the degree of safety. The improvement of the technical condition of the infrastructure may result in increase in speed with negative effects on incidence of traffic accidents. This argument is also supported by the results of evaluation depending on the implementation of projects in urban and rural areas and the results of projects evaluation in relation to the relative number of crossroads in the individual projects. The goals declared for the project were an almost complete elimination of all accidents with a fatality and accidents with serious or light injuries, but these goals were missed.

The ex-post verification of investment costs brought very surprising results, and these results were different than in past investigations. Contrary to all expectations, the actual investment cost was lower in total terms by 11%.

The impact on the CBA result is significant, the weighted profitability index for the entire set of projects dropped from 16.7% to −2.8%. However NPV did not change dramatically in 75% of the projects. The results bring a clear recommendation for the reconstruction and modernisation of existing regional roads: impact on the incidence of traffic accidents shall not be taken into account in the CBA. The exceptions should be fully justified by safety audit or similar assessment.

In general, the results demonstrate the importance of ex-post evaluation of projects in order to improve the decision-making process in the future. At the application level, it was not possible to prove the positive effect of the modermisation of regional roads on the reduction of accidents, although this impact was ex-ante widely expected.

The potential users of the evaluation results are mainly local and regional authorities responsible for the maintenance and modernisation of the regional roads. An important group of users of the results of the evaluation are also the managing authorities responsible for the implementation of national and international resources in the field of road infrastructure development or regional development in general. The methodology could also be adapted for the evaluation at the national level and can provide a stimulus for further research on the correlation between road infrastructure modernization and number of accidents. The research could be applied in the national or general methodologies for the evaluation of investment projects in the field of transport infrastructure. The findings can improve the telling value of CBAs, make the decision-making process easier and improve the efficiency of road maintenance and modernisation cost.

Author Contributions: Conceptualization, P.H. and M.R.; methodology, P.H., R.M. and M.R.; validation, P.H., R.M. and M.R.; formal analysis, P.H.; investigation, resources, data curation, P.H., R.M. and M.R.; writing—original draft preparation, P.H.; writing—review and editing, R.M. and M.R.; visualization, P.H.; supervision, R.M. and M.R.; project administration, R.M. All authors have read and agreed to the published version of the manuscript.

Funding: This research received no external funding.

Acknowledgments: We would like to thank the Regional South-East Cohesion Government Office for their data.

Conflicts of Interest: The authors declare no conflict of interest.

Appendix A

Table A1. List of projects evaluated.

Project	Section Length (km)	NPV, Ex Ante (CZK Million)	NPV, Ex-Post (CZK Million)	MPE (NPV)	MAPE (NPV)
III/3657 Letovice, bridge 3657-3	0.01	0.0	−1.9	−5521.2%	5521.2%
III/4135 Rybníky bridge 4135-3	0.01	−11.7	−12.9	10.6%	10.6%
II/379 Lipůvka—Blansko, bridge 379-023	0.03	−29.7	−18.8	−36.7%	36.7%
III/3771 Předklášteří bridge 3771-3	0.03	−42.3	−42.5	0.4%	0.4%
II/129 Humpolec—bridge 129-011	0.06	52.5	55.4	5.5%	5.5%
III/4206 Pouzdřany, bridge 4206-2	0.07	−43.5	−10.8	−75.1%	75.1%
II/408 roundabout Kuchařovice II/399	0.10	2.2	9.8	355.8%	355.8%
II/397 and III/3974, roundabout Čejkovice	0.10	34.4	21.4	−37.7%	37.7%
II/130 Miletín—bridge 130-011	0.10	14.5	21.2	46.5%	46.5%
III/13035 Hořice—bridge 13035-2	0.10	−13.9	−3.4	−75.2%	75.2%
II/398 Horní Dunajovice, bridge 398-009	0.12	−7.9	−9.8	24.7%	24.7%
III/3993 Naloučany—bridge	0.14	−10.1	−5.3	−47.4%	47.4%
III/15280 Modřice bridge	0.18	−74.5	−34.7	−53.4%	53.4%
II/421 Zaječí bridge 421-010	0.25	−47.8	−40.1	−16.2%	16.2%
III/00221 Ladná, bridge 00221-2	0.26	−16.1	−13.3	−17.5%	17.5%
II/430 Vyškov through road, bridge 430-017	0.27	−2.5	−8.8	253.6%	253.6%
II/391 Žd'árec bridges 391-003, 391-004	0.28	−8.9	2.2	−124.8%	124.8%
II/399 Stropešín—bridge 399-002	0.30	−47.7	−103.0	116.1%	116.1%
II/380 Sokolnice roundabout	0.33	−9.8	4.5	−145.9%	145.9%
II/602 hr.kraje—Pelhřimov, 5. phase	0.35	−64.0	−68.7	7.4%	7.4%
II/409 Panské Dubenky—bridge 409-016	0.38	14.9	6.7	−55.2%	55.2%
II/602 Veselka roundabout	0.40	−18.5	−19.0	3.1%	3.1%
II/430 Podolí, roundabout	0.47	35.2	44.3	26.0%	26.0%
II/373 Ochoz through road 2. phase	0.49	3.5	4.7	33.8%	33.8%
III/03810 Hesov—bridges 03810-006, 007, 008	0.53	136.6	134.9	−1.2%	1.2%
II/374 Cetkovice through road	0.55	−5.0	4.4	−187.6%	187.6%
II/379 Lipůvka through road	0.56	−5.4	0.7	−112.4%	112.4%
II/408 Dyje through road	0.57	−8.1	7.8	−196.4%	196.4%
II/387 Bořínov—Nedvědice, border bridge 387-012	0.65	−30.3	−35.6	17.7%	17.7%
II/399 Tavíkovice bridge 399-006	0.66	−18.6	−19.1	2.8%	2.8%
II/400 Chlupice bridge 400-007	0.66	−14.7	−16.1	9.6%	9.6%
Lužice through road IV. and V. phase	0.77	30.3	12.7	−58.1%	58.1%
II/413 Prosiměřice bridge 413-014	0.80	−19.5	−0.6	−97.2%	97.2%
II/412 Znojmo, bridge 412-001, 412-002	0.85	−55.2	−43.7	−20.9%	20.9%
II/431 Vyškov—through road	0.92	9.4	22.1	133.8%	133.8%
II/422 Čejkovice through road III. phase	0.94	4.3	3.5	−18.7%	18.7%
II/377 Brt'ov through road	0.94	−15.3	−21.4	40.1%	40.1%
II/602 Domašov bridge 602-013	0.97	−33.4	−20.6	−38.2%	38.2%
Connection I/38 Znojmo	1.11	25.6	41.7	62.9%	62.9%
III/37418 Boskovice, Chrudichromská	1.12	1.4	−11.4	−909.5%	909.5%
II/427 Moravský Písek II. and III. phase	1.15	−33.9	−53.2	57.2%	57.2%
III/4133 Moravský Krumlov through road	1.17	−22.2	35.5	−260.3%	260.3%
II/394 Neslovice through road	1.19	27.7	12.9	−53.5%	53.5%
II/405 Zašovice—through road	1.26	0.6	−7.5	−1355.1%	1355.1%
II/394 Tetčice through road	1.28	1.6	18.1	1059.7%	1059.7%
II/422 Čejkovice—through road, phase 1	1.40	7.2	7.9	9.8%	9.8%
Reconstruction of road III/4301 Bukovany—through road	1.46	15.1	4.5	−70.5%	70.5%

Table A1. *Cont.*

Project	Section Length (km)	NPV, Ex Ante (CZK Million)	NPV, Ex-Post (CZK Million)	MPE (NPV)	MAPE (NPV)
II/398 Vranov nad Dyjí through road—Onšov	1.55	81.0	−42.2	−152.1%	152.1%
II/393 Oslavany—II. and IV. phase	1.60	19.2	29.5	53.6%	53.6%
II/416, 417 Křenovice through road	1.61	−18.6	−39.4	111.5%	111.5%
II/602 Ostrovačice 1. phase	1.62	36.0	−20.0	−155.6%	155.6%
II/381 Velké Němčice through road	1.65	12.0	26.9	124.1%	124.1%
II/432 Ratíškovice through road	1.66	12.5	12.5	0.2%	0.2%
II/421 Kobylí through road	1.70	9.0	7.8	−12.7%	12.7%
II/408 Dyjákovice through road	1.73	−40.6	−20.2	−50.4%	50.4%
III/01926, III/01928, III/01929 in Nová Cerekev	1.75	−46.3	−45.9	−0.9%	0.9%
II/400 Hostěradice—Višňové, I. phase	1.76	9.7	−10.2	−205.5%	205.5%
II/360 Štěpánovice—Vacenovice	1.78	6.4	−10.6	−266.1%	266.1%
II/425 Nosislav through road	1.78	−6.9	13.9	−301.0%	301.0%
II/152 Jamolice—border of the region Vysočina	1.79	−3.8	−7.7	102.7%	102.7%
II/398 Mikulovice through road	1.80	−51.9	−53.5	3.0%	3.0%
II/150 Havlíčkův Brod—Okrouhlice	1.81	7.2	22.5	213.5%	213.5%
II/405 Zašovice—Okříšky	1.85	−124.9	−138.0	10.5%	10.5%
II/383 Pozořice—Sivice	1.88	−15.9	−22.3	39.9%	39.9%
II/360 ul. Rafaelova—Pocoucov	1.88	−60.3	−62.4	3.5%	3.5%
II/379 Tišnov—Lipůvka, section Nuzířov—Lipůvka	1.90	−0.6	−34.4	5261.7%	5261.7%
II/405 Příseka—Brtnice	1.92	−48.8	−142.9	193.1%	193.1%
II/385 Kuřim through road	1.92	102.8	111.0	7.9%	7.9%
II/152 Jamolice—Polánka up to crossroad III/15250	2.00	−10.6	−12.5	17.7%	17.7%
II/353 D1—Rytířsko—Jamné, 2. phase	2.02	22.7	11.2	−50.5%	50.5%
II/422 Svatobořice—Mistřín, phase 1	2.07	56.7	83.8	47.7%	47.7%
II/602 Říčany-Říčky	2.10	0.5	15.8	3231.4%	3231.4%
II/432 Hodonín roundabout	2.18	29.5	13.1	−55.7%	55.7%
III/42117 Bulhary through road	2.19	−6.8	35.4	−620.9%	620.9%
II/422 Svatobořice—Mistřín through road, II. phase	2.20	3.0	42.6	1311.0%	1311.0%
II/416 Pohořelice through road	2.21	65.4	54.6	−16.6%	16.6%
III/4194 Vážany n/Litavou—Hrušky	2.26	−30.4	−27.9	−8.3%	8.3%
II/345 Chotěboř—through road, 2. phase	2.28	17.7	−6.1	−134.3%	134.3%
II/374 Blansko through road	2.30	283.6	303.2	6.9%	6.9%
II/379 Lažánky through road	2.41	80.7	70.9	−12.1%	12.1%
II/431, III/4301 Ždánice, Nechvalín through roads	2.55	−20.4	29.8	−246.5%	246.5%
II/380 Těšany through road	2.87	14.8	80.7	444.7%	444.7%
III/3773 Brumov—Bedřichov, bridge 3773-17	2.88	−18.3	−10.8	−40.7%	40.7%
III/40819 Hradiště through road	2.89	−15.2	−46.6	206.0%	206.0%
II/425 Hustopeče through road, roundabout	3.00	142.3	135.1	−5.1%	5.1%
II/405 Příseka bypass	3.09	60.4	63.4	4.9%	4.9%
II/421 Bořetice through road—Kobylí	3.34	1.6	27.0	1599.1%	1599.1%
II/400 Přeskače—through road	3.36	−15.5	15.7	−201.3%	201.3%
II/152 Modřice bridges	3.53	55.2	72.1	30.5%	30.5%
II/395 Zastávka u Brna—border of the region, III. section	3.58	4.8	15.1	212.6%	212.6%
II/422 Čejč—Čejkovice	3.67	−18.4	−72.3	293.9%	293.9%
II/377 Rájec—Bořitov	3.76	−34.3	−52.3	52.5%	52.5%
II/347 Světlá n. S.—D1, 2. phase, section 1	3.76	−69.2	−54.9	−20.7%	20.7%
II/395 Odrovice through road	3.84	9.0	3.7	−58.5%	58.5%

Table A1. *Cont.*

Project	Section Length (km)	NPV, Ex Ante (CZK Million)	NPV, Ex-Post (CZK Million)	MPE (NPV)	MAPE (NPV)
II/425, III/4217 Hustopeče—Horní Bojanovice	3.96	−16.7	80.2	−579.3%	579.3%
II/425 Hustopeče roundabout—Horní Bojanovice	3.96	5.2	40.3	673.1%	673.1%
II/602 border of the region—Pelhřimov, 4. phase, section 1	4.08	−88.6	−105.2	18.7%	18.7%
II/602 border of the region—Pelhřimov, 7. phase, section 1	4.09	45.5	−102.0	−323.8%	323.8%
II/385 Tišnov—Hradčany—Čebín	4.10	165.1	182.0	10.2%	10.2%
II/420 Hustopeče—Kurdějov	4.15	−18.2	10.9	−160.0%	160.0%
II/432 Ratíškovice—Hodonín, I/55 rural area	4.28	31.6	−15.6	−149.5%	149.5%
II/365 Letovice—Horní Poříčí, residential area	4.45	−10.1	14.5	−242.8%	242.8%
III/39613 crossroad Pasohlávky—Drnholec	4.49	−29.5	−12.1	−59.1%	59.1%
II/402 Třešť—crossroad I/38	4.50	3.1	−28.8	−1019.0%	1019.0%
III/4301 Ždánice—Bukovany, phase 1	4.77	−35.7	−25.6	−28.3%	28.3%
II/602 border of the region—Pelhřimov, 6. phase	4.89	−57.6	−45.5	−20.9%	20.9%
II/347 Světlá n. S.—D1, 1. stavba	5.00	−145.2	−203.2	39.9%	39.9%
II/602 Popůvky—Ostrovačice	5.03	64.1	84.5	31.9%	31.9%
II/408 Suchohrdly u Znojma—Přímětice—I/38 (rural areas)	5.07	21.4	20.3	−5.3%	5.3%
II/152 Ivančice—Polánka	5.26	66.6	67.4	1.2%	1.2%
II/398 Podmyče—Šafov	5.36	−21.6	−29.2	35.5%	35.5%
II/402 Batelov—Třešť	5.50	−9.8	−33.9	245.7%	245.7%
II/365 Horní Poříčí—Letovice, rural areas	5.55	13.1	−28.2	−315.1%	315.1%
II/430 Rousínov—Tučapy	5.55	169.3	202.0	19.4%	19.4%
II/373, III/37357 Benešov—Žďárná rural area	5.69	7.3	−33.0	−552.4%	552.4%
III/3783 Holštejn, bridge 3783-1	5.73	−10.8	29.1	−370.4%	370.4%
II/408 Vranov nad Dyjí—Znojmo, rural area	5.87	14.3	−158.6	−1211.5%	1211.5%
II/425 Rajhrad—Židlochovice	5.97	28.5	150.6	427.7%	427.7%
II/374 Adamov—Bílovice nad Svitavou	6.97	−25.1	−60.4	140.5%	140.5%
II/413 Prosiměřice—Suchohrdly, extravilány	7.23	17.0	−49.9	−393.4%	393.4%
II/422 Čejkovice—Velké Bílovice	7.35	−42.8	−45.9	7.3%	7.3%
II/339 Ledeč nad Sázavou—border of the region	7.61	12.7	−13.4	−205.3%	205.3%
II/380 Těšany—Borkovany—Kašnice	7.89	126.8	143.6	13.3%	13.3%
II/602 hr. kraje—Pelhřimov, 3. phase	7.91	−66.3	−143.0	115.6%	115.6%
II/150 Boskovice—Valchov—Žďárná	8.13	37.7	−21.8	−157.8%	157.8%
II/424 Mor.Nová Ves—Tvrdonice—Lanžhot	8.53	4.2	28.0	572.2%	572.2%
II/523 Jihlava—Větrný Jeníkov	9.19	−32.5	−238.0	632.7%	632.7%
II/348, II/131 Štoky—Petrovice—Větrný Jeníkov	9.40	231.0	11.8	−94.9%	94.9%
II/523 Větrný Jeníkov—Humpolec	9.95	−36.9	7.1	−119.4%	119.4%
II/398 Horní Dunajovice—Mikulovice—Pavlice rural zone	10.66	−53.6	−82.0	52.8%	52.8%
II/425 Nosislav—Velké Němčice—Starovičky	10.73	210.7	169.3	−19.6%	19.6%
II/411 Moravské Budějovice—border of the region	10.94	−12.5	−52.4	320.1%	320.1%
III/03810 Havlíčkův Brod—Přibyslav	11.36	23.4	28.5	21.6%	21.6%
II/430 Vyškov through road	11.41	333.4	276.3	−17.1%	17.1%
II/152 Jaroměřice—Hrotovice	11.58	−65.3	−111.7	71.0%	71.0%
II/410 from I/23—Želetava	12.60	−48.5	−89.3	84.2%	84.2%
II/639 Horní Cerekev—Kostelec	12.95	81.1	50.3	−38.1%	38.1%
II/133 Horní Cerekev—crossroad II/602	14.59	−67.3	−85.8	27.5%	27.5%
II/151, III/15113 from I/38—Budeč+Štěpkov-Budkov	15.81	−48.9	−9.7	−80.1%	80.1%
II/379 Podomí—Drnovice	16.16	73.7	61.9	−16.0%	16.0%

Table A1. *Cont.*

Project	Section Length (km)	NPV, Ex Ante (CZK Million)	NPV, Ex-Post (CZK Million)	MPE (NPV)	MAPE (NPV)
II/354 Nové Město na Moravě—Svratka	16.51	29.6	−26.2	−188.7%	188.7%
II/129 Cetoraz—Jiřičky	19.71	−77.1	−143.7	86.3%	86.3%
II/351 from II/602—Třebíč	22.88	68.0	−126.0	−285.2%	285.2%
II/360 Jimramov—Moravec	23.76	24.3	−2.8	−111.7%	111.7%
total	564.62	1081.2	−245.0	3.8%	267.2%

References

1. Boardman, A.E.; Mallery, W.L.; Vining, A.R. Learning from ex ante/ex post cost-benefit comparisons: The Coquihalla highway example. *Socio-Econ. Plan. Sci.* **1994**, *28*, 69–84. [CrossRef]
2. Odeck, J.; Kjerkreit, A. The accuracy of benefit-cost analyses (BCAs) in transportation: An ex-post evaluation of road projects. *Transp. Rev. Part A Policy Pract.* **2019**, *120*, 277–294. [CrossRef]
3. European Commission. Transport/Task 3—Major Project Case Studies Work Package 5/Ex Post Evaluation of Cohesion Policy Programmes 2007–2013, Focusing on the European Regional Development Fund (ERDF) and the Cohesion Fund (CF). Available online: https://ec.europa.eu/regional_policy/sources/docgener/evaluation/pdf/expost2013/wp5_task3_en.pdf (accessed on 10 February 2020).
4. eCBA s.r.o. Regional South-East Cohesion Government Office. System for Financial and Economic Evaluation of Regional Development Projects. 2008. Available online: https://cba.jihovychod.cz/login.aspx?ReturnUrl=%2f (accessed on 1 December 2019).
5. European Commission, Directorate-General for Regional and Urban Policy. Guide to Cost-Benefit Analysis of Investment Projects—Economic Appraisal Tool for Cohesion Policy 2014–2020. Available online: https://ec.europa.eu/regional_policy/sources/docgener/studies/pdf/cba_guide.pdf (accessed on 10 December 2019).
6. Anguera, R. The Channel Tunnel—An ex post economic evaluation. *Transp. Rev. Part A Policy Pract.* **2006**, *40*, 291–315. [CrossRef]
7. Börjesson, M.; Jonsson, R.; Lundberg, D. An ex-post CBA for the Stockholm Metro. *Transp. Rev. Part A Policy Pract.* **2014**, *70*. [CrossRef]
8. Jong, G.; Vignetti, S.; Pancotti, C. Ex-post evaluation of major infrastructure projects. *Transp. Res. Procedia* **2019**, *42*, 75–84. [CrossRef]
9. Welde, M.; Volden, G. Measuring efficiency and effectiveness through ex-post evaluation: Case studies of Norwegian transport projects. *Res. Transp. Bus. Manag.* **2018**. [CrossRef]
10. Welde, M. In search of success: Ex-post evaluation of a Norwegian motorway project. *Case Stud. Transp. Policy* **2018**, *6*, 475–482. [CrossRef]
11. Meunier, D.; Welde, M. Ex-post evaluations in Norway and France. *Transp. Res. Procedia* **2017**, *26*. [CrossRef]
12. Nicolaisen, M.; Driscoll, P. An International Review of Ex-Post Project Evaluation Schemes in the Transport Sector. *J. Environ. Assess. Policy Manag.* **2016**, *18*. [CrossRef]
13. Sudiana, I.P. How Effective is Cost-Benefit Analysis in Assisting Decision Making by Public Sector Managers? Case Studies of Two Australian Departments. 2010. Available online: http://www.crawford.anu.edu.au/degrees/pogo/discussion_papers/PDP10-01.pdf (accessed on 12 February 2020).
14. Czech Republic Police, Centrum Dopravního Výzkumu V.V.I.; Czech Statistical Office. Statistical Evaluation of Accidents in the Map. Available online: http://maps.jdvm.cz/cdv2/apps/nehodyvmape/Search.aspx (accessed on 28 January 2020).
15. Florio, M.; Vignetti, S. The Use of Ex Post Cost-Benefit Analysis to Assess the Long-Term Effects of Major Infrastructure Projects. *SSRN Electron. J.* **2013**. [CrossRef]
16. Lord, D.; Mannering, F. The statistical analysis of crash-frequency data: A review and assessment of methodological alternatives. *Transp. Rev. Part A Policy Pract.* **2010**, *44*, 291–305. [CrossRef]
17. Czech Republic Police. Statistics of the Incidence of Traffic Accidents. 2019. Available online: https://www.policie.cz/clanek/statistika-nehodovosti-900835.aspx?q=Y2hudW09MTE%3d (accessed on 16 February 2020).
18. Czech Statistical Office. Traffic Accidents and Their Consequences in the Czech Republic in the Long Term. 2014. Available online: https://www.czso.cz/documents/10180/20534694/32025414a.pdf/57d484eb-1939-47ad-8fef-f38d6dd2c19e?version=1.0 (accessed on 16 February 2020).
19. Knudsen, D.C. Shift-Share Analysis: Further Examination of Models for the Description of Economic Change. *Socio-Econ. Plan. Sci.* **2000**, *34*, 177–198. [CrossRef]
20. Střeleček, F.; Zdeněk, R.; Lososová, J. Shift-Share Analysis of Labour Productivity. In *Inproforum 2009*; University of South Bohemia in České Budějovice: České Budějovice, Czech Republic, 2009; ISBN1 978-80-7394-173-4. Available online: http://ocs.ef.jcu.cz/index.php/inproforum/INP2009/paper/viewFile/240/233 (accessed on 15 January 2020)ISBN2 978-80-7394-173-4.
21. Ministerstvo dopravy ČR a SUDOP PRAHA a.s. Methodology for the Economic Effectiveness Evaluations of Transportation Projects. 2017. Available online: https://www.sfdi.cz/soubory/obrazky-clanky/metodiky/2017_02_rezortni_metodika-komplet.pdf (accessed on 20 November 2019).

22. Centrum Dopravního Výzkumu, v.v.i. What Was the Total Cost of Traffic Accidents in 2015? 2016. Available online: https://www.czrso.cz/clanek/dopravni-nehody-nas-v-roce-2015-staly-68-miliard-zemrelo-737-osob/?id=1664 (accessed on 1 December 2019).
23. Regional South-East Cohesion Government Office. Evaluation Tables ROP JV Support Area 1.1 Regional Infrastructure Development—Roads. 2010. Available online: http://www.jihovychod.cz/pro-prijemce/aktualni-vyzvy/dokumentace-k-vyzve-1-1/1-1-vyzva-k-20-12-2010 (accessed on 10 February 2020).
24. European Commission. Directorate General Regional Policy. Guide to Cost-Benefit Analysis of Investment Projects. 2008. Available online: https://ec.europa.eu/regional_policy/sources/docgener/guides/cost/guide2008_en.pdf (accessed on 26 February 2020).
25. Regional South-East Cohesion Government Office. *Form R17—Funding Draw-Down Projection*; Regional South-East Cohesion Government Office: Brno, Czech Republic, 2019.
26. Ambros, J. Jak měřit bezpečnost?—2. část [How to measure road traffic safety?—Part 2]. *Road Superv.* **2012**, *73*, 103–105.
27. Odeck, J. Cost overruns in road construction—What are their sizes and determinants? *Transp. Policy* **2004**, *11*, 43–53. [CrossRef]
28. Odeck, J. Do reforms reduce the magnitudes of cost overruns in road projects? Statistical evidence from Norway. *Transp. Rev. Part A Policy Pract.* **2014**, *65*. [CrossRef]
29. Flyvbjerg, B.; Skamris Holm, M.K.; Buhl, S.L. What Causes Cost Overrun in Transport Infrastructure Projects? *Transp. Rev.* **2004**, *24*, 3–18. [CrossRef]
30. Subramani, T.; Sruthi, P.S.; Kavitha, M. Causes of Cost Overrun In Construction. *IOSR J. Eng.* **2014**, *4*. [CrossRef]
31. Czech Statistical Office. Míra růstu reálného HDP. 2020. Available online: http://apl.czso.cz/pll/eutab/html.h?ptabkod=tec00115 (accessed on 26 February 2020).

MDPI

Article

Stabilising Rural Roads with Waste Streams in Colombia as an Environmental Strategy Based on a Life Cycle Assessment Methodology

Alejandra Balaguera [1,2,*], Jaume Alberti [1], Gloria I. Carvajal [2,*] and Pere Fullana-i-Palmer [1,*]

1 UNESCO Chair in Life Cycle and Climate Change ESCI-UPF, Pompeu Fabra University, Passeig Pujades 1, 08003 Barcelona, Spain; jaume.alberti@esci.upf.edu
2 Facultad de Ingenierías, Universidad de Medellín, Medellín 050010, Colombia
* Correspondence: abalaguera@udem.edu.co (A.B.); gicarvajal@udem.edu.co (G.I.C.); pere.fullana@esci.upf.edu (P.F.-i.-P.)

Abstract: Roads with low traffic volume link rural settlements together and connect them with urban centres, mobilising goods and agricultural products, and facilitating the transportation of people. In Colombia, most of these roads are in poor conditions, causing social, economic, and environmental problems, and significantly affecting the mobility, security, and economic progress of the country and its inhabitants. Therefore, it is essential to implement strategies to improve such roads, keeping in mind technical, economic, and environmental criteria. This article shows the results of the application of the environmental life cycle assessment—LCA—to sections of two low-traffic roads located in two different sites in Colombia: one in the Urrao area (Antioquia), located in the centre of the country; and another in La Paz (Cesar), located in the northeast of the country. Each segment was stabilised with alternative materials such as brick dust, fly ash, sulfonated oil, and polymer. The analysis was carried out in three stages: the first was the manufacture of the stabiliser; the second included preliminary actions that ranged from the search for the material to its placement on site; and the third was the stabilisation process, which included the entire application process, from the stabiliser to the road. The environmental impacts are mainly found in the manufacture of stabilisers (60% of the total), for sulfonated oil or polymer, due to the different compounds used during production, before their use as stabilisers. The impact categories with the greatest influence were abiotic depletion potential (ADP), global warming potential (GWP) and terrestrial ecotoxicity potential (TETP). For the stabilisation stage (impact between 40% and 99%), ash and brick dust have the highest impacts. The impact categories most influenced in this stage were: acidification potential (AP), freshwater aquatic ecotoxicity potential (FAETP), human toxicity potential (HTP), marine aquatic ecotoxicity potential (MAETP) and photochemical ozone creation potential (POCP).

Keywords: life cycle assessment; waste management; circular economy; alternative materials; construction; road stabilisation

Citation: Balaguera, A.; Alberti, J.; Carvajal, G.I.; Fullana-i-Palmer, P. Stabilising Rural Roads with Waste Streams in Colombia as an Environmental Strategy Based on a Life Cycle Assessment Methodology. *Sustainability* **2021**, *13*, 2458. https://doi.org/10.3390/su13052458

Academic Editor: Edoardo Bocci

Received: 22 September 2020
Accepted: 16 December 2020
Published: 25 February 2021

1. Introduction

1.1. Low Traffic Roads in Colombia and Its Stabilisation

The Colombian road network has 204,855 km, of which approximately 70% corresponds to tertiary roads [1]. They are essential to facilitate the integration of rural areas with their respective urban headwaters, and boost the economy through the development of agricultural, mining and tourism activities [2]. About 96% of tertiary roads are in bad condition [3]. High levels of deterioration are found, consistent with the great topographic variability, soil susceptibility and hydrological regime to which they are exposed, hindering their proper functioning, especially in rainy seasons [4]. This fact, added to the financial impossibility of paving the entire tertiary network of the country, implies a need for rehabilitation and maintenance, implementing techniques that contribute to its stability and proper functioning.

In Colombia, the stabilisation of tertiary roads includes the use of materials such as lime, Portland cement, rock dust, colloidal transport materials, and organic bases, with procedures governed mainly by the Urban Development Institute of Bogotá (IDU), collected in its guide of design and construction of structural layers of pavements through chemical processes [5].

Traditionally, the implementation of soil stabilisation activities for roads includes the use of materials such as cement and lime. Some studies have found improvements in the mechanical properties of the soil under the addition of such materials [6–10].

The general specifications for road construction of the National Institute of Roads [11] also establish guidelines for road soil stabilisation, by using cement or lime as stabilising materials.

1.2. Assessing Environmental Impacts and Life Cycle Assessment (LCA)

The extensive use of traditional materials in roads stabilisation, such as natural aggregates, cement, and lime, is causing a gradual depletion of natural resources [12,13]. At the same time, the industry sector produces large quantities of different types of waste, which have to be managed, with the consequent environmental impacts [14]. Following the circular economy principles [15], some of this waste may become alternative materials to be reused in road stabilisation, as a means to (partially) solve both problems described above: resource depletion and environmental impacts.

Of course, there is a need to objectively quantify these improvements in terms of environmental impact. The LCA methodology is known as an objective assessment tool [16], and its results may facilitate the decision-making process [17–22]. In our case, LCA may allow builders and government decision-makers to measure the consumption and emissions generated during the life cycle of low traffic roads stabilised with alternative materials.

Since 2001, different applications have been developed for the use of the LCA methodology to assess the impacts of construction in general [23], to communicate them through environmental product declarations [24], and specifically to apply them for different materials in the construction of roads. Mroueh et al. [25] found that the use of fly ash, steel slag, and crushed concrete, as substitutes for natural aggregates in road construction in Finland, reduces the contributions for some impact categories. In the U.S., Rajendran & Gambatese [26] developed a life cycle-based comparative analysis of energy consumption and the generation of solid waste associated with concrete and asphalt pavements. Birgisdóttir et al [27] analysed, in Denmark, the environmental impacts associated with road construction and the use of ash produced by municipal waste incineration. Chowdhury et al. [20], in the United States, compared by-products, such as fly ash and recycled asphalt (RAP), against natural aggregates by evaluating some impacts, such as energy consumption, acidification potential and toxicity potentials.

Celauro et al. [28] developed a study in Italy, in which different percentages of RAP were applied both to the asphalt and to the base layers, and the stabilisation of clay soil was performed with lime. They showed a reduction in energy consumption and emissions thanks to the avoided use and transport of traditional materials. In addition, the use of lime in the stabilisation and a greater amount of RAP lead to a significant reduction in CO_2 by 48.79%, CO by 41.11% and human toxicity potential (cancer) by 36.97%. Finally, in Paraguay, the environmental impacts of using clay–lime mixtures on soil stabilisation were analysed. They used different doses, which enabled the obtaining of different levels of stiffness and resistance. This guaranteed the desirable mechanical properties of the soil. In this study, lime production represented more than 75% of the total energy consumption, greenhouse gas emissions and photochemical oxidation for each of the analysed mixtures [12].

A literature review [29] has shown that not much research has been done related to LCA on low-traffic roads in developing countries, nor to stabilizing agents, nor to other more simplified life-cycle-based indicators that other sectors are using, such as the carbon footprint [30,31] or energy demand [32]. The aim of this paper is to present the results of a pilot study in which different stabilising agents coming from waste streams have been technically tested in segments of low traffic roads in Colombia, and for which an LCA has been performed to compare their environmental impact and to show if they perform better than traditional virgin materials.

2. Materials and Methods

2.1. Roads Being Studied

The experimentation was performed in two different tertiary towns of Colombia. At the international level, roads that are known as tertiary in Colombia are associated with low traffic volumes and are called low volume roads (LVR). Low volume roads are characterised by not being paved and having an average daily traffic equivalent of fewer than 200 vehicles, as well as slow modes of travel that are mostly made up of pedestrians and non-motorised traffic [33].

The first town was Urrao, which is located in the southwest sub-region of Antioquia; and the second town was La Paz, which is located in Cesar region (between San José de Oriente and Filomachete region), in the northwest of Colombia (see Figure 1a), and close to the border with Venezuela (see Figure 1b). The reason to choose them was the relationship between those towns, their administrations, and their participation in the Red Innovial. The Red Innovial was a network that joined different public and private institutions in Colombia, such as universities and state organisations to investigate and perform new materials and constructive techniques for low traffic roads taking into account lower environmental impact and technical and economic viability.

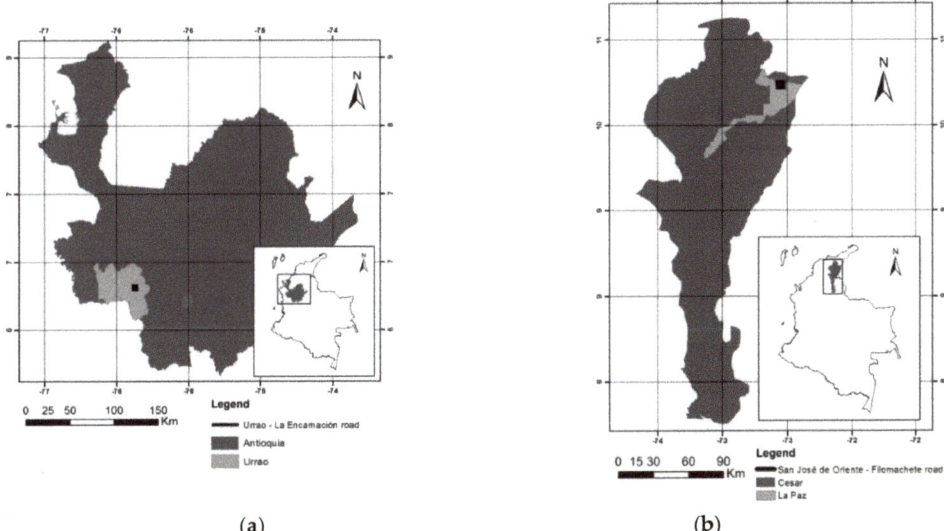

(a) (b)

Figure 1. Location of the roads and cross section.

These regions soils have clay (Urrao) and sandy-clay (La Paz) characteristics. Table 1 provides an overview of the characteristics of the studied soils considering the standards defined by the National Institute of Roads in Colombia (INVIAS). The dimensions of each road were: width 5.0 m, length 1 km and thickness 0.20 m. The tests were performed by the authors within Red Innovial during 2016 (See Figure 2).

Table 1. Characterisation of the soils under study.

Properties	Standard	Urrao	La Paz
Type of soil		Lime	Sandy-clay
Unified Soil Classification System (USCS)		MH [1]	SC [2]
AASHTO classification		A-7-5	A-6
Natural humidity (%)	INV E [3]-122	27	9
Specific gravity (Gs)	INV E-128	2.71	2.70
Liquid limit (%)	INV E-25	66	36
Plastic limit (%)	INV E-126	48	20
Plastic index (%)	INV E-126	18	16
Maximum size of particle (mm)		19	9.5
Clay (%)		23	14
Dry unit weight (kN/m^3)	INV E-142	14.8	19.5
Optimal humidity (%)		25.8	11.8

Source: INVIAS (2012).[1] Unified Soil Classification System (USCS): MH means lime type; [2] Unified Soil Classification System (USCS): MH means Sandy-clay type [3] INV E: Test Standard of the National Highway Institute (INVIAS) of Colombia.

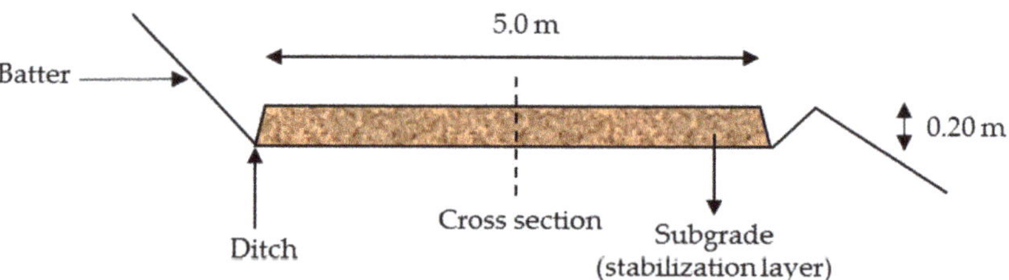

Figure 2. Cross section.

The process of extraction of a soil sample in Urrao, the process of stabilisation, and the different testing methods descriptions may be found in the literature for different configurations of fly ash as stabilising material [34]. The methods used for the La Paz region were the same as those described for Urrao.

2.2. Stabilising Materials

As explained above, roads must be stabilised to correctly perform their function. This is required because of the characteristics of the current soil properties existing on the locations considered (see Table 1). This can be performed using natural aggregates or, as an alternative, some types of waste.

Four alternative waste materials have been tested, some of them needing a pre-treatment in order to be able to act as a stabilising agent (see Table 2). Before developing the LCA, laboratory tests were performed for each waste. Leaching potential was analysed; in the case of presenting any level of concentration in the soil it was discarded, being a decision criterion for application as stabilisers on the selected roads. Material 1 was fly ash waste from the combustion of coal in a thermoelectric process of a Colombian textile company located in Medellin city. The ash must follow an activation process with lime to become a stabilising agent. The description of this process and the testing methods are described in a previous paper [34].

Table 2. Design of the stabilising mixture.

Material	Quantity of Material for Urrao (t/m^3)	Quantity of Soil for Stabilisation Urrao (t/m^3)	Quantity of Material for La Paz (t/m^3)	Quantity of Soil for Stabilisation La Paz (t/m^3)
Fly ash	97.4	997.6	155	1272.8
Lime (Fly ash)	65		51.8	
Brick dust	113	1049.2	162	1272.8
Lime (brick dust)	534		45.6	
Sulphonated oil	0.426	1280	1.12	1480
Polymer	33.3	1280	50	1480

Material 2 was brick dust, which comes from the waste generated in the process of making bricks cooked at temperatures of 950 °C, and collected when emptying the storage wagons that are inside the brickyard. This material must follow, as described for the ash, an activation process with lime in order to become a stabilising agent [31].

Material 3 was a waste oil, which had a chemical transformation (a sulphonation process). This oil is a catalyst agent that produces ion exchange. Chemically, it is an organic compound derived from combined sulphides and acids. The most important function of this stabiliser is the reduction in the water contained between the soil particles, increasing the number of voids that allow the rearrangement of the particles, by attraction among them or by compaction [35]. The main effects of sulphonated oil on clay-clad soils are a reduction in interstitial spaces, reduction in permeability, increase in sedimentation, improvement of the response to compaction, and increase in the soil density. The studies carried out with sulphonated oils and the evidence obtained through field tests showed that the electrochemical stabilisation system is a competitive alternative to reduce the expansive potential of clay soils [36].

Finally, Material 4 was a polymer waste used as an emulsion. Polymer emulsions are a dispersed system in which the phases are immiscible or partially miscible liquids, one of which is dispersed in the other and whose structure is stabilised by a surfactant called an emulsifier.

2.3. LCA Methodology

The environmental assessment was made by using the LCA methodology following the ISO 14040 and ISO14044 standards [37,38] which involve four phases: "(i) goal and scope definition, (ii) inventory analysis, (iii) impact assessment, and (iv) interpretation". The software used was GaBi version 6.115 and most databases were also from thinkstep GaBi.

The scope of the study was restricted to cradle-to-gate, collecting data up to the stabilisers. This scope consideration is common in the construction sector, where environmental product declarations (EPDs) are widely used following the EN 15804 standard. In addition, because this is a first pilot scale study, technical environmental data collection is a challenge and data will have a high uncertainty. However, a simplified LCA is considered a better option than any other type of environmental assessment. ISO 14044 states that "The scope, including system boundary and level of detail, of an LCA depends on the subject and the intended use of the study. The depth and the breadth of LCA can differ considerably depending on the goal of a particular LCA." Knowing the limitations in obtaining environmental information from this pilot system, it is not the intention of the study to advance further than a simplified LCA.

A complete LCA is not always needed to identify where the main impacts could be, and practice suggests different ways of applying the LCA methodology [39]. For instance, LCA is used to find the items (stages, processes, materials, etc.) which account for most of the impact in a system under an environmental product declaration programme [40] and it is recommended by the UNEP/SETAC life cycle initiative to help perform hot-spot analyses [41]. Depending on the goal of the environmental analysis, a simplified LCA can be used and a selective assessment may be performed, taking into consideration only generic data and/or covering the life cycle in a restricted way (e.g., from cradle to gate), but without abandoning rigour [16,30]. The European Commission [42] introduced life cycle thinking (not only LCA) as essential to the sustainable use of resources. Some examples of this may be found in the literature [43].

The impact assessment method of CML [44], updated to 2016, was applied to evaluate the impacts, as recommended by the "Building Research Establishment Product Category Rules (PCR)" of construction products [12,44,45]. CML 2001 is an impact assessment method which restricts quantitative modelling to early stages in the cause–effect chain to limit uncertainties. Results are grouped in midpoint categories according to common mechanisms (e.g., climate change) or commonly accepted groupings (e.g., ecotoxicity) [44]. It is recommended because it restricts quantitative modelling to the early stages of the cause–effect chain to reduce uncertainties. As in other studies [34,39,46–49] the analysed impact categories were: abiotic depletion potential (both ADP elements and ADP fossil), acidification potential (AP), eutrophication potential (EP), freshwater aquatic ecotoxicity (FAETP), global warming potential (GWP 100 years) excluding biogenic carbon, human toxicity potential (HTP), marine aquatic ecotoxicity potential (MAETP), ozone depletion potential (ODP), photochemical ozone creation potential (POCP), and terrestrial ecotoxicity potential (TETP).

3. Life Cycle Assessment Methodology

3.1. Goal and Scope Definition

The functional unit was chosen as: "1 km of low traffic road".

This functional unit enabled the assessment of environmental impact values and their comparison with other LCA studies applied to roads [28,50–52].

The scope of the study was defined as from cradle to gate. Therefore, according to CEN 15804 (2012) [45], a better naming might be "reference unit" instead of "functional unit", but we will keep the latter, more commonly used. The scope included the following processes:

- The transformation of the polymer and sulphonated oil into stabilisers.
- The ash and brick dust taken as delivered by the industrial facility where they were produced as waste, without any additional processing of the material. Lime was used as an alkali activator in both cases, during stabilisation process.
- The preliminary activities stage included all the necessary processes before the stabilisation process. Among these activities, we included the transport of machinery and materials necessary for the stabilisation of the road, identifying its origin, time (in hours) and distances travelled (in km) from its origin to the study area.
- In the stabilisation process, the following activities are considered: scarification, stabiliser application, compaction, wetting, and curing depending on the case; as well as diesel consumption by each machine that was used in the process (see Figure 3).

Other processes were not included, considered as outside the limits of the system:

- Machinery manufacture, because the impact assigned to the time used for this system is very small in relation to the useful life of the machines.
- The maintenance stage, because it was beyond the scope of the Red Innovial, and is a stage that takes longer to be performed and evaluated than the duration of the project. As indicated above, this is an LCA from "cradle to gate" and, therefore, the use stage was not considered.
- The "end of life" stage, because no significant types or quantities of waste were generated during the whole process, as equally considered in previous studies [25,28].

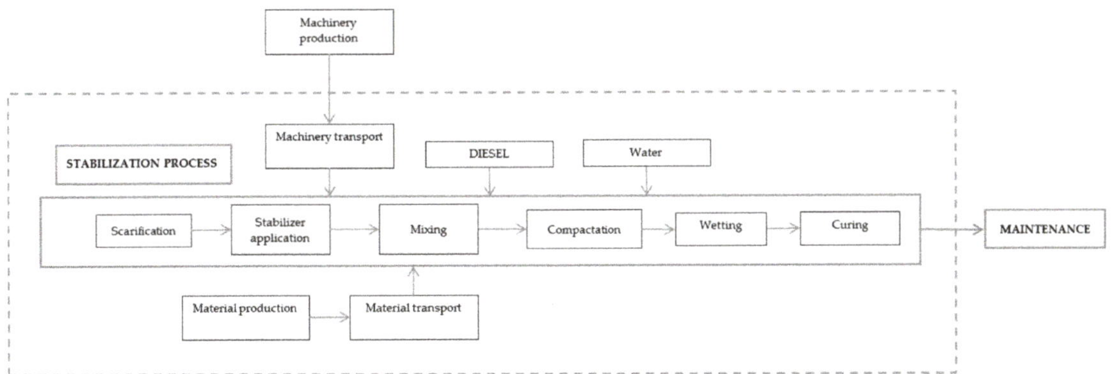

Figure 3. Flow chart of the stabilisation of the roads.

3.2. Inventory Analysis: Inputs and Outputs

3.2.1. Stabilisers

The following assumptions were taken:

- The "cut-off rule" [38] was applied and, therefore, wastes such as ash and brick dust entered the system with no environmental impact other than their transformation, if needed, and their transport to site.
- The polymer and the sulphonated oil were of waste origin, but they were mixed with virgin chemical compounds to be transformed into stabilisers. For instance, for the sulphonated oil manufacture, the following processes were used to model the substances used and obtained from the GaBi database. The needed amounts are presented in Table 3.
    - Silicone–resin plaster; technology mix; production mix, at plant; based on mineral fillers and a silicone bonding agent (en) from Germany; 100 mL;
    - Deionised water; highly pure, via ion exchange, from hydrochloric acid and caustic soda; single route, at plant from the U.S.; 1 kg/L (en);
    - Phenol; hock process, oxidation of cumene; single route, at plant; 1.07 g/cm^3, 94 g/mol (en) from Germany;
    - Propylene glycol; via Propylene oxide (PO)-hydrogenation; single route, at plant; 1.04 g/cm^3, 76.10 g/mol (en) from Germany;
    - Polyvinyl alcohol (from vinyl acetate) (PVAL); technology mix; production mix, at plant; without additives from the U.S.;
    - Antistatic agent (quaternary ammonium compound); technology mix; production mix, at plant; quaternary ammonium compound (en) Global;
    - Sulphuric acid aq. (96%); concentrated, sulphur dioxide route; single route, at plant; 96%, 1.84 g/cm^3 (en) from the U.S.
- On the other hand, the environmental data for the chemical products added to the polymer were obtained from the GaBi database of the United States, and the processes for the compounds used in the production of the stabiliser are mentioned below (see Table 4 for quantities):
    - Acrylic acid (Propene); oxidation of propene; production mix, at plant; 1.05 g/cm^3, 72.06 g/mol (en; Dipropylene glycol by product propylene glycol via PO hydrogenation; hydration of propylene oxide; single route, at plant; 1.02 g/cm^3, 134 g/mol (en);
    - Dipropylene glycol by product propylene glycol via PO hydrogenation; hydration of propylene oxide; single route, at plant; 1.02 g/cm^3, 134 g/mol (en) from the U.S.;

- Polycarbonate–acrylonitrile–butadiene–styrene compound (80% PC, 20% ABS); mixing, pelletising and compounding; single route, at plant; 80% polycarbonate, 20% acrylonitrile–butadiene–styrene (en);
- Ethylene vinylacetate copolymer (E/VA) (72% ethylene, 28% vinylacetate); copolymerisation of ethylene and vinyl acetate; production mix, at plant; without additives, 72% ethylene, 28% vinyl acetate (en);
- Sodium chloride (rock salt); salt mining and leaching; production mix, at plant; 2.17 g/cm^3, 58.44 g/mol (en).

- For the alkaline activator (lime) model, the process used was: "Limestone flour (0.115 mm), production mix, at producer; grain size 0.115 mm", from the GaBi database.

Table 3. Substances used for the production of kg of sulphonated oil.

Silicone–Resin Plaster (kg)	Deionised Water (kg)	Phenol (kg)	Propylene Glycol (kg)	Antistatic Agent (Quaternary Ammonium Compound) (kg)	Polyvinyl Alcohol (kg)	Sulphuric Acid Aq. (96%) (kg)
0.2	0.093	0.06	0.01	0.02	0.6	0.0004

Table 4. Substances used for the production of 1.0 kg of polymer.

Acrylic Acid (kg)	Dipropylene Glycol (kg)	Polycarbonate-Acrylonitrile-Butadiene-Styrene (kg)	Sodium Chloride (kg)	Ethylene Vinylacetate Copolymer (kg)
0.08	0.004	0.5	0.002	0.5

3.2.2. Material Transportation

To calculate consumptions related to materials transportation and their corresponding environmental impacts, the following data were used in the model:

- The distance travelled by each material from the production site to the corresponding road (Table 5);
- Material weight (Table 2), which depended on the type of soil at each road;
- In addition, for ash and brick dust, lime was used as an activator (Table 2);
- For any material transportation, the type of truck was chosen from the GaBi database, based on the amounts of materials needed to be transported: Truck-Heavy Heavy-duty Diesel Truck/53.333 lb payload-8b; Unit process, not pre-allocated; consumption mix;
- The weight of the motor grader (9844 kg), of the roller (7144.1 kg), and of each material used (see Table 2) was considered. The vehicle that transported the machinery was low-key, while the truck that transported the materials was a diesel-based truck (see Figure 3).

Table 5. Weights and distances travelled to transport the materials for each intervened road.

Material	Urrao		La Paz	
	Distance (km)	Time (h)	Distance (km)	Time (h)
Ash	159	3.5	234	5.9
Lime (ash)	159	3.5	25	0.7
Brick dust	159	3.5	322	8.1
Lime (brick dust)	159	3.5	25	0.7
Sulphonated oil	159	3.5	25	0.7
Polymer	159	3.5	25	0.7

3.2.3. Machinery Transportation

To calculate the consumptions of machinery transportation and their environmental impacts, the following data were used in the model:

- The distances for machinery acquisition were calculated based on the closer city to each stabilised road: 2 km for Urrao (Antioquia region) and 14 km for La Paz (Cesar region);
- The machines weights were [53,54]: 7144 kg for the roller and 9844 kg for the motor grader;
- The machinery was transported in a low bed truck, taken from GaBi database, and chosen based on the capacity to support the weight of each machine: "Flatbed, platform, etc./49,000 lb payload-8b; Unit process, not pre-allocated; consumption mix".

3.2.4. Stabilisation Stage

For the stabilisation stage, diesel machines were used. The consumptions were calculated taking into account functional unit; the calculations (Tables 6 and 7) considered the following conditions:

- Performances obtained from the manufacturers of: the motor grader (2080 L/h); the roller (1510 L/h); and the water tanker (3581 km/L);
- Distances used to perform each activity within the stabilisation stage (including scarification, application of stabiliser (depending on the case), wetting/curing and compaction): calculated based on the number of times the relevant machinery was to pass over each cell, multiplied by the length of the cell;
- The time to carry the materials to each place was calculated taking the distance travelled in km and the speed used by each type of transport, for each material or machinery;
- Amount of water in the mixtures of lime + ash and lime + brick powder, calculated according to the proportion of the material used in each of the sections (See Table 8);
- Environmental impacts of diesel production: calculated using the GaBi data set for the United States (USA), because no Colombian diesel data were available (Diesel at refinery; from crude oil; production mix, at refinery; 15 ppm sulphur);
- Water consumption, which was measured in the field, adding up the amounts used for the curing and wetting activities, which depended on each type of soil and its moisture requirements (Table 8);
- Environmental impacts of water production calculated using the U.S. GaBi dataset (Tap water from groundwater; filtration, disinfection, ion removal, etc.; production mix, at plant; 1000 kg/m^3, 18 g/mol from the U.S.).

Table 6. Diesel consumption for the stabilisation stage in Urrao.

Activity	Diesel Consumption per Material (kg)			
	Sulphonated Oil	**Polymer**	**Brick Dust**	**Ash**
Scarification	12,300	67,700	49,100	106,000
Stabiliser application	4.08	14.6	*	*
Mixing	27,800	57,300	40,900	63,900
Compaction	16,700	16,900	13,300	15,000
Wetting/curing	2.04	1.36	6.12	8.16

Table 7. Diesel consumption for the stabilisation stage in La Paz.

Activity	Diesel Consumption per Material (kg)			
	Sulphonated Oil	Polymer	Brick Dust	Ash
Scarification	82,900	27,200	53,000	21,600
Stabiliser application	4.42	2.04	*	*
Mixing	29,200	10,300	19,200	45,800
Compaction	26,400	41,400	15,000	10,000
Wetting/curing	**	**	1.02	2.38

* These were applied manually, without a machine. ** Because of soil and material type, water addition was not needed.

Table 8. Water consumption during the stabilisation stage.

Material	Urrao kg	La Paz kg
Ash	57,000	15,400
Ash + lime	38,000	10,200
Brick dust	11,800	17,500
Brick dust + lime	5570	8220
Sulphonated oil	23,200	102,000
Polymer	20,300	50,300

* These were applied manually, without a machine.

3.3. Impact Assessment

This set of categories ensures the quantification of global impacts (ADP elements, ADP fossils, GWP, ODP), regional impacts (AP, EP), and local impacts (FAETP, HTP, MAETP, POCP, TETP). It also ensures the consideration of impacts on the terrestrial environment (ADP elements, ADP fossils, AP, EP, HTP, POCP, TETP), impacts on the aquatic environment (AP, EP, FAETP, MAETP), impacts on the air environment (GWP, ODP, POCP), and impacts on human health (HTP, ODP, POCP).

The reasoning for why to choose those categories is as follows. The global warming potential (or carbon footprint) generally has a more intense social perception than the other categories. A great interest in Colombia is focused on the impact on air quality due to the vehicle fleet and the industry [55]. On the other hand, supporting the circular economy and the need to save natural resources is of increasing interest worldwide. When emissions contributing to HTP and FAETP impact categories increase, they directly affect population health in the rural and urban areas. Finally, water consumption has been chosen because most of the Colombian water system is in the process of alteration, due to the detrimental effects caused by the transport of sediments, organic load, and toxic substances, with a high incidence in the industrial corridors located in the corresponding basins. In addition, the average consumption of urban households with drinking water service is 200 L/inhabitant per day, and 120 L/inhabitant per day for rural households. These figures exceed the minimum volume of 80 L necessary to guarantee life quality [56].

4. Results and Discussion

The results of the environmental impacts are presented in Tables 9–12, showing the differences of choosing the alternative materials used in the stabilisation process in the two road sections.

The stabilisation alternative which presented a higher environmental impact in all impact categories (with at least a difference of two orders of magnitude) was the sample using sulphonated oil. This difference is due to the need for significant additional resources for it to become a stabiliser [12]. If the environmental impact of the polymer (the second in the list) is normalised to 100%, for each road, the oil would have a contribution from 1500% up to 9500%, depending on the impact category.

If sulfonated oil, due to its high environmental impact, is discarded as a viable alternative for road stabilisation, the relative impacts of the other three options can be better analysed, without the interference of the high results produced by the oil. The results are presented below for the two roads studied.

4.1. Results for Urrao

For the soil of Urrao (see Tables 9 and 10 and Figure 4), the manufacture of the polymer and its application as a stabiliser were found to have greater impacts than the other two alternatives, which offer very similar results.

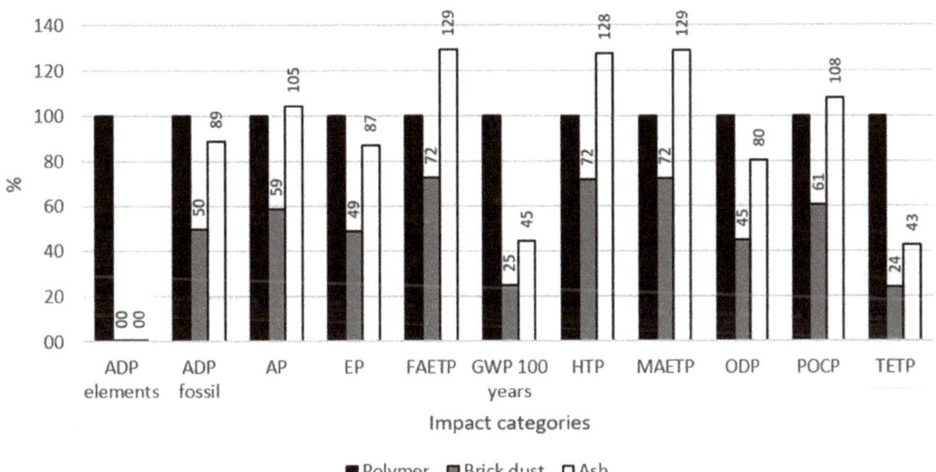

Figure 4. Environmental impacts for each stabiliser applied in Urrao.

Taking the polymer values (100%) as a reference, it was found that they presented higher impacts for the ADP elements, and GWP and TETP impact categories. This is due to the manufacture of the chemical compounds used during its transformation into a stabiliser. Some authors such as da Rocha et al.; Muench; and Siracusa et al.; [12,57,58] also found that, for the traditional materials such as natural aggregates and cement, the environmental impacts were generated during extraction and manufacturing.

For the other impact categories, the ash had the greatest impacts, because of the stabilisation stage: AP 104.5%, FAETP 129.4%; HTP 127.7%; MAETP 129.0%; and POCP 107.9%. Due to the type of soil in Urrao (clay soil), the machinery needed more time performing the processes of scarification, mixing, profiling, and wetting; therefore, more diesel consumption was needed. Larrea-Gallegos et al. [13] also found that stabilisation of soil had the highest impact due to fuel consumption. However, other authors Johnson et al.,; Karavalakis et al.; Wu, Zhang, Lou, Li, & Chen [52,59,60] found that, for traditional materials, GWP was higher due to the combustion of fuels during the process of stabilisation.

Table 9. Environmental impacts for liquid stabilisers at each life cycle stage for the Urrao system.

Impact Categories	Polymer				Sulphonated Oil			
	Material Production	Preliminaries Activities	Stabilisation	Total	Material Production	Preliminaries Activities	Stabilisation	Total
ADP elements [4] kg Sb-Equiv.	0.362	7.21×10^{-5}	24,100	0.362	0.00565	6.10×10^{-7}	9.76×10^{-5}	0.00575
ADP fossil [5] MJ	3.41×10^{6}	5930	7.30×10^{6}	1.07×10^{7}	1.64×10^{8}	50.2	2.92×10^{6}	1.67×10^{8}
AP [6] kg SO2-Equiv.	259	0.212	1060	1320	58.000	0.00179	424	58.400
EP [7] kg Phosphate-Equiv.	36.2	0.0705	73.2	110	4880	5.96×10^{-4}	29.3	4910
FAETP [8] kg DCB-Equiv.	822	3.54	163,000	1.63×10^{5}	3.69×10^{6}	0.0299	65.000	3.76×10^{6}
GWP 100 years [9] kg CO2-Equiv.	1.50×10^{5}	53.0	78,000	2.28×10^{5}	1.32×10^{7}	0.448	31.300	1.32×10^{7}
HTP [10] kg DCB-Equiv.	7650	44.7	459,000	4.66×10^{5}	1.19×10^{7}	0.379	1.84×10^{5}	1.21×10^{7}
MAETP [11] kg DCB-Equiv.	3.63×10^{6}	6720	6.16×10^{8}	6.20×10^{8}	1.40×10^{10}	56.9	2.47×10^{8}	1.42×10^{10}
ODP [12] kg R11-Equiv.	1.35×10^{-5}	2.73×10^{-9}	2.18×10^{-5}	3.53×10^{-5}	4.92×10^{-4}	2.31×10^{-11}	8.72×10^{-6}	5.01×10^{-4}
POCP [13] kg Ethene-Equiv.	34.6	0.0474	170	204	5490	4.01×10^{-4}	68.0	5560
TETP [14] kg DCB-Equiv.	60.5	0.159	28.7	89.4	8430	0.00135	11.5	8440

[4] ADP: Abiotic Depletion Potential, [5] AP: Acidification Potential, [6] EP: Eutrophication Potential, [7] FAETP: Freshwater Aquatic Ecotoxicity, [8] GWP: Global Warming Potential, [10] HTP: Human Toxicity Potential, [11] MAETP: Marine Aquatic Ecotoxicity, [12] ODP: Ozone Layer Depleticn Potential, [13] POCP: Photochemical Ozone Creation Potential, [14] TETP: Terrestrial Ecotoxicity Potential.

Table 10. Environmental impacts for solid stabilisers at each life cycle stage for the Urrao system.

Impact Categories	Brick Dust + Lime				Ash + Lime			
	Material Production	Preliminaries Activities	Stabilisation	Total	Material Production	Preliminaries Activities	Stabilisation	Total
ADP elements [4] kg Sb-Equiv.	6.44×10^{-7}	2.31×10^{-4}	1.76×10^{-4}	4.07×10^{-4}	6.44×10^{-7}	2.31×10^{-4}	3.18×10^{-4}	5.49×10^{-4}
ADP fossil [5] MJ	15.5	19,000	5.32×10^{6}	5.34×10^{6}	15.5	19,000	9.48×10^{6}	9.50×10^{6}
AP [6] kg SO2-Equiv.	0.00429	0.676	773	773	0.00429	0.676	1380	1380
EP [7] kg Phosphate-Equiv.	9.90×10^{-4}	0.226	53.3	53.5	9.90×10^{-4}	0.226	95.2	95.4
FAETP [8] kg DCB-Equiv.	0.00415	11.3	118,000	118,000	0.00415	11.3	211,000	211,000
GWP 100 years [9] kg CO2-Equiv.	1.50	170	56,900	57,000	1.50	170	102,000	102,000
HTP [10] kg DCB-Equiv.	0.0624	143	334,000	334,000	0.0624	143	596,000	596,000
MAETP [11] kg DCB-Equiv.	135	21.500	4.49×10^{8}	4.49×10^{8}	135	21,500	8.00×10^{8}	8.00×10^{8}
ODP [12] kg R11-Equiv.	5.41×10^{-12}	8.74×10^{-9}	1.59×10^{-5}	1.59×10^{-5}	5.41×10^{-12}	8.74×10^{-9}	2.83×10^{-5}	2.83×10^{-5}
POCP [13] kg Ethene-Equiv.	3.79×10^{-4}	0.152	124	124	3.79×10^{-4}	0.152	221	221
TETP [14] kg DCB-Equiv.	0.00494	0.510	21.0	21.5	0.00494	0.510	37.5	38.0

[4] ADP: Abiotic Depletion Potential, [5] AP: Acidification Potential, [6] EP: Eutrophication Potential, [7] FAETP: Freshwater Aquatic Ecotoxicity, [8] GWP: Global Warming Potential, [10] HTP: Human Toxicity Potential, [11] MAETP: Marine Aquatic Ecotoxicity, [12] ODP: Ozone Layer Depletion Potential, [13] POCP: Photochemical Ozone Creation Potential, [14] TETP: Terrestrial Ecotoxicity Potential

Table 11. Environmental impacts for liquid stabilisers at each life cycle stage for the La Paz system.

Impact Categories	Polymer				Sulphonated Oil			
	Material Production	Preliminaries Activities	Stabilisation	Total	Material Production	Preliminaries Activities	Stabilisation	Total
ADP elements [4] kg Sb-Equiv.	0.542	2.25×10^{-5}	1.34×10^{-4}	0.542	0.0150	6.31×10^{-6}	2.41×10^{-4}	0.0152
ADP fossil [5] MJ	5.11×10^{6}	1850	4.04×10^{6}	9.15×10^{5}	4.36×10^{8}	518	7.12×10^{6}	4.43×10^{8}
AP [6] kg SO2-Equiv.	389	0.0661	587	976	154,000	0.0185	1030	155,000
EP [7] kg Phosphate-Equiv.	54.4	0.0220	40.6	95.0	12,900	0.00616	71.5	13,000
FAETP [8] kg DCB-Equiv.	1230	1.11	90,000	91,300	9.78×10^{6}	0.309	159,000	9.94×10^{6}
GWP 100 years [9] kg CO2-Equiv.	225,000	16.5	43,200	268,000	3.51×10^{7}	4.63	76,200	3.52×10^{7}
HTP [10] kg DCB-Equiv.	11,500	14.0	254,000	266,000	3.16×10^{7}	3.91	448,000	3.20×10^{7}
MAETP [11] kg DCB-Equiv.	5.45×10^{6}	2100	3.42×10^{8}	3.47×10^{8}	3.73×10^{10}	587	6.02×10^{8}	3.79×10^{10}
ODP [12] kg R11-Equiv.	2.03×10^{-5}	8.54×10^{-10}	1.21×10^{-5}	3.24×10^{-5}	0.0013	2.38×10^{-10}	2.13×10^{-5}	0.00132
POCP [13] kg Ethene-Equiv.	51.8	0.0148	94.2	146	14,600	0.00414	166	14,800
TETP [14] kg DCB-Equiv.	90.8	0.0498	16.0	107	22,400	0.0139	28.2	22,400

[4] ADP: Abiotic Depletion Potential, [5] AP: Acidification Potential, [6] EP: Eutrophication Potential, [7] FAETP: Freshwater Aquatic Ecotoxicity, [8] GWP: Global Warming Potential, [10] HTP: Human Toxicity Potential, [11] MAETP: Marine Aquatic Ecotoxicity, [12] ODP: Ozone Layer Depletion Potential, [13] POCP: Photochemical Ozone Creation Potential, [14] TETP: Terrestrial Ecotoxicity Potential.

Table 12. Environmental impacts for solid stabilisers at each life cycle stage for the La Paz system.

Impact Categories	Brick Dust + Lime				Ash + Lime			
	Material Production	Preliminaries Activities	Stabilisation	Total	Material Production	Preliminaries Activities	Stabilisation	Total
ADP elements [4] kg Sb-Equiv.	6.44×10^{-7}	7.26×10^{-4}	1.49×10^{-4}	8.76×10^{-4}	6.44×10^{-7}	5.32×10^{-4}	1.33×10^{-4}	6.65×10^{-4}
ADP fossil [5] MJ	15.5	59,800	4.48×10^{6}	4.54×10^{6}	15.5	43,700	3.98×10^{6}	4.03×10^{6}
AP [6] kg SO2-Equiv.	0.000429	2.13	651	653	0.000429	1.56	578	580
EP [7] kg Phosphate-Equiv.	9.90×10^{-4}	0.710	45.0	45.7	9.90×10^{-4}	0.520	40.0	40.5
FAETP [8] kg DCB-Equiv.	0.00415	35.6	99,800	99,900	0.00415	26.1	88,700	88,800
GWP 100 years [9] kg CO2-Equiv.	1.50	533	48,000	48,500	1.50	390	42,600	43,000
HTP [10] kg DCB-Equiv.	0.0624	451	282,000	282,000	0.0624	330	250,000	251,000
MAETP [11] kg DCB-Equiv.	135	67,700	3.78×10^{8}	3.78×10^{8}	135	49,500	3.36×10^{8}	3.36×10^{8}
ODP [12] kg R11-Equiv.	5.41×10^{-12}	2.75×10^{-8}	1.34×10^{-5}	1.34×10^{-5}	5.41×10^{-12}	2.02×10^{-8}	1.19×10^{-5}	1.19×10^{-5}
POCP [13] kg Ethene-Equiv.	3.79×10^{-4}	0.478	104	105	3.79×10^{-4}	0.349	92.8	93.1
TETP [14] kg DCB-Equiv.	0.00494	1.61	17.7	19.3	0.0494	1.17	15.7	16.9

[4] ADP: Abiotic Depletion Potential, [5] AP: Acidification Potential, [6] EP: Eutrophication Potential, [7] FAETP: Freshwater Aquatic Ecotoxicity, [8] GWP: Global Warming Potential, [10] HTP: Human Toxicity Potential, [11] MAETP: Marine Aquatic Ecotoxicity, [12] ODP: Ozone Layer Depletion Potential, [13] POCP: Photochemical Ozone Creation Potential, [14] TETP: Terrestrial Ecotoxicity Potential

4.2. Results for La Paz

The environmental comparison was applied to the road in La Paz as well (see Tables 11 and 12 for absolute values, and Figure 5 for relative values). Results for the sulfonated oil are not presented in the figure, because its impacts exceeded the rest of the stabilisers by more than 1000%. The polymer results are taken as a reference value at 100%.

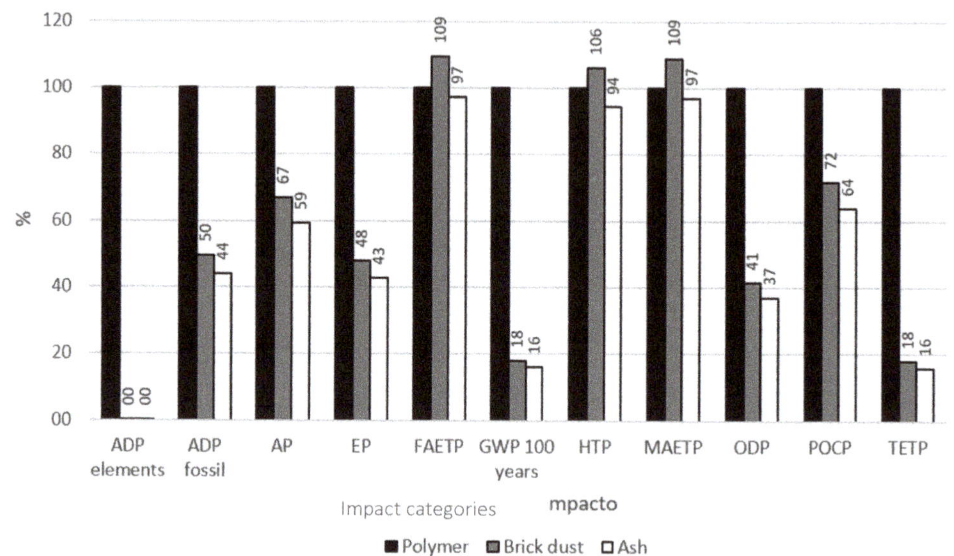

Figure 5. Environmental impacts for each stabiliser applied in La Paz.

For La Paz, the results obtained showed that the polymer impacts exceeded the rest of the stabilisers in even more impact categories than for Urrao (ADP elements, fossil ADP, AP, EP, GWP, ODP, POCP and TETP). For this road, the material that had the second place in impact was not the ash but the brick dust, because of the diesel consumption required for the stabilisation stage, offering greater impacts for three categories: FAETP (109.4%), HTP (106.2%) and MAETP (109.0%).

Some very significant data influencing the environmental results were obtained from processes included in databases corresponding to other countries (for example, the impacts associated with the generation of energy, the production of diesel and the manufacture of some raw materials), which may represent a significant uncertainty. During the project discussions, it was clearly acknowledged that a strong recommendation was to be sent to public authorities to start the construction of an environmental database of LCA data in Colombia, including energy, materials, and construction processes.

An extended analysis and comparison (obtaining environmental, social and economic information) with traditional materials used to stabilise soils, such as cement, lime and natural aggregates, would be very useful to produce policy changes in the sustainability direction.

Finally, although less common in Colombia, it would be also interesting to study roads with a high traffic volume, and promote environmental impact reduction in these more complex and exemplifying systems.

5. Conclusions

Due to the much higher environmental impact produced by the sulphonated oil for two types of soil, policy-makers may generally discard its use as stabiliser.

Regarding the three non-discarded stabilisation options (polymer, brick dust and ash), different results were obtained for the two tested roads, and even the order of priority varied from one road to the other. The decision process on the stabiliser should consider the soil type and the impact categories considered relevant, together with economic and social variables, which have not been assessed in this study.

If the focus is how the different life cycle stages contribute to the total impact of each alternative material option, it can be concluded that, for both regions, the environmental impacts are mostly caused by the stabiliser manufacture (60% of the total). This is due to the production of different chemicals needed during this stage.

Within the most contributing life cycle stage, stabilisation, the percentages of the impacts ranged from 40% to 90%, depending on the type of stabiliser (solid or liquid). Solids (brick dust and ash) had the greatest impacts because of the processes required to perform this stage. For instance, its impact on climate change was due to fuel combustion. This stage is (by far) the most relevant.

As presented in the introduction, the most relevant categories for Colombia seem to be GWP, HTP, and MAETP. In this sense, the best alternative material found for the Urrao type of soil is brick dust, because its impact on GWP is 25% lower. The results for La Paz indicated that the best material is fly ash, with 18.1% of the impact in GWP impact category compared to the other options.

Having an alternative material instead of a conventional raw material can be a great advantage in environmental terms, because resources are kept in a more circular economy. Further research will be needed to compare those types of materials with traditional substances such as natural aggregates or cement. This is essential to ensure that the overall performance along the life cycle of the alternative stabilisers is better (and by how much) from an environmental point of view.

A more holistic sustainability assessment, taking into account social and economic indicators, would help improve sound decision-making by public authorities when choosing among road stabilising processes. To develop and test methodology on this matter is recommended, because tertiary roads highly influence the rural settlements and their economic systems, both during construction and use.

Author Contributions: Conceptualization, A.B. and G.I.C.; methodology, P.F.-i.-P.; software, J.A.; validation, P.F.-i.-P., G.I.C.; formal analysis, A.B.; investigation, A.B.; writing—original draft preparation, A.B.; J.A.; P.F.-i.-P.; G.I.C.; writing—review and editing A.B.; J.A.; P.F.-i.-P.; G.I.C.; supervision, P.F.-i.-P., G.I.C.; project administration, G.I.C. All authors have read and agreed to the published version of the manuscript.

Funding: This research was funded by COLCIENCIAS and ESCI-UPF.

Institutional Review Board Statement: Not applicable.

Informed Consent Statement: Not applicable.

Data Availability Statement: Not applicable.

Acknowledgments: Thanks are given to Universidad de Medellín, Red Innovial and ESCI-UPF, for providing the funds of this research. One of the authors (A.B.) wishes to thank the UNESCO Chair in Life Cycle and Climate Change ESCI-UPF for hosting her during the 2016–2017 academic year, when research was carried out and this paper was started. The authors are responsible for the choice and presentation of information contained in this paper, as well as for the opinions expressed therein, which are not necessarily those of UNESCO and do not commit this organisation.

Conflicts of Interest: The authors declare no conflict of interest. The funders had no role in the design of the study; in the collection.

References

1. DNP. *Política para la Gestión de la Red Terciaria: Presupuesto Informado por Resultados*; Departamento Nacional de Planeación: Bogotá, Colombia, 2017.
2. Muñoz, F.; Arias, Y.P.; Hidalgo, C. *Evaluación del polvo de ladrillo como estabilizante de suelo perteneciente a vías terciarias. Proceedings from: Simposio Colombiano Sobre Ingeniería de Pavimentos*; Santa Marta, Colombia, 2015; pp. 1–10.

3. La República. Del Total de la red vial terciaria con la que cuenta Colombia, 96% está en mal estado. 2017. Available online: https://www.larepublica.co/infraestructura/del-total-de-la-red-vial-terciaria-con-la-que-cuenta-colombia-96-esta-en-mal-estado-2828335 (accessed on 27 September 2019).
4. Muñoz, F. *Evaluación de las Propiedades Hidráulicas de Suelos Estabilizados con Polvo de Ladrillo Utilizados en vías de Bajos Volúmenes de Tránsito*; Universidad de Medellín: Medellín, Colombia, 2016.
5. IDU. *Guía para el Diseño y la Construcción de Capas Estructurales de Pavimentos Estabilizados Mediante Procesos Químicos*; Instituto de Desarrollo Urbano: Bogotá, Colombia, 2005.
6. Cadavid, L.; Restrepo, F. *Evaluación del Efecto de la Erosión por Escorrentía Superficial en Suelos Residuales Arenosos Estabilizados con Cemento o cal*; Universidad de Medellín: Medellín, Colombia, 2007.
7. Cardona, C.; Ruiz, N. *Determinación de la Dosificación de cal para la Estabilización de Suelos Granulares de Dioritas, bajo Análisis de Pruebas Estadísticas*; Universidad de Medellín: Medellín, Colombia, 2005.
8. Echavarria, J.; Arroyave, N. *Diseño de Mezclas de Suelo-Cemento para Estructuras Ingenieriles, Utilizando Suelo Residual Proveniente de Diorite*; Universidad de Medellín: Medellín, Colombia, 2006.
9. Gómez, C.; Osorio, J. *Análisis del Comportamiento Estático de un Suelo Tipo Granular Adicionado con Cal*; Universidad de Medellín: Medellín, Colombia, 2004.
10. Ortiz, C.; Barreto, J. *Mejoramiento de Suelos con cal Fase III*; Universidad de Medellín: Medellín, Colombia, 2005.
11. INVIAS. *Especificaciones Generales de Construcción de Carreteras*; Instituto Nacional de Vías: Bogotá, Colombia, 2012.
12. Da Rocha, C.G.; Passuello, A.C.; Consoli, N.C.; Samaniego, R.A.Q.; Kanazawa, N.M. Life cycle assessment for soil stabilization dosages: A study for the Paraguayan Chaco. *J. Clean. Prod.* **2016**, *139*, 309–318. [CrossRef]
13. Larrea-Gallegos, G.; Vázquez-Rowe, I.; Gallice, G. Life cycle assessment of the construction of an unpaved road in an undisturbed tropical rainforest area in the vicinity of Manu National Park, Peru. *Int. J. Life Cycle Assess.* **2016**, *22*, 1109–1124. [CrossRef]
14. Kneese, A.V.; Ayres, R.U.; D'Arge, R.C. *Economics and the Environment: A Materials Balance Approach*; RFF Resources for the Future: New York, NY, USA, 2015.
15. Ghisellini, P.; Cialani, C.; Ulgiati, S. A review on circular economy: The expected transition to a balanced interplay of environmental and economic systems. *J. Clean. Prod.* **2016**, *114*, 11–32. [CrossRef]
16. Fullana-i-Palmer, P.; Puig, R.; Bala, A.; Baquero, G.; Riba, J.; Raugei, M. From Life Cycle Assessment to Life Cycle Management. *J. Ind. Ecol.* **2011**, *15*, 458–475. [CrossRef]
17. De Larriva, R.A.; Rodríguez, G.C.; López, J.M.C.; Raugei, M.; I Palmer, P.F. A decision-making LCA for energy refurbishment of buildings: Conditions of comfort. *Energy Build.* **2014**, *70*, 333–342. [CrossRef]
18. Bloom, E.F.; Horstmeier, G.J.; Pakes Ahlman, A.; Edil, T.B.; Whited, G. Assessing the Life Cycle Benefits of Recycled Material in Road Construction. GeoChicago-Conference-Paper-I-94-and-Beltline-LCA. 2017. Available online: http://rmrc.wisc.edu/wp-content/uploads/2017/05/GeoChicago-Conference-Paper-I-94-and-Beltline-LCA.pdf (accessed on 30 December 2020).
19. Chiu, C.-T.; Hsu, T.-H.; Yang, W.-F. Life cycle assessment on using recycled materials for rehabilitating asphalt pavements. *Resour. Conserv. Recycl.* **2008**, *52*, 545–556. [CrossRef]
20. Chowdhury, R.; Apul, D.; Fry, T. A life cycle based environmental impacts assessment of construction materials used in road construction. *Resour. Conserv. Recycl.* **2010**, *54*, 250–255. [CrossRef]
21. Hung, M.-L.; Ma, H.-W.; Chao, C.-W. Screening the Life Cycle Impact Assessment Methods and Modifying Environmental Impact Model to Determine Environmental Burdens. *J. Environ. Eng. Manag.* **2009**, *19*, 155–164. Available online: http://ser.cienve.org.tw/download/19-3/jeeam19-3_155-164.pdf (accessed on 30 December 2020).
22. Aldaco, R.; Margallo, M.; Fullana-i-Palmer, P.; Bala, A.; Gazulla, C.; Irabien, A.; Aldaco, R. Introducing life cycle thinking to define best available techniques for products: Application to the anchovy canning industry. *J. Clean. Prod.* **2017**, *155*, 139–150. [CrossRef]
23. Lasvaux, S.; Gantner, J.; Wittstock, B.; Bazzana, M.; Schiopu, N.; Saunders, T.; Gazulla, C.; Mundy, J.A.; Sjöström, C.; Fullana-i-Palmer, P.; et al. Achieving consistency in life cycle assessment practice within the European construction sector: The role of the EeBGuide InfoHub. *Int. J. Life Cycle Assess.* **2014**, *19*, 1783–1793. [CrossRef]
24. Benveniste, G.; Gazulla, C.; Fullana-i-Palmer, P.; Celades, I.; Ros, T.; Zaera, V.; Godes, B. Análisis de ciclo de vida y reglas de categoría de producto en la construcción. El caso de las baldosas cerámicas. *Informes de la Construcción* **2011**, *63*, 71–81. [CrossRef]
25. Mroueh, U.-M.; Eskola, P.; Laine-Ylijoki, J. Life-cycle impacts of the use of industrial by-products in road and earth construction. *Waste Manag.* **2001**, *21*, 271–277. [CrossRef]
26. Rajendran, S.; Gambatese, J. Solid Waste Generation in Asphalt and Reinforced Concrete Roadway Life Cycles. *J. Infrastruct. Syst.* **2007**, *13*, 88–96. [CrossRef]
27. Birgisdottir, H.; Bhander, G.; Hauschild, M.; Christensen, T. Life cycle assessment of disposal of residues from municipal solid waste incineration: Recycling of bottom ash in road construction or landfilling in Denmark evaluated in the ROAD-RES model. *Waste Manag.* **2007**, *27*, S75–S84. [CrossRef] [PubMed]
28. Celauro, C.; Corriere, F.; Guerrieri, M.; Casto, B.L. Environmentally appraising different pavement and construction scenarios: A comparative analysis for a typical local road. *Transp. Res. Part D Transp. Environ.* **2015**, *34*, 41–51. [CrossRef]
29. Balaguera, A.; Carvajal, G.I.; Albertí, J.; Fullana-i-Palmer, P. Life cycle assessment of road construction alternative materials: A literature review. *Resour. Conserv. Recycl.* **2018**, *132*, 37–48. [CrossRef]
30. Bala, A.; Raugei, M.; Benveniste, G.; Gazulla, C.; Fullana-i-Palmer, P.; Bala, A. Simplified tools for global warming potential evaluation: When 'good enough' is best. *Int. J. Life Cycle Assess.* **2010**, *15*, 489–498. [CrossRef]

31. Navarro, A.; Puig, R.; Fullana-I-Palmer, P. Product vs corporate carbon footprint: Some methodological issues. A case study and review on the wine sector. *Sci. Total Environ.* **2017**, *581–582*, 722–733. [CrossRef]
32. Puig, R.; Fullana-i-Palmer, P.; Baquero, G.; Riba, J.-R.; Bala, A. A Cumulative Energy Demand indicator (CED), life cycle based, for industrial waste management decision making. *Waste Manag.* **2013**, *33*, 2789–2797. [CrossRef]
33. Mackie, P.; Nellthorp, J.; Laird, J. Low Volume Rural Roads. 1 January 2005. pp. 1–9. Available online: http://documentos.bancomundial.org/curated/es/286811468339867486/Low-volume-rural-roads (accessed on 30 December 2020).
34. Balaguera, A.; Carvajal, G.I.; Jaramillo, Y.P.A.; Albertí, J.; Fullana-i-Palmer, P. Technical feasibility and life cycle assessment of an industrial waste as stabilizing product for unpaved roads, and influence of packaging. *Sci. Total. Environ.* **2018**, *651*, 1272–1282. [CrossRef]
35. Paez, D. Efectos de la estabilización electroquímica de suelos finos. *Rev. Ing. UPTC* **2005**, *18*, 83–96.
36. Camacho-Tauta, J.; Ortiz, O.J.R.; Mayorga, C. Efecto de la radiación UV en arcillas expansivas tratadas con aceite sulfonado. *Ingeniería y Competitividad* **2011**, *12*, 41–50. [CrossRef]
37. ISO. ISO 14044:2006(E)—Environmental management—Life cycle assessment—Requirements and guidelines. English, 2006. p. 42. Available online: https://www.iso.org/standard/38498.html (accessed on 28 January 2021).
38. ISO 14040. Environmental management—Life cycle assessment—Principles and framework. In *Environmental Management—Life Cycle Assessment—Principles and Framework*; ISO: Geneva, Switzerland, 2006. [CrossRef]
39. Baitz, M.; Albrecht, S.; Brauner, E.; Broadbent, C.; Castellan, G.; Conrath, P.; Fava, J.; Finkbeiner, M.; Fischer, M.; I Palmer, P.F.; et al. LCA's theory and practice: Like ebony and ivory living in perfect harmony? *Int. J. Life Cycle Assess.* **2013**, *18*, 5–13. [CrossRef]
40. ISO. ISO 14025—Environmental Labels and Declarations—Type III Environmental Declarations—Principles and Procedures. 2006. Available online: https://www.normativa-iso-une-pdf.com/descarga/pdf/une-en-iso-140252010/ (accessed on 28 January 2021).
41. Barthel, M.; Fava, J.; James, K.; Khan, S. Hotspots Analysis an Overarching Methodological Framework and Guidance for Product and Sector Level Application. 2017. Available online: www.lifecycleinitiative.org (accessed on 30 December 2020).
42. Chomkhamsri, K.; Wolf, M.-A.; Pant, R. International Reference Life Cycle Data System (ILCD) Handbook: Review Schemes for Life Cycle Assessment. In *Towards Life Cycle Sustainability Management*; Springer: Cham, Switzerland, 2011; pp. 107–117. [CrossRef]
43. Delgado-Aguilar, M.; Tarrés, Q.; Pèlach, M.À.; Mutjé, P.; Fullana-i-Palmer, P. Are Cellulose Nanofibers a Solution for a More Circular Economy of Paper Products? *Environ. Sci. Technol.* **2015**, *49*, 12206–12213. [CrossRef]
44. Guinée, J.B. *Handbook on Life Cycle Assessment: Operational Guide to the ISO Standards*; Kluwer Academic Publishers: New York, NY, USA, 2002.
45. BRE. Product Category Rules for Type III Environmental Product Declaration of Construction Products to EN 15804: 2012. 2013. Available online: http://www.bre.co.uk/filelibrary/materials/bre_en_15804_pcr.pn514.pdf (accessed on 30 December 2020).
46. Rashid, A.F.A.; Idris, J.; Sumiani, Y. Environmental Impact Analysis on Residential Building in Malaysia Using Life Cycle Assessment. *Sustainability* **2017**, *9*, 329. [CrossRef]
47. Quintero, A.B.; Cano, D.G.; Peláez, G.C.; Arias, Y.P. Technical and Environmental Assessment of an Alternative Binder for Low Traffic Roads with LCA Methodology. In *Proceedings of the 3rd Pan American Materials Congress*; Springer International Publishing: Cham, Switzerland, 2017.
48. Bueno, C.; Hauschild, M.Z.; Rossignolo, J.; Ometto, A.R.; Crespo-Mendes, N. Sensitivity analysis of the use of Life Cycle Impact Assessment methods: A case study on building materials. *J. Clean. Prod.* **2016**, *112*, 2208–2220. [CrossRef]
49. Huarachi, D.A.R.; Gonçalves, G.; De Francisco, A.C.; Canteri, M.H.G.; Piekarski, C.M. Life cycle assessment of traditional and alternative bricks: A review. *Environ. Impact Assess. Rev.* **2020**, *80*, 106335. [CrossRef]
50. Araújo, J.P.C.; Oliveira, J.R.; Da Silva, H.M.R.D. The importance of the use phase on the LCA of environmentally friendly solutions for asphalt road pavements. *Transp. Res. Part D Transp. Environ.* **2014**, *32*, 97–110. [CrossRef]
51. Santos, J.; Ferreira, A.; Flintsch, G.W. A life cycle assessment model for pavement management: Road pavement construction and management in Portugal. *Int. J. Pavement Eng.* **2014**, *16*, 315–336. [CrossRef]
52. Wu, D.; Zhang, F.; Lou, W.; Li, D.; Chen, J. Chemical characterization and toxicity assessment of fine particulate matters emitted from the combustion of petrol and diesel fuels. *Sci. Total Environ.* **2017**, *605–606*, 172–179. [CrossRef]
53. Ingersoll Rand. Ficha Técnica Vibrocompactador. 2015. Available online: http://maquqam.com/tecnicas/carreteras-3337/ingersoll-rand/sd70d.html (accessed on 30 December 2020).
54. John Deere. Ficha Técnica Motoniveladora. 2015. Available online: http://maquqam.com/tecnicas/construccion-6735/john-deere/570b.html (accessed on 30 December 2020).
55. Ministerio de Ambiente y Desarrollo Sostenible. Estrategia Nacional de Economía Circular 2018–2022. 2018. Available online: https://id.presidencia.gov.co/Paginas/prensa/2019/190614-Presidente-Duque-Estrategia-Nacional-Economia-Circular-primera-politica-ambiental-de-este-tipo-en-America-Latina.aspx#:~{}:text=La%20Estrategia%20Nacional%20de%20Econom%C3%ADa%20Circular%20transformar%C3%A1%20las%20cadenas%20de,Am%C3%A9rica%20Latina%20para%20el%20203057 (accessed on 30 December 2020).
56. CEPAL. Agua para el siglo XXI para América del Sur. De la División a la Acción. 2016. Available online: https://www.cepal.org/samtac/noticias/documentosdetrabajo/5/23345/InCo00200.pdf (accessed on 30 December 2020).
57. Siracusa, V.; Ingrao, C.; Giudice, A.L.; Mbohwa, C.; Rosa, M.D. Environmental assessment of a multilayer polymer bag for food packaging and preservation: An LCA approach. *Food Res. Int.* **2014**, *62*, 151–161. [CrossRef]

58. Muench, S.T. Roadway Construction Sustainability Impacts. *Transp. Res. Rec. J. Transp. Res. Board* **2010**, *2151*, 36–45. [CrossRef]
59. Johnson, D.; Heltzel, R.; Nix, A.C.; Clark, N.; Darzi, M. Greenhouse gas emissions and fuel efficiency of in-use high horsepower diesel, dual fuel, and natural gas engines for unconventional well development. *Appl. Energy* **2017**, *206*, 739–750. [CrossRef]
60. Karavalakis, G.; Hajbabaei, M.; Jiang, Y.; Yang, J.; Johnson, K.C.; Cocker, D.R.; Durbin, T.D. Regulated, greenhouse gas, and particulate emissions from lean-burn and stoichiometric natural gas heavy-duty vehicles on different fuel compositions. *Fuel* **2016**, *175*, 146–156. [CrossRef]

Article

Comparison of the Bearing Capacity of Pavement Structures with Unbound and Cold Central-Plant Recycled Base Courses Based on FWD Data

Audrius Vaitkus, Judita Gražulytė *, Igoris Kravcovas and Rafal Mickevič

Road Research Institute, Vilnius Gediminas Technical University, Linkmenų Str. 28, LT-08217 Vilnius, Lithuania; audrius.vaitkus@vilniustech.lt (A.V.); igoris.kravcovas@vilniustech.lt (I.K.); rafal.mickevic@vilniustech.lt (R.M.)
* Correspondence: judita.grazulyte@vilniustech.lt

Abstract: Bearing capacity changes over the year, depending on the water content in a pavement structure: the higher the water content, the lower the bearing capacity. As expected, the highest water content in a pavement structure is observed in the early spring as the ice lenses melt. Thus, spring is a critical period for pavement performance, because a decrease in bearing capacity results in faster pavement deterioration. The bearing capacity of pavement structures with an unbound base course and the negative effect of spring thawing on pavement performance have been analyzed by a considerable number of researchers. However, very little is known about the bearing capacity of pavement structures with a cold-recycled base course despite the significantly increasing usage of cold-recycled mixtures. This paper focuses on the bearing capacity of both unbound and cold central-plant recycled base courses at different seasons and their stability. A cold central-plant recycled (CCPR) base course was constructed from a mixture of 38.8% reclaimed asphalt pavement (RAP), 3.1% foamed bitumen and 2.3% cement. A virgin aggregate was added to achieve desirable aggregate gradation. The bearing capacity of the unbound and CCPR base layers, as well as the whole pavement structure, was evaluated by back-calculated E moduli from falling weight deflectometer (FWD) data. In addition to this, the residual pavement life was calculated using mechanistic-empirical pavement design principles. The results showed that the durability of pavement structures with a CCPR base course is more than seven times lower compared to that of pavement structures with an unbound base course, irrespective of season. Nevertheless, the bearing capacity (surface modulus E0) of the pavement structure with a CCPR base course gradually increases due to the curing processes of bituminous and hydraulic binders (in this study, within four years of operation, it increased by 28–47%, depending on the side of the road).

Keywords: bearing capacity; unbound base course; cold central-plant recycled base course; falling weight deflectometer (FWD), cold recycling in-plant

Citation: Vaitkus, A.; Gražulytė, J.; Kravcovas, I.; Mickevič, R. Comparison of the Bearing Capacity of Pavement Structures with Unbound and Cold Central-Plant Recycled Base Courses Based on FWD Data. *Sustainability* **2021**, *13*, 6310. https://doi.org/10.3390/su13116310

Academic Editor: Edoardo Bocci

Received: 26 April 2021
Accepted: 28 May 2021
Published: 2 June 2021

1. Introduction

Bearing capacity changes over the year depending on the water content in the pavement structure: the higher water content, the lower the bearing capacity [1–5]. As expected, the highest water content in the pavement structure is observed in the early spring [6,7]. The reason for that is the inability to drain water accumulated from melted ice and snow on the shoulders and embankment slopes within the pavement structure, because the bottom part of the pavement is still frozen. Consequently, the pavement part above the frost zone (typically subgrade and base courses) becomes soaked. This leads to a decrease in the bearing capacity of both individual layers and the whole pavement structure and afterward, results in faster pavement deterioration. This kind of phenomenon typically occurs in the regions where pavement structures are exposed to and affected by frost, ice, and snow and especially, where they periodically freeze and thaw.

The bearing capacity of pavement structures with an unbound base course and the negative effect of spring thawing on pavement performance have been analyzed by a considerable number of researchers [8–15]. Salour and Erlingsson [8] concluded that the back-calculated E modulus of the unbound base course in spring is 48% lower than in summer. Doré and Savard [12] found that pavement distresses, such as fatigue cracks, alligator cracks and permanent deformation, significantly increase during the spring thaw. Zhang and Macdonald [13] also confirmed that. Their study showed that about 60–75% of permanent deformation occurs during the spring thaw.

However, only a few studies have attempted to investigate the bearing capacity of pavement structures with a cold-recycled base course, despite the significantly increasing usage of cold-recycled mixtures. Vaitkus et al. [16] summarized the results from the plate load test on a cold-recycled base course in nine different road sections and concluded that mixtures bound with both a bituminous emulsion and cement have a higher bearing capacity than those with foamed bitumen and cement. Meocci et al. [17] analyzed measurements from a falling weight deflectometer (FWD) on the cold recycled base course after 29 days and 90 days of curing. They concluded that the bearing capacity increases within time and that their study confirms that the evolution of the curing process relates to the bituminous and hydraulic binders. A similar tendency was observed by [18]. However, they analyzed a much longer operation time (8 years), and measurements were done on the wearing course (the thickness of all asphalt layers was 19 cm). The back-calculated stiffness modulus based on the measured deflections revealed that the bearing capacity of the whole pavement structure increased within 3–5 years of construction and afterwards, decreased because of traffic. In addition to this, it was determined that the cold-recycled mixture with foamed bitumen and cement has 29% higher initial-bearing capacity than bituminous emulsion and cement, but after 3 years, both mixtures perform similarly.

Arimilli et al. [19] tried to mathematically assess the structural performance of pavement structures with a cold-recycled base course. In this case, cold-recycled mixtures were bound with either a bituminous emulsion or foamed bitumen. A stress-strain analysis showed that both mixtures were suitable for constructing the base course, but the use of the emulsion provided higher vertical compressive strain and deflection. Gu et al. [20] also used mathematical models to predict the performance of pavement structures with both cold central-plant and in-place recycled mixtures. In addition to this, they compared the predicted performance with the actual one and concluded that similar trends existed only in the first two years. Later, the rut depth was higher than predicted. It has to be noted that the prediction was performed using the Mechanistic-Empirical Pavement Design Guide, taking into account the viscoelasticity of cold-recycled mixtures with either bituminous emulsion or foamed bitumen.

Taking these studies together, there has been little discussion about the bearing capacity of pavement structures with a cold-recycled base course during different seasons and about the spring-thaw effect on pavement performance. Consequently, this paper analyzes the bearing capacity of pavement structures with both unbound and cold central-plant recycled (CCPR) base courses using a back-calculated surface modulus E_0 from the FWD data measured at different seasons within four years of operation and directly measured surface deflections, as well as the calculated surface curvature index (SCI), base damage index (BDI) and base curvature index (BCI). In addition to this, the remaining pavement life was calculated to evaluate the effect of a CCPR base course on pavement durability in comparison with the unbound base course.

This study is very important for the further usage of a CCPR base course with a thin asphalt surface course on it, instead of the unbound base course with both asphalt base and surface courses, which is a typical pavement structure for newly constructed and reconstructed roads. The alternative solution to the typical one uses recycled asphalt from the existing old pavement and helps to save natural resources. In addition to this, a much thinner asphalt layer is constructed on the base course, which results in a lower cost for the pavement structure. It is worth highlighting that, according to Lithuanian normative documents, both pavement structures are assumed equal and are used when the number of equivalent single-axle loads over a 20-year period is from 0.3 million to 1.0 million. However, as mentioned earlier, little is known about the bearing capacity of pavement structures with a CCPR base course at different seasons and the spring-thaw effect on that pavement performance. In addition to this, it is unknown if alternative pavement structures respond to load in the same manner as the typical structures and how this response changes depending on the season. This study will help to address this knowledge gap. Furthermore, the calculation of the remaining pavement life will show if there exist differences in durability.

2. Experiment

2.1. Test Site

The experiment to evaluate the bearing capacity of pavement structures with unbound and cold central-plant recycled base courses was conducted on a two-lane Lithuanian national road, No. 209 Joniškis-Žeimelis-Pasvalys, from 16.836 km to 27.900 km, which was reconstructed in 2016 using three types of pavement structures (Figure 1). Types A and B were designed with 20 cm of a cold central-plant recycled base course. On top of the base course was laid an asphalt surface course with a thickness of 4.5 cm. The difference between these two pavement structures is that type A was constructed on top of the existing sub-base using a 0–15 cm thick regulating course with a frost-resistant layer only where the road was widened, while in type B, the existing pavement structure was entirely removed, and a new 30 cm thick frost-resistant layer was constructed. Type C was designed as a typical pavement structure for newly constructed and reconstructed roads. It consisted of a frost-resistant layer (31 cm), an unbound base course (20 cm) and asphalt pavement (14 cm). The asphalt pavement consisted of an asphalt base course (10 cm) and an asphalt surface course (4 cm). In all types, the subgrade was improved with lime.

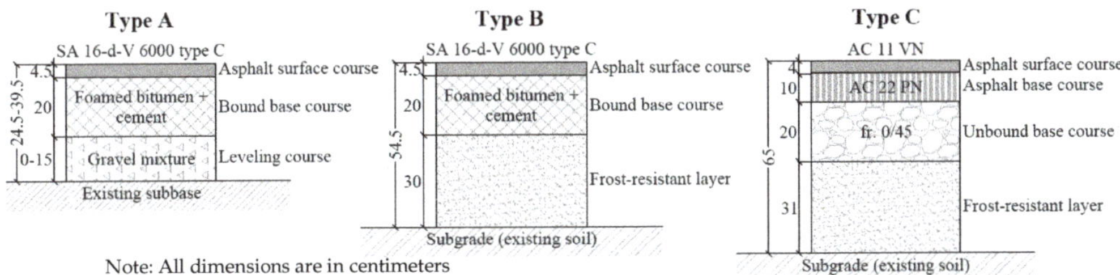

Figure 1. Pavement structures used in the test site.

The sections, in which each type of pavement structure was used, are represented in Figure 2.

All pavement structures were constructed in an area where the average annual temperature in the 1990–2019 period was 7.2 °C. The average temperature in January was −3.7 °C, and the average temperature in July was 18.4 °C. The average annual precipitation during the same period was 669 mm; the average depth of snow was 17.6 cm, and the average frost line was 50 cm. During the measurement period 2016–2019, the annual parameters were as follows: average annual temperature—7.6 °C, average temperature in January—−4.7 °C, average temperature in July—17.9 °C, average precipitation—638 mm, average snow depth—14.3 cm and average frost line—42.3 cm.

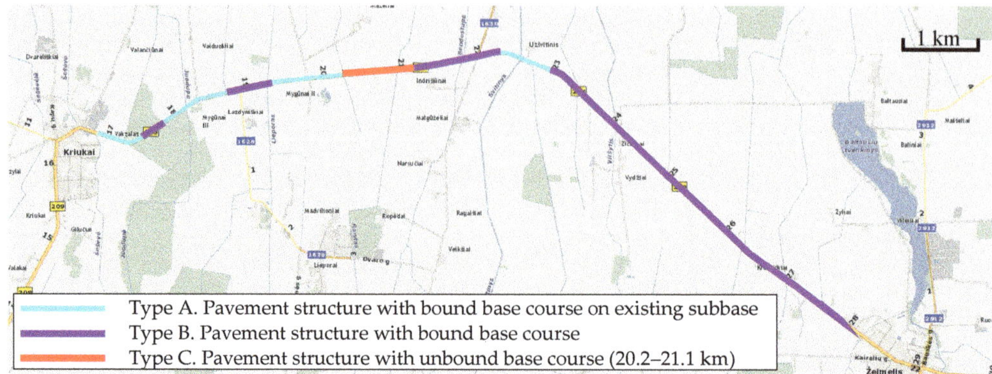

Figure 2. Test site location.

The annual average daily traffic (AADT) on road No. 209 for the period 2016–2019 is given in Figure 3. The numbers are derived from a stationary measurement post located at 8.181 km of this road, corresponding to the road section from 3.09 km to 29.14 km.

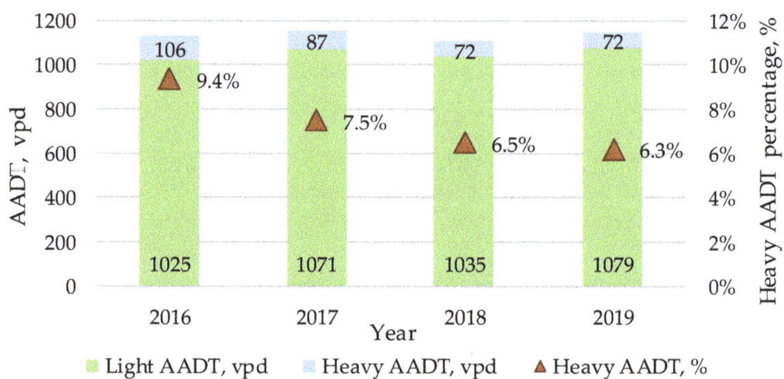

Figure 3. AADT on road No. 209 from 2016 to 2019.

2.2. Materials

The bound base course was constructed from a cold central-plant recycled mixture. For this purpose, the old pavement from the whole road section (16.836–21.200 km) was milled, broken, crushed and transported to a central plant, where 38.8% of it was mixed with 55.8% of crushed dolomite (0–32 mm) and bound with 2.3% of cement (CEM II/A-LL 42.5 N), as well as 3.1% of bitumen 50/70 (Figure 4). The properties of cold central-plant recycled mixture properties are represented in Table 1, and the gradation of the whole mixture, as well as its components, is depicted in Figure 5.

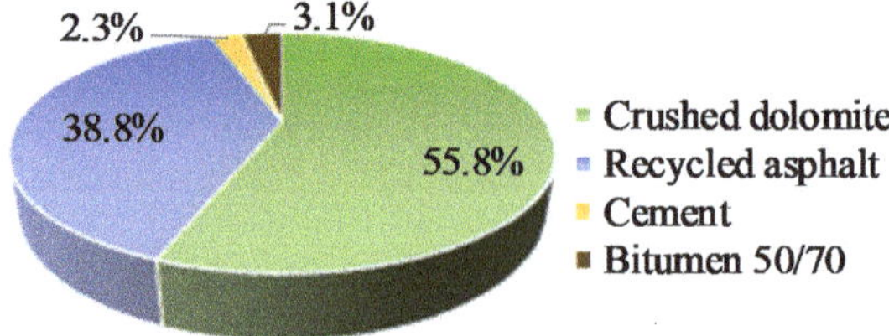

Figure 4. Cold-recycled mixture composition.

Table 1. Properties of cold-recycled mixture.

Parameter	Value	
	Actual	Required
Indirect tensile strength after 7 days, MPa	0.6	0.6–0.8
Indirect tensile strength after 28 days, MPa	0.7	0.7–1.0
Air voids content, %	9.2	5–15
Optimal water content, %	6.4	–
Proctor density, Mg/m^3	2.12	–

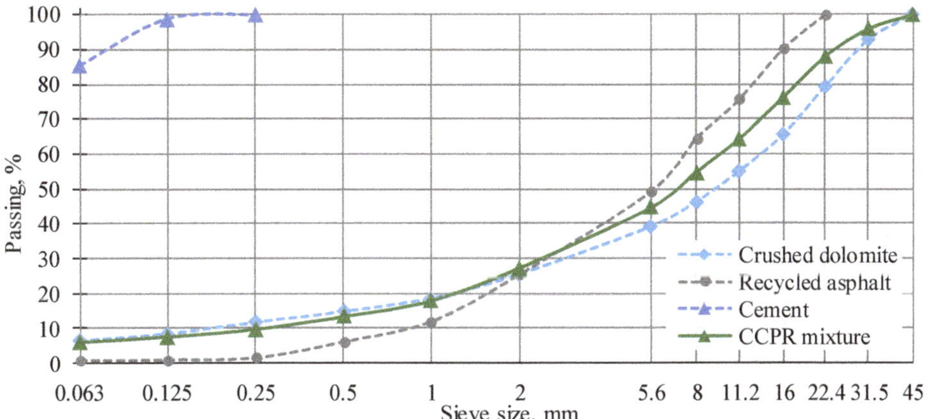

Figure 5. Cold-recycled mixture gradation.

2.3. Field Testing

A falling weight deflectometer (FWD), which is one of the most popular nondestructive testing (NDT) devices, was used to evaluate the bearing capacity of the constructed pavement structures. A FWD transfers a 50 kN load to the road pavement through a 300 mm diameter circular plate, which results in 707 MPa pressure. The generated haversine pulse lasts about 30 ms. Dynamic deflections on the road surface due to applied loads are captured by sensors (geophones), which are positioned at different distances from the center of the loading plate (0, 200, 300, 450, 600, 900, 1200, 1500 and 1800 mm).

Measurements with a FWD enable the evaluation of the bearing capacity of the whole pavement structure (on the surface level), as well as at different depths (according to the position of the geophones). Over the years of using NDTs, many different parameters have been introduced to evaluate different layers or parts of pavement structures. Talvik and Aavik [21] compiled a table with a list of widely used deflection basin parameters. As a result, in addition to the surface deflection d_0 and surface modulus E_0, this research also focuses on three basic deflection basin parameters:

- The surface curvature index (SCI) depicts the condition of top pavement layers;
- The base damage index (BDI) depicts the condition of base layers;
- The base curvature index (BCI) depicts the condition of the subgrade.

FWD measurements were carried out in October 2016 (immediately after the reconstruction of the road section), April 2017 and 2018 (as ice lenses melted), and May 2020. Each time, the same measurement procedure was applied, i.e., measurements were taken in the middle of the lane in 50 m intervals on the road section with an unbound base layer (20.2–21.2 km) and at 100 m intervals for the rest of the test road sections.

In addition to deflections, air, surface and asphalt-layer temperatures were recorded during FWD measurements. The asphalt layer temperature was measured in the middle of all asphalt layers. The lowest, highest and average asphalt surface and layer temperatures registered during all measurement periods are presented in Table 2.

Table 2. Asphalt surface and layer temperatures during all measurement periods.

Measurement Period	Temperature of Asphalt Surface, °C			Temperature of Asphalt Layer, °C		
	Min	Max	Average	Min	Max	Average
Oct 2016	2	12	8.5	5	8	5.9
April 2017	11	29	23.4	13	17	16.0
April 2018	18	24	22.2	17	19	18.2
May 2020	18	31	25.9	11	25	17.9

Since the measured deflections cannot be directly compared to each other due to the varying temperatures of the asphalt surface layer and loads, they were adjusted (normalized) to a reference (standard) load (50 kN) and temperature (20 °C). The adjustment (normalization) of the deflection was done by multiplying the measured deflection by the load and temperature correction factors. The load correction factor was derived by comparing the reference pressure, which is equal to 707,355 kPa and is caused by the 30 cm diameter plate and 50 kN load, with the measured one. The load correction factor was calculated by Equation (1):

$$k_L = \frac{P_{meas}}{P_{ref}} \tag{1}$$

where P_{meas} is the pressure under the pressure plate measured during the test, and P_{ref} is the reference pressure under the pressure plate.

The temperature correction factor was calculated by Equation (2) [22]:

$$k_T = 10^{-0.000221 \cdot h_{asf}^{1.0229} \cdot (T-20)} \tag{2}$$

where h_{asf} is the thickness of the asphalt layers, and T is the temperature of the asphalt layer measured during the test.

The surface modulus E_0 was calculated from the normalized surface deflections using Boussinesq's equation [23]:

$$E_0 = \frac{f \cdot (1 - \nu^2) \cdot \sigma_0 \cdot a}{d_0} \tag{3}$$

where f is a stress distribution factor (f = 2, as the stress was distributed uniformly); ν is Poisson's ratio; σ_0 is the stress (pressure) at the surface under the plate; a is the plate radius, and d_0 is the deflection at the center under the plate.

The surface curvature index (SCI) can be calculated at depths of 200 mm or 300 mm. The 200 mm depth was chosen to represent the condition of top bound layers because the top bound layer thickness is closer to this number (245 mm and 140 mm) than it is to 300 mm. The SCI is calculated by Equation (4). The base damage index (BDI) was calculated for the base layers using Equation (5). The base curvature index (BCI) was calculated for the subgrade by Equation (6). The lower values of these parameters mean the better condition of the layer.

$$\text{SCI} = d_0 - d_{200} \tag{4}$$

$$\text{BDI} = d_{300} - d_{600} \tag{5}$$

$$\text{BCI} = d_{600} - d_{900} \tag{6}$$

where d_0 is the deflection at the center under the plate, and d_r is the deflection measured at r distance from the center of the plate

2.4. Pavement Life Calculation

The pavement life calculation includes an estimation of the predicted number of ESALs using Equation (6) or Equation (7).

Pavement life was calculated by the mechanistic-empirical method using MN LAYER software [24] and accepting these assumptions (Yoder and Witczak, 1975 [25]):

- the thickness of the layers of the pavement structure is even;
- the layers are unrestricted in the horizontal direction;
- the properties of the materials in the layers are homogeneous, and isotropic asphalt layers and unbound layers have full adhesion, but there is partial adhesion between asphalt and the unbound layer;
- Poisson's ratio is constant, for asphalt layers—0.35 and for unbound layers—0.45;
- the reaction of the pavement structure calculated from a single wheel subjected to a force of 50 kN with a 15 cm radius of the contact area.

The fatigue function of asphalt was calculated using Equation (7) [26]:

$$N_{rib.} = \frac{k_1(T)}{F_{(\varepsilon 6)}} \cdot \left(\frac{S_{mix}(T)}{\sigma_v \cdot \gamma_{AC}} \right)^{k_2(T)} \quad (7)$$

where $S_{mix}(T)$ is the temperature-dependent modulus of asphalt E; σ_v is the vertical stress due to load; γ_{AC} is the safety factor of the asphalt layer; $F_{(\varepsilon 6)}$ is the fatigue safety factor, and $k_1(T)$ and $k_2(T)$ are the temperature coefficients.

The limit number of loads of unbound base layers and earth bed was calculated according to Equation (8) [26]:

$$N_{rib.} = 10^{\frac{1}{0.7}\left(\frac{0.00875 \cdot E_{V2} \cdot 1.0}{\sigma_{ZZ} \cdot \gamma}\right)} \quad (8)$$

where σ_{zz} is the vertical stresses resulting from the effect of the load; γ is the safety factor of the hydraulically bound layer, and E_{v2} is deformation modulus.

Pavement structure Type C was selected as a reference for the calculation of theoretical surface deflection. Data for theoretical surface deflection calculations are represented in Table 3. Layers' modulus E values are, according to LST EN 12697-26:2018 [27], 4-point bending (4 PB) at 20 °C and 10 Hz.

Table 3. Data for theoretical surface deflection calculation.

Pavement Structure	Thickness, cm	E modulus, MPa
Asphalt surface course AC 11 VN	4	4000
Asphalt base course AC 22 PN	10	5600
Unbound base course	20	350
Frost-resistant layer	31	120
Subgrade	–	45

In order to determine the remaining life of pavement structures Type B and Type C, calculations were made using MN LAYER software [24] and back-calculated E moduli. The E modulus of each layer was back-calculated with ELMOD software using the deflection basin fit method. A back-calculated *E* modulus shows the real time condition and the bearing capacity of pavement structural layers. After back-calculation, statistical indicators such as minimum, maximum, average and standard deviation values were calculated. One standard deviation below the average of back-calculated E modulus values was used in the calculations of the remaining pavement life. The use of one standard deviation below the average according to advisory circular 150/5320-6F, "Airport Pavement

Design and Evaluation" [28] allows a more accurate prediction of the remaining life of pavement structures.

The remaining pavement life was calculated by analyzing data from October 2016 and April 2018. That data were selected since measurements in October 2016 represent the performance of the pavement structure directly after construction, while data from April 2018 reveals pavement durability in the spring thaw.

3. Results and Discussion

3.1. Surface Deflection and Modulus

The normalized surface deflections (d_0) measured with a FWD on each side are represented in Figure 6. The statistical analysis for this parameter is presented in Table 4. The directly correlating parameter with deflections is the surface modulus E_0, which is represented in Figure 7. The statistical analysis for this parameter is presented in Table 5.

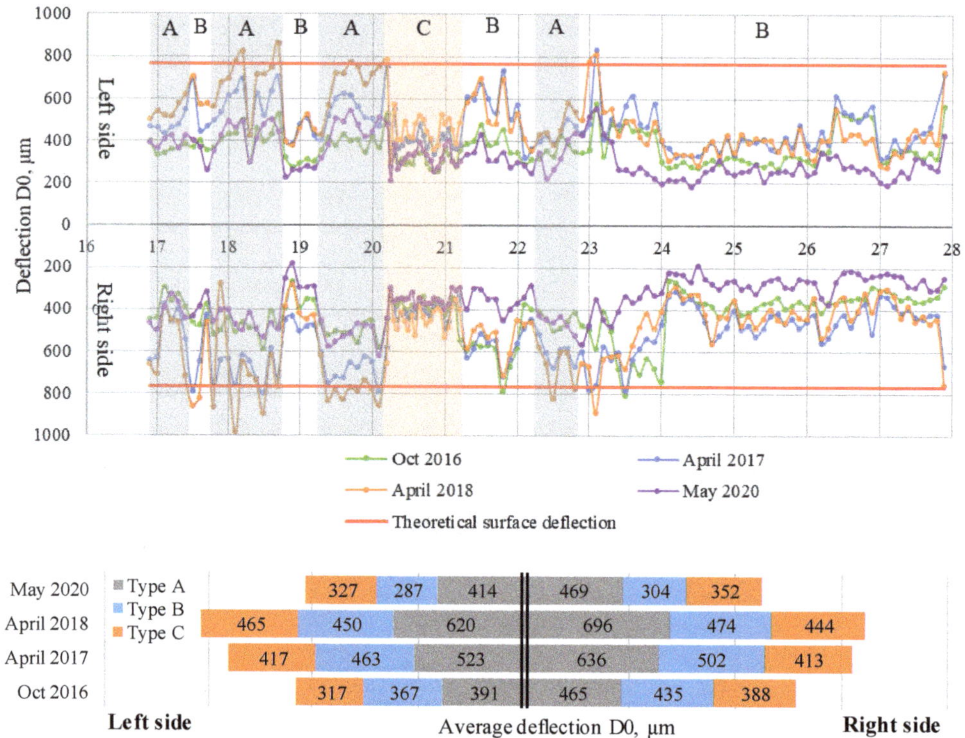

Figure 6. Deflection on each side of the road and their averages.

Table 4. Statistical analysis of deflections.

Param.	Test Date	Side	Pavement Structure Type A					Pavement Structure Type B					Pavement Structure Type C				
			Min	Max	Avg.	St. dev.	CV, %	Min	Max	Avg.	St. dev.	CV, %	Min	Max	Avg.	St. dev.	CV, %
Deflection, μm	Oct 2016	Left	297	526	391	50	12.8	268	577	367	83	22.5	254	530	317	57	17.9
		Right	295	587	465	65	14.0	262	807	435	133	30.5	338	546	388	45	11.5
	April 2017	Left	395	704	523	86	16.4	320	833	463	112	24.2	281	752	417	92	22.0
		Right	373	798	636	101	15.9	312	786	502	111	22.0	331	579	413	54	13.0
	April 2018	Left	382	863	620	132	21.4	281	812	450	116	25.9	297	787	465	96	20.7
		Right	278	983	696	152	21.9	279	889	474	132	27.8	312	656	444	72	16.2
	May 2020	Left	223	543	414	77	18.5	184	556	287	73	25.3	213	479	327	58	17.7
		Right	328	618	469	59	12.6	183	566	304	74	24.3	294	441	352	35	10.1

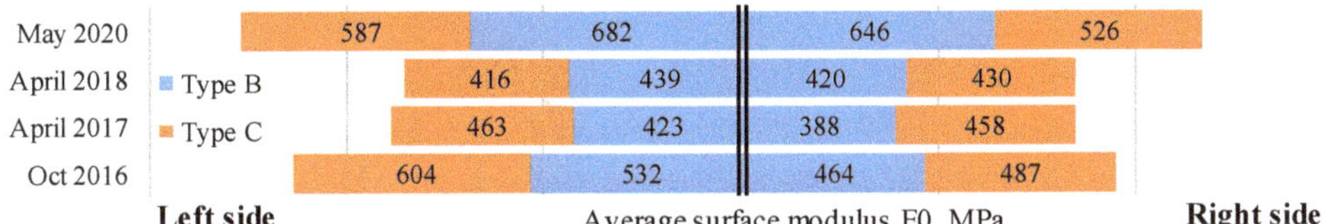

Figure 7. Surface moduli on each side of the road.

Table 5. Statistical analysis of surface moduli.

Param.	Test date	Side	Pavement Structure Type B					Pavement Structure Type C				
			Min	Max	Avg.	St. dev.	CV, %	Min	Max	Avg.	St. dev.	CV, %
Surface modulus, MPa	Oct 2016	Left	324	698	532	103	19.4	354	738	604	85	14.1
		Right	231	716	464	117	25.3	342	551	487	48	9.9
	April 2017	Left	222	584	423	90	21.4	247	661	463	79	17.1
		Right	237	596	388	81	20.9	322	564	458	55	11.9
	April 2018	Left	229	664	439	99	22.5	236	627	416	78	18.8
		Right	209	666	420	103	24.5	283	598	430	67	15.5
	May 2020	Left	334	984	682	142	20.8	385	858	587	104	17.7
		Right	327	1025	646	142	22.0	408	627	526	55	10.5

As seen from Figure 6, deflection varied from 183 to 983 μm irrespective of the measurement time, while the surface moduli varied from 189 to 1025 MPa. The highest average deflection was determined on the pavement structure of Type A (with a bound base course on top of the existing sub-base) in April 2018 on the right side of the road—696 μm, while Type B (with a bound base with a frost-resistant layer) and Type C (with an unbound base layer) performed quite similarly—474 μm and 444 μm, respectively. The surface moduli were 287, 420 and 430 MPa, respectively, for Type A, Type B and Type C. Thus, the pavement structure of Type A performed 31.7% and 33.3% worse than Types B and C, respectively, during these conditions. In fact, during each measurement on each side of the road, the pavement structure of Type B and C always outperformed Type A, with an improvement varying from 8.6% to 38.0%. One possible reason for this is that the FWD measurement point across the road could have been right above the edge of the existing sub-base, on top of which the rest of the pavement structure Type A was installed and that the existing sub-base could have been weak, which influenced the new pavement performance; thus, specific attention has to be paid to the construction in those places to ensure the desirable bearing capacity. This type of pavement structure (Type A) is unreliable in comparison with the other two structures and, as a result, it is omitted from further discussion about the results.

The lowest average deflection was measured on pavement structure Type B in May 2020—287 and 304 μm (left and right side, respectively), with resulting surface moduli of 682 and 646 MPa, respectively. At that period, the average deflection on pavement structure Type C was 327 and 352 μm (left and right side, respectively), and the surface moduli—587 and 526 MPa, respectively. This resulted in the average surface modulus for pavement structure Type B to be 16.3–22.8% higher than Type C, depending on the side of the road at that measurement period. When compared to the measurement period right after reconstruction (October 2016), the average surface modulus on pavement structure Type B improved by 28.3–39.3% (for the left and right side of the road, respectively) over the first 4.5 years, while on pavement structure Type C on the right side of the road, it increased by only 8.1%, and on the left side of the road, it actually decreased by 2.9%.

The theoretical deflection calculated for standard asphalt pavement with unbound base layers is 764 μm. During all four measurement periods, the measured deflection on pavement Type C exceeded the theoretical deflection only at one point: the thaw period in April 2018 on the left side of the road at 20.2 km. This station is the first station on the border with pavement Type A, which is most likely the reason for the high deflections. The average deflection on this pavement type was 1.64–2.41 times lower than the theoretical deflection. During all four measurement periods, the deflection on pavement Type B exceeded the theoretical deflection only on several measurement points during thaw periods in 2017 and 2018 (up to three stations per measurement period per road side, which is up to 4.3% of all measurements). Most of the stations are also on a border with pavement structure Type A, which is the probable reason for the high deflections at those stations. The average deflection on this pavement type was 1.52–2.66 times lower than the theoretical deflection, which is similar to pavement structure Type C. The resulting numbers show a reserve for both pavement structures when comparing the measured deflections to the theoretical deflection of a standard pavement even during thaw periods. Based on an average deflection, the reserve ranges from 34.3% to 62.4% of the theoretical deflection.

In general, the average surface modulus on pavement structure Type B was 16.3–20.5% higher right after reconstruction in October 2016 when compared to the thaw period next year (April 2017) and 35.0–35.7% higher during the latest measurement period, May 2020, when compared to the thaw period in April 2018. For pavement structure Type C, the respective numbers are 5.9–23.3% (comparing 2016 to 2017) and 18.3–29.1% (comparing 2020 to 2018). The decrease in performance during thaw seasons is more substantial for pavement structure Type B than Type C, which could indicate that a pavement structure with a CCPR base course is more susceptible to hydrothermal conditions than the pavement structure with an unbound base course.

The surface modulus for the left side of the road was higher irrespective of different pavement structures and measurement time (period) than the right side of the road. For example, in October 2016, the average moduli of the left side for pavement structures Type B and Type C were, respectively, 14.6% and 24.1% higher than on the right side. The difference gradually decreased, and, in May 2020, the average moduli of the left side for pavement structures Type B and Type C were, respectively, only 5.6% and 11.5% higher than on the right side.

3.2. Deflection Basin Parameters (SCI, BDI, BCI)

The surface curvature indices (SCI) calculated for each side are represented in Figure 8. The statistical analysis for this parameter is presented in Table 6. The SCI varied from 31 to 251 μm irrespective of the measurement time. The lowest average SCI was calculated on pavement structure Type C immediately after reconstruction in October 2016—41 and 51 μm (left and right side of the road, respectively). The SCI on pavement structure Type B was two times higher than on Type C—83 and 108 μm, respectively. In May 2020, the average SCI increased to 51 and 57 μm for Type C and decreased to 60 and 71 μm for Type B. Thus, from 2016 to 2020, CCPR-based pavement structures' disadvantage decreased from 100.4–109.7% to 18.3–25.2% compared to regular pavement regarding top layer performance. What is noticeable is that the average SCI on pavement structure Type B improved by 26.8–34.2% (depending on the road side) over the first 4.5 years, while on pavement structure Type C, the performance actually decreased by 10.2–24.0%.

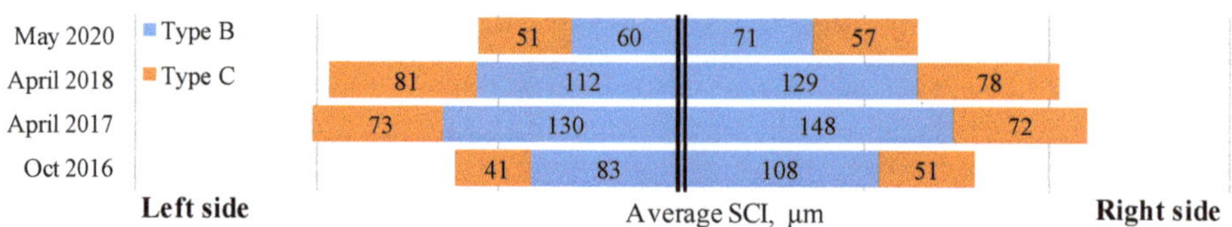

Figure 8. SCI on each side of the road.

Table 6. Statistical analysis of SCI.

Param.	Test Date	Side	Pavement Structure Type B					Pavement Structure Type C				
			Min	Max	Avg.	St. dev.	CV, %	Min	Max	Avg.	St. dev.	CV, %
SCI, µm	Oct 2016	Left	51	168	83	28	34.1	31	91	41	16	38.8
		Right	52	247	108	48	44.8	38	169	51	27	51.6
	April 2017	Left	79	229	130	45	34.6	53	151	73	20	26.9
		Right	85	251	148	42	28.5	63	83	72	6	7.7
	April 2018	Left	57	209	112	34	30.7	57	133	81	20	24.1
		Right	65	246	129	44	33.8	63	125	78	13	16.3
	May 2020	Left	39	113	60	15	24.4	37	72	51	7	14.6
		Right	42	125	71	21	30.1	50	63	57	4	6.6

The average SCI on pavement structure Type B was 37.6–57.8% higher during the thaw period in April 2017 when compared to the measurement period right after reconstruction in October 2016. The average SCI was 81.4–85.5% higher during the thaw period in April 2018 when compared to the latest measurement period, May 2020. For pavement structure Type C, the respective numbers are 40.9–76.4% (comparing 2017 to 2016) and 36.8–59.3% (comparing 2018 to 2020). This indicates that the top bound layer is almost equally susceptible to hydrothermal conditions irrespective of the pavement structure type.

As with the surface moduli, the condition on the left side of the road in regard of SCI was better than on the right side of the road. The average SCI for the left side of the road was lower than for the right side of the road irrespective of pavement structures and measurement time. For example, in October 2016, the average left side SCIs of the left sides of pavement structures Type B and Type C were, respectively, 23.4% and 19.9% lower than on the right side. The difference gradually decreased and, in May 2020, when the average SCIs of the left side of pavement structures Type B and Type C were, respectively, only 14.9% and 9.9% lower than on the right side. Although it is noticeable that, during thaw seasons, the SCI on the left side of the road was higher for pavement structure Type C (0.3% and 4.9%), while, for Type B, the left side condition was still better, lower by 12.2% and 13.0%, respectively.

The base damage index is presented in Figure 9. The statistical analysis for this parameter is presented in Table 7. The BDI varied from 32 to 236 µm irrespective of the measurement time. The lowest average BDI was determined to be on pavement structure Type B during the latest filed test, on May 2020—69 and 70 µm (left and right side of the road, respectively). In October 2016, the average BDIs were 85 and 100 µm. This means that, over 4.5 years on average, the base condition improved by 18.4–30.2% for Type B, depending on the side of the road. At the same time, the base condition on pavement structure Type C improved only on the right side of the road (by 5.6%) and decreased by 8.4% on the left side of the road. During the latest measurement period in May 2020, the average BDI for pavement structure Type B was 15.1–20.9% lower than for Type C, depending on the side of the road.

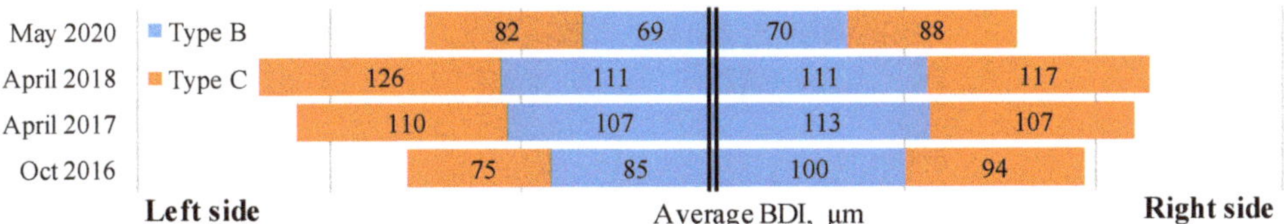

Figure 9. BDI on each side of the road.

Table 7. Statistical analysis of BDI.

Param.	Test Date	Side	Pavement Structure Type B					Pavement Structure Type C				
			Min	Max	Avg.	St. dev.	CV, %	Min	Max	Avg.	St. dev.	CV, %
BDI, μm	Oct 2016	Left	60	164	85	20	23.6	61	122	75	12	16.3
		Right	58	200	100	32	31.5	79	129	94	10	10.2
	April 2017	Left	73	196	107	26	24.5	72	195	110	23	20.7
		Right	73	204	113	26	23.3	80	130	107	11	9.9
	April 2018	Left	67	215	111	32	28.9	86	210	126	24	19.2
		Right	59	229	111	35	31.5	86	155	117	16	14.1
	May 2020	Left	44	138	69	19	28.1	58	102	82	12	14.5
		Right	32	149	70	19	27.1	78	102	88	7	8.1

The average BDI on pavement structure Type B was 13.1–26.9% higher during the thaw period in April 2017 when compared to the measurement period right after reconstruction in October 2016. The average BDI was 59.0–60.3% higher during the thaw period in April 2018 when compared to the latest measurement period, in May 2020. For pavement structure Type C, the respective numbers are 14.5–46.8% (comparing 2017 to 2016) and 32.1–54.9% (comparing 2018 to 2020). The results are inconsistent, as the first comparison shows a more substantial decrease in base performance for pavement structure Type C, while the second comparison shows a more substantial decrease in base performance for pavement structure Type B. This could be the result of an ongoing decrease of the average BDI for pavement structure Type B.

The average BDI for the left side of the road was lower than for the right side, irrespective of the measurement period, for pavement structure Type B, ranging from 0.2% to 15.4%. For pavement structure Type C, on the other hand, the left side of the road performed better than the right side only during the measurement periods October 2016 and May 2020 (by 7.8–19.8%), while, during the thaw seasons, the condition on the left side was worse than the right side (by 2.8–3.1%).

The base curvature index is presented in Figure 10. The statistical analysis for this parameter is presented in Table 8. The BCI varied from 25 to 134 μm irrespective of the measurement time. The lowest average BCI was determined on pavement structure Type B during the latest filed test, in May 2020—43 and 44 μm (left and right side of the road, respectively). In October 2016, the average BCI was 49 and 54 μm. This means that, over 4.5 years on average, the subgrade condition improved by 12.5–18.3% for Type B, depending on the side of the road. At the same time, the subgrade condition on pavement structure Type C improved only on the right side of the road (by 14.7%) and decreased by 0.5% on the left side of the road. During the latest measurement period of May 2020, the average BCI for pavement structure Type B was 15.9–19.0% lower than Type C depending on the side of the road.

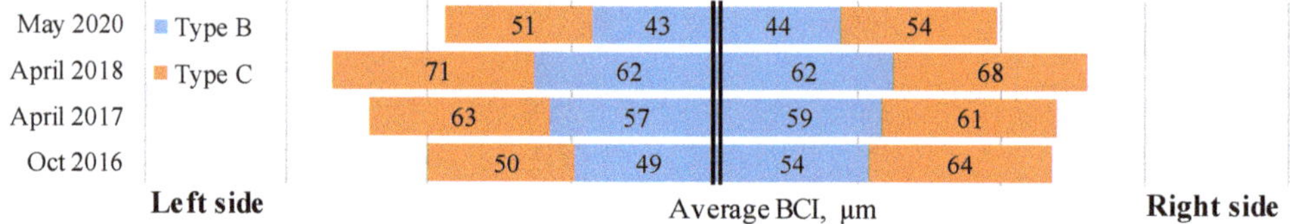

Figure 10. BCI on each side of the road.

Table 8. Statistical analysis of all relevant parameters.

Param.	Test Date	Side	Pavement Structure Type B					Pavement Structure Type C				
			Min	Max	Avg.	St. dev.	CV, %	Min	Max	Avg.	St. dev.	CV, %
BCI, µm	Oct 2016	Left	36	88	49	10	19.8	41	71	50	7	14.7
		Right	36	86	54	11	19.8	56	74	64	4	5.9
	April 2017	Left	36	117	57	14	25.0	38	110	63	15	23.7
		Right	38	107	59	13	22.6	46	87	61	9	14.8
	April 2018	Left	36	109	62	17	27.0	43	118	71	15	21.5
		Right	38	118	62	15	24.8	44	111	68	14	21.1
	May 2020	Left	25	87	43	12	27.9	30	76	51	10	20.3
		Right	25	88	44	11	24.7	43	69	54	7	12.6

The average BCI on pavement structure Type B was 9.1–17.5% higher during the thaw period in April 2017 when compared to the measurement period right after reconstruction in October 2016. The average BCI was 41.0–45.8% higher during the thaw period in April 2018 when compared to the latest measurement period, May 2020. For pavement structure Type C, an anomaly was detected, as the average BCI on the right side of the road was lower during the thaw period in April 2017 when compared to October 2016. While, on the right side during the thaw period, the subgrade performed 24.2% worse than right after reconstruction. The average BCI on pavement structure Type C was 26.0–39.9% higher during the thaw period in April 2018 when compared to the latest measurement period, May 2020.

The average BCI for the left side of the road was lower right after reconstruction for both pavement structures—Type B and Type C during the measurement period of October 2016 by 9.4% and 20.7%, respectively. During both thaw measurement periods, the difference was not significant, ranging from 3.7% lower performance on the left side to 2.4% higher performance on the left side when compared to the right side, irrespective of the pavement structure. The average BCI again improved for the left side for both pavement structures—Type B and Type C during the latest measurement period, May 2020, by 2.9% and 6.6% respectively.

3.3. Backcalculation E Modulus and Remaining Pavement Life

The back-calculated E moduli with ELMOD software for pavement structures of Type B and Type C are represented in Table 9. During the measurement period October 2016, the average E modulus of the bound layer of pavement structure Type B was 1.88 times greater on the left side of the road (1.32 times greater on the right side of the road) than in the thaw season in April 2018. The average E modulus of the frost-resistant layer in October 2016 was 1.57 times greater on the left side of the road (1.29 times greater on the right side of the road) than in the thaw season in April 2018. The average E modulus of the subgrade in October 2016 was 1.23 times greater on the left side of the road (1.18 times greater on the right side of the road) than in the thaw season in April 2018.

The average back-calculated E modulus of the bound layer of pavement structures of Type C during the measurement period October 2016 was 1.91 times greater on the left side of the road (2.42 times greater on the right side of the road) than during the thaw season in April 2018. The average E modulus of the unbound layer in October 2016 was 1.37 times greater on the left side of the road (1.75 times greater on the right side of the road) than in the thaw season in April 2018. The average E modulus of the frost-resistant layer in October 2016 was 1.10 times greater on the left side of the road (1.45 times greater on the right side of the road) than in the thaw season in April 2018. The average E modulus of the subgrade in October 2016 was 1.08 times greater on the left side of the road (1.29 times greater on the right side of the road) than in the thaw season in April 2018.

Table 9. Statistical indicators of back-calculated E moduli.

Parameter	Test Date	Side	Layer	Pavement Structure Type B				Pavement Structure Type C			
				Min	Max	Avg.	St. dev.	Min	Max	Avg.	St. dev.
E modulus, MPa	Oct 2016	Left	Bound	836	6385	1895	643	5190	8029	6599	836
		Right		616	7488	1602	716	2864	9546	8128	1537
		Left	Unbound	–	–	–	–	276	579	448	79
		Right		–	–	–	–	250	673	528	104
		Left	Frost-resistant	12	730	394	121	149	321	225	37
		Right		10	753	310	161	213	504	346	90
		Left	Subgrade	28	158	92	23	58	82	68	7
		Right		18	184	86	31	53	114	80	13
	April 2018	Left	Bound	490	2015	1007	368	2897	4555	3457	416
		Right		621	2403	1217	360	1773	4335	3356	641
		Left	Unbound	–	–	–	–	244	426	327	45
		Right		–	–	–	–	138	420	301	58
		Left	Frost-resistant	30	558	251	110	86	429	247	86
		Right		91	541	241	98	113	397	239	75
		Left	Subgrade	25	167	75	27	37	106	63	14
		Right		30	138	73	26	26	101	62	16

The average back-calculated E modulus of the bound layer during the measurement period October 2016 of pavement structures Type C was 3.48 times greater on the left side of the road (5.07 times greater on the right side of the road) than pavement structure Type B. The average E modulus of the frost-resistant layer of pavement structure Type B was 1.75 times greater on the left side of the road (0.90 times smaller on the right side of the road) than pavement structure Type C. The average E modulus of the subgrade of pavement structure Type B was 1.35 times greater on the left side of the road (1.08 times greater on the right side of the road) than pavement structure Type C.

The average back-calculated E modulus of the bound layer during the measurement period of April 2018 of pavement structures Type C was 3.44 times greater on the left side of the road (2.76 times greater on the right side of the road) than pavement structure Type B. The average E modulus of the frost-resistant layer of pavement structure Type B was 1.02 times greater on the left side of the road (1.01 times greater on the right side of the road) than pavement structure Type C. The average E modulus of the subgrade of pavement structure Type B was 1.19 times greater on the left side of the road (1.18 times greater on the right side of the road) than pavement structure Type C.

Calculated pavement surface deflection, E_0 and the predicted number of ESALs are represented in Table 10. Calculated pavement surface deflection of October 2016 for pavement structure Type B is 0.90 times smaller than the thaw season in April 2018 and, for pavement structure Type C—0.60 times smaller than the thaw season in 2018. E_0 modulus calculated for October 2016 for pavement structure Type B is 1.11 times greater than in the thaw season in April 2018 and, for pavement structure Type C, it is 1.65 times greater than the thaw season of April 2018. The predicted number of ESALs calculated of October 2016 for pavement structures B is 3 times greater than the thaw season of April 2018, for pavement structures C—45.43 times greater than the thaw season of April 2018.

Table 10. Results of pavement life calculations.

Test Date	Side	Pavement Structure Type	Parameter		
			Deflection, μm	E_0, MPa	Predicted Number ESALs, mln.
Oct 2016	Right	B	886	210	0.03
April 2018	Right	B	983	190	0.01
Oct 2016	Right	C	505	369	3.18
April 2018	Right	C	839	222	0.07

Calculated pavement surface deflection of October 2016 for pavement structure Type B is 1.75 times greater than for pavement structure Type C, for the thaw season in April 2018 is 1.17 times greater. E_0-modulus-calculated pavement surface deflection in October 2016 for pavement structure Type B is 0.57 times smaller than pavement structure Type C, for the thaw season of April 2018—0.86 times smaller. The predicted number of ESALs in October 2016 on pavement structure Type C is 106 times greater than pavement structure Type B and, for the thaw season of April 2018, 7 times greater.

4. Conclusions

The analysis of FWD data and calculated deflection basin parameters (SCI, BDI, BCI), as well as a comparison of measured and theoretically calculated deflection, in addition to the predicted pavement life on the basis of back-calculated E moduli, led to the following conclusions:

- In all cases, the pavement with a CCPR base course constructed on the existing sub-base (Type A) performs 9–38% worse than fully reconstructed pavements (Types B and C) in terms of surface modulus E_0, irrespective of the base course type (CCPR or unbound). The reason for this may be that the measurements' position on pavement structure Type A coincided with the edge of the existing sub-base, since the road was widened in those sections. Thus, to ensure a desirable bearing capacity, specific attention has to be paid to the construction in such conditions.
- The theoretical deflection on the pavement surface calculated using mechanistic-empirical pavement design principles is exceeded only in some individual measurement points in the spring thaw (Type B and C). The average deflection on the pavement surface was lower than the theoretical value even in the thaw of 2017 and 2018 when it increased by 9–26% (Type B) and by 6–47% (Type C), respectively, compared to in October. The same tendency was determined regarding SCI (increased by 35–57% in Type B and by 41–98% in Type C) and BDI (increased by 13–31 in Type B and by 14–68% in Type C).
- The use of a CCPR base course (Type B) leads to much more scattered FWD data (the coefficient of variation varied from 22% to 31%) than from the pavement structure with the unbound base course (Type C, the coefficient of variation varied from 10% to 22%). The reason for this may be related to the inhomogeneity of the CCPR base course (38.8% reclaimed asphalt). Therefore, for the determination of the residual life of pavement structures with a CCPR base course, we recommend the use of input data that reflects the pavement performance with a deflection value equal to the average deflection calculated based on the measured FWD data and decreased by one standard deviation (average deflection minus one standard deviation). Such an approach will prevent the pavement structure from premature failures.
- Based on the pavement design with back-calculated E moduli, the durability of the pavement structure with a CCPR base course (Type B) is more than seven times lower compared to that of the pavement structure with an unbound base course (Type C), irrespective of season. This result may be explained by the fact that soft asphalt with a very thin layer (4.5 cm) was constructed on the CCPR base course, and its E modulus is more than five times lower than that of asphalt concrete mixtures used in typical pavement structures with an unbound base course. Nevertheless, it must be emphasized that it has already been proven by a number of researchers that the bearing capacity of pavement structures with a CCPR base course gradually increases over time due to the curing processes of bituminous and hydraulic binders (in this study, the surface modulus E0 increased by 28–47%, depending on the side of the road) and significantly prolongs pavement life. Unfortunately, this increase in bearing capacity cannot be assessed by typical mechanistic-empirical pavement design principles.
- Seeking to design a pavement structure based on the back-calculated E moduli from FWD data, it is vital to adjust (normalize) all measured data to a reference (standard)

load and climate conditions. However, in this study, an applied temperature correction factor is determined based on the FWD measurements on the test road in the summer when hydrothermal conditions did not affect the stiffness of unbound layers. While in practice, measurements are done irrespective of the season, for example, in spring when the bearing capacity (E modulus) of unbound layers are affected by water content. Thus, this effect should be also incorporated in the correction factor. Further studies, which take this into account, will be undertaken.

Author Contributions: Conceptualization, A.V.; methodology, J.G.; software, R.M.; formal analysis, I.K. and R.M.; writing—original draft preparation, I.K. and R.M.; writing—review and editing, J.G.; supervision, A.V. All authors have read and agreed to the published version of the manuscript.

Funding: This research received no external funding.

Institutional Review Board Statement: Not applicable.

Informed Consent Statement: Informed consent was obtained from all subjects involved in the study.

Data Availability Statement: The data presented in this study are available on request from the corresponding author.

Conflicts of Interest: The authors declare no conflict of interest.

References

1. Simonsen, E.; Isacsson, U. Thaw weakening of pavement structures in cold regions. *Cold Reg. Sci. Technol.* **1999**, *29*, 135–151. [CrossRef]
2. Yi, J.; Doré, G.; Bilodeau, J.-P.; Prophète, F. Monitoring the structural behaviour variation of a flexible pavement structure during freeze and thaw. In Proceedings of the 2014 Conference of the Transportation Association of Canada, Montreal, QC, Canada, 28 September–1 October 2014; pp. 1–12.
3. Ishikawa, T.; Lin, T.; Kawabata, S.; Kameyama, S.; Tokoro, T. Effect evaluation of freeze-thaw on resilient modulus of unsaturated granular base course material in pavement. *Transp. Geotech.* **2019**, *21*, 100284. [CrossRef]
4. Saevarsdottir, T.; Erlingsson, S. Effect of moisture content on pavement behaviour in a heavy vehicle simulator test. *Road Mater. Pavement Des.* **2013**, *14*, 274–286. [CrossRef]
5. Saevarsdottir, T.; Erlingsson, S. Water Impact on the Behaviour of Flexible Pavement Structures in an Accelerated Test. *Road Mater. Pavement Des.* **2013**, *14*, 256–277. [CrossRef]
6. Charlier, R.; Hornych, P.; Srsen, M.; Hermansson, A.; Bjarnason, G.; Erlingsson, S.; Pavsic, P. Water Influence on Bearing Capacity and Pavement Performance. In *Water in Road Structures*; Dawson, A., Ed.; Springer: Dordrecht, The Netherlands, 2009; pp. 175–192.
7. Wen, Z.; Zhang, M.; Ma, W.; Wu, Q.; Niu, F.; Yu, Q.; Fan, Z.; Sun, Z. Thermal–moisture dynamics of embankments with asphalt pavement in permafrost regions of central Tibetan Plateau. *Eur. J. Environ. Civ. Eng.* **2015**, *19*, 387–399. [CrossRef]
8. Salour, F.; Erlingsson, S. Investigation of a pavement structural behaviour during spring thaw using falling weight deflectometer. *Road Mater. Pavement Des.* **2013**, *14*, 141–158. [CrossRef]
9. Vaitkus, A.; Laurinavičius, A.; Oginskas, R.; Motiejūnas, A.; Paliukaitė, M.; Barvidienė, O. The road of experimental pavement structures: Experience of five years operation. *Balt. J. Road Bridg. Eng.* **2012**, *7*, 220–227. [CrossRef]
10. White, T.D.; Coree, B.J. Treshold Pavement Thickness to Survive Spring Thaw. In *Proceedings of the Third International Conference on Bearing Capacity of Roads and Airfields*; Norwegian University of Science and Technology, Ed.; Tapir Publishers: Trondheim, Norway, 1990; pp. 41–51.
11. St-Laurent, D.; Roy, M. Évaluation Structurale des Chausses Souples dans un Contexte Climatique Nordique: Une Étude Avec le FWD (Structural Evaluation of Flexible Pavements in a Northern Context: A Study Using the FWD.). In Proceedings of the 30th Annual Conference of AQTR, Association Québécoise du Transport et des Routes, Quebec, QC, Canada, 1995. Available online: https://corpus.ulaval.ca/jspui/handle/20.500.11794/45150 (accessed on 20 May 2021).
12. Doré, G.; Savard, Y. Analysis of Seasonal Pavement Deterioration. In Proceedings of the Transportation Research Board; Preprint no. 981046, Transportation Research Board of the National Academies, Washington, DC, USA, 1998.
13. Zhang, W.; Macdonald, R.A. Response and Performance of a Test Pavement to Freeze-Thaw Cycles in the Danish Road Testing Machine. In Proceedings of the Fifth International Conference on Unbound Aggregates in Roads, Nottingham, UK, 21–23 June 2000; pp. 77–86.
14. Lekarp, F.; Isacsson, U.; Dawson, A. State of the Art. I: Resilient Response of Unbound Aggregates. *J. Transp. Eng.* **2000**, *126*, 66–75. [CrossRef]
15. Lekarp, F.; Isacsson, U.; Dawson, A. State of the Art. II: Permanent Strain Response of Unbound Aggregates. *J. Transp. Eng.* **2000**, *126*, 76–83. [CrossRef]

16. Vaitkus, A.; Gražulytė, J.; Juknevičiūtė-Žilinskienė, L.; Andrejevas, V. Review of Lithuanian Experience in Asphalt Pavements Cold Recycling. In Proceedings of the 10th International Conference "Environmental Engineering", VGTU Technika, Vilnius, Lithuania, 27–28 April 2017.
17. Meocci, M.; Grilli, A.; La Torre, F.; Bocci, M. Evaluation of mechanical performance of cement–bitumen-treated materials through laboratory and in-situ testing. *Road Mater. Pavement Des.* **2017**, *18*, 376–389. [CrossRef]
18. Godenzoni, C.; Graziani, A.; Bocci, E.; Bocci, M. The evolution of the mechanical behaviour of cold recycled mixtures stabilised with cement and bitumen: Field and laboratory study. *Road Mater. Pavement Des.* **2018**, *19*, 856–877. [CrossRef]
19. Arimilli, S.; Nagabhushana, M.N.; Jain, P.K. Comparative mechanistic-empirical analysis for design of alternative cold recycled asphalt technologies with conventional pavement. *Road Mater. Pavement Des.* **2018**, *19*, 1595–1616. [CrossRef]
20. Gu, F.; Ma, W.; West, R.C.; Taylor, A.J.; Zhang, Y. Structural performance and sustainability assessment of cold central-plant and in-place recycled asphalt pavements: A case study. *J. Clean. Prod.* **2019**, *208*, 1513–1523. [CrossRef]
21. Talvik, O.; Aavik, A. Use of FWD Deflection Basin Parameters (SCI, BDI, BCI) for Pavement Condition Assessment. *Balt. J. Road Bridg. Eng.* **2009**, *4*, 196–202. [CrossRef]
22. Motiejūnas, A.; Paliukaitė, M.; Vaitkus, A.; Čygas, D.; Laurinavičius, A. Research on the Dependence of Asphalt Pavement Stiffness Upon the Temperature of Pavement Layers. *Balt. J. Road Bridg. Eng.* **2010**, *5*, 50–54. [CrossRef]
23. Ullidtz, P. *Pavement Analysis*; Elsevier: New York, NY, USA, 1987.
24. Khazanovich, L.; Wang, Q. MnLayer: High-performance layered elastic analysis program. *Transp. Res. Rec.* **2008**, *2038*, 63–75. [CrossRef]
25. Yoder, E.J.; Witczak, M.W. *Principles of Pavement Design*, 2nd ed.; Willey: New York, NY, USA, 1975.
26. FGSV. *Richtlinien für die Rechnerische Dimensionierung des Oberbaus von Verkehrsflächen mit Asphaltdeckschicht. RDO—Asphalt 09*; FGSV: Köln, Germany, 2009.
27. LST EN 12697-26:2018 Bituminous Mixtures—Test Methods—Part 26: Stiffness (Bituminiai Mišiniai Bandymo Metodai. 26 Dalis. Standis). 2018. Available online: https://standards.iteh.ai/catalog/standards/cen/99e7a977-c432-4028-b14d-4e1cb9f84ddc/en-12697-26-2018 (accessed on 20 May 2021).
28. FAA. Airport Pavement Design and Evaluation. AC No. 150/5320-6F. 2016. Available online: https://www.faa.gov/documentlibrary/media/advisory_circular/150-5320-6f.pdf (accessed on 20 May 2021).

Review

A Review on Bitumen Aging and Rejuvenation Chemistry: Processes, Materials and Analyses

Emiliano Prosperi [1,*] and Edoardo Bocci [2]

1 Department of Construction, Civil Engineering and Architecture, Marche Polytechnic University, 60131 Ancona, Italy
2 Faculty of Engineering, eCampus University, 22060 Novedrate, Italy; edoardo.bocci@uniecampus.it
* Correspondence: e.prosperi@pm.univpm.it

Abstract: During the last decades, extensive research has been carried out on using reclaimed asphalt pavement (RAP) material in the production of hot recycled mix asphalt. Unfortunately, the aged, stiff, and brittle binder in the RAP typically increases the mixture stiffness and can therefore cause fatigue and low-temperature damages. In the scientific literature, there are many studies concerning the aging and rejuvenation of bitumen, but there is a lack of up-to-date reviews that bring them together, especially those facing the phenomena from a chemical point of view. In this paper, a recap of the chemical aspects of virgin, aged, and rejuvenated bitumen is proposed in order to provide a useful summary of the state of the art, with the aim of both encouraging the use of an increasing quantity of RAP in hot mix asphalt and trying to give indications for further research.

Keywords: bitumen; aging; rejuvenation; reclaimed asphalt; recycling

Citation: Prosperi, E.; Bocci, E. A Review on Bitumen Aging and Rejuvenation Chemistry: Processes, Materials and Analyses. *Sustainability* **2021**, *13*, 6523. https://doi.org/10.3390/su13126523

Academic Editor: Rui Micaelo

Received: 5 May 2021
Accepted: 4 June 2021
Published: 8 June 2021

1. Introduction

Nowadays, the world is living through the most severe environmental crisis ever, due to biodiversity loss; air, soil, and water pollution; resource exhaustion; and disproportionate land use. To face these issues, the concepts of circular economy and sustainability are getting more and more attention by administrations, technicians, and researchers. Indeed, these topics have become important themes of research with a steep increase in the number of papers and journals [1]. One of the sectors that greatly contributes to the environmental impact is road construction, which entails the exploitation of raw materials, the emission of pollutants, and the generation of wastes [2].

In the sector of road pavement engineering, the easiest way to promote the circular economy is to encourage the use of Reclaimed Asphalt (RAP). RAP is defined as removed and/or reprocessed pavement material composed of bitumen and aggregates. Four methods—hot in-plant recycling, hot in-place recycling, cold in-plant recycling, and cold in-place recycling—allow RAP recycling in new bituminous mixtures, determining the saving of virgin resources and the progressive cyclic reuse of this waste material, in a sustainable and circularly economic way [3].

The most widespread method to recycle RAP is the hot in-plant technique, which allows both the aggregates and binder contained in the RAP to be exploited. Unfortunately, the aged, stiff, and brittle binder from the RAP increases the mixture stiffness and can therefore cause fatigue and low-temperature damages [4,5]. For this reason, road agencies usually limit the maximum amount of RAP that can be recycled in new hot mixes [6]. With the aim of solving this issue, many products worldwide have been used with the function of rejuvenating agents, which allow restoring (fully or partly) the mechanical properties that the RAP binder loses with aging [7].

In the scientific literature, there are several studies concerning the aging and rejuvenation of bitumen. In particular, most of the studies are focused on how "improved" (in terms of performance or even environmental friendliness) bituminous binders behave with aging and when blended with aged bitumen from RAP. For instance, the use of polymer-modified bitumen is one of the hottest topics in this field [8]. Bitumen including elastomeric or plastomeric polymers experiences different changes with aging with respect to neat bitumen, as a function of the polymer type and content [9–13]. Moreover, it is a theme of research whether RAP including polymer-modified bitumen is as recyclable as the RAP containing neat bitumen, in terms of final mix performance and emission during mix production [14,15]. Another technique to increase the bitumen performance, which is actually investigated in terms of aging behavior, deals with the use of nanomaterials as modifiers, such as fumed silica, clay, diatomite, titanium dioxide, graphene, and carbon nanotubes [16–22]. However, there are still some gaps of knowledge on the understanding of neat-bitumen-aging phenomena, particularly from a chemical point of view. Many researchers have been trying to fill these gaps in the last five years. As shown in Table 1 and Figure 1, the search for the keywords "bitumen" (or "asphalt", in the American dictionary) and "aging" or "rejuvenation" in the Scopus database has provided several occurrences when associated with chemical analyses. More than half of these publications come from Chinese authors (first author), but many studies have been carried out also in North America (the United States and Canada) and Europe.

Table 1. Number of papers indexed by Scopus in the period 2016–2021 on the theme of bitumen aging/rejuvenation chemistry (date of the research 27 May 2021).

Keywords	Papers Published Since 2016
"Bitumen/Asphalt", "Aging", "AFM"	114
"Bitumen/Asphalt", "Aging", "Chemistry"	71
"Bitumen/Asphalt", "Aging", "Chromatography"	115
"Bitumen/Asphalt", "Aging", "FTIR"	395
Total number of papers of bitumen aging chemistry	589
"Bitumen/Asphalt", "Rejuvenation", "AFM"	24
"Bitumen/Asphalt", "Rejuvenation", "Chemistry"	15
"Bitumen/Asphalt", "Rejuvenation", "Chromatography"	29
"Bitumen/Asphalt", "Rejuvenation", "FTIR"	73
Total number of papers on bitumen rejuvenation chemistry	121

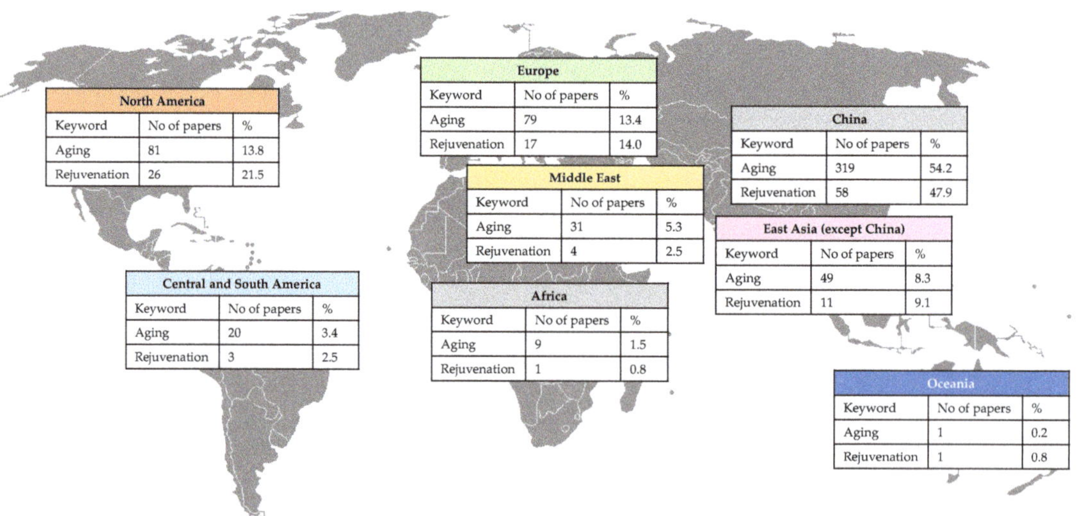

Figure 1. Origin of the papers indexed by Scopus in the period 2016–2021 on the theme of bitumen aging/rejuvenation chemistry.

This paper tries to recap the most recent research and innovation in the field of chemical analyses on virgin, aged, and rejuvenated bitumen. With respect to other reviews on aging and rejuvenation chemistry, the present state of the art is more updated (the most recent paper was published in 2009 [23]) and includes discussion about both processes (while the review by Loise et al. [24] only deals with rejuvenation, and specifically on additives). The aim is to provide a useful summary and some inputs for further research on this topic, in order to encourage the use of an increasing quantity of RAP in Hot Mix Asphalt (HMA).

2. Bitumen Chemistry

Bitumen is the visco-elasto-plastic material obtained through crude oil distillation. Basically, during this process, the various phases of the crude oil are separated due to the differences in their boiling and condensing temperatures [25]. A typical distillation process involves a first step in which the lighter components are separated, subjecting the crude oil to a temperature of about 350 °C at atmospheric pressure. The residue of the first step is subjected to a higher temperature, around 350–425 °C, under a controlled pressure ranging from 1 kPa to 10 kPa [26]. The residue of the second process is called straight-run bitumen [27]. Moreover, if the residue of this second process is subjected to another process of thermal distillation at temperatures between 455 °C and 510 °C, visbreaker bitumen is produced [28]. Vis-breaking allows refineries to reduce the amount of the residue produced, as it allows the further recovery of lighter products such as diesel and gas. This penalizes the quality of the bitumen that is obtained, which is more rigid, brittle, and susceptible to aging [29,30].

2.1. Basic Characterization

The bitumen composition is strictly related to the characteristics of the starting crude oil, particularly on its age and the depth from which it is extracted. The study of bitumen chemical composition is very tricky because it contains many chemical elements. Although bitumen is mainly composed of hydrogen (8–12% by weight) and carbon (80–88% by weight), which together give a hydrocarbon content of about 90%, heteroatoms as nitrogen (0–2% by weight), oxygen (0–2% by weight), and sulfur (0–9% by weight) are also present. Moreover, there can also be traces of heavy metals such as nickel and vanadium, in the order of hundreds of parts per million [31].

The hydrocarbons (C+H) can be classified, according to the type of bond between the carbon atoms, into:

- saturated: only simple bonds are present between the carbon atoms;
- unsaturated: double or triple bonds are present.

The heteroatoms (N, S, O) can be found in the correspondence of the unsaturated bonds. When combined with carbon, these can cause an imbalance of the electrochemical forces that gives polarity to the molecule. Therefore, the heteroatoms, despite being present in small percentages, have the ability to make the unsaturated molecules more active, influencing the bitumen rheological properties [32].

2.2. Chemical-Structural Analysis—SARA Analysis

To interpret bitumen properties from its chemistry, it is necessary to consider it on different scales and not only at a global level. In fact, bitumen can be described as several central structures consisting of polyaromatic assemblies containing a various number of molten rings, saturated polycyclic structures, and combinations. Saturated hydrocarbon side chains, which are characterized by different dimensions and patterns, are linked to these central assemblies. Therefore, the number of possible isomers is almost unlimited. This is the reason why bitumen is characterized by millions of different molecules and none of these are present in such a high quantity to isolate and characterize them. Therefore, a chemical-structural analysis is more useful and appropriate to understand bitumen composition and mechanical behavior [33].

The chemical-structural analysis has progressed with fractionation techniques, through which it is possible to separate the bitumen molecules into chemical groups according to the dimensions or the soluble properties in various kinds of solvents (polar, apolar, or aromatic). Over the years, the fractionation techniques applied to bitumen have undergone a series of advances that have led to increasingly interesting results, clarifying more and more the structure of the material and therefore allowing a deeper understanding of the bitumen chemical composition.

In 1836, Boussingault separated two components of bitumen by distillation. He obtained two fractions named "petrolenes" (85% by weight) and "asphaltene" (15% by weight). Given the similar H/C ratio of the two fractions, he thought that asphaltene could derive from the oxidation of petrolene [34]. A few decades later, Richardson made his contribution, defining "asphaltenes" as the part of the bitumen insoluble in naphtha but soluble in carbon tetrachloride (CCl_4). Moreover, he introduced the "carbeni" and the "carboids", which are, respectively, soluble and insoluble in CS_2 [35].

Kayser, in 1897, used three solvents, chloroform, ether, and alcohol, to obtain three bitumen fractions [36]. Hoiberg achieved greater success in 1939 when he achieved the separation of the maltenes in resins (precipitate) and oils (soluble part) [37]. Corbett proposed a method to further split the maltenes and obtain three categories: saturated, aromatics, and resins [38]. Therefore, he managed to separate the bitumen into the four fractions that are still considered today in the chromatographic analysis: Saturated, Aromatic, Resins, and Asphaltenes (SARA); hence the SARA terminology is obtained by joining the initials of each fraction. Figure 2 summarizes the processes for SARA fraction separation.

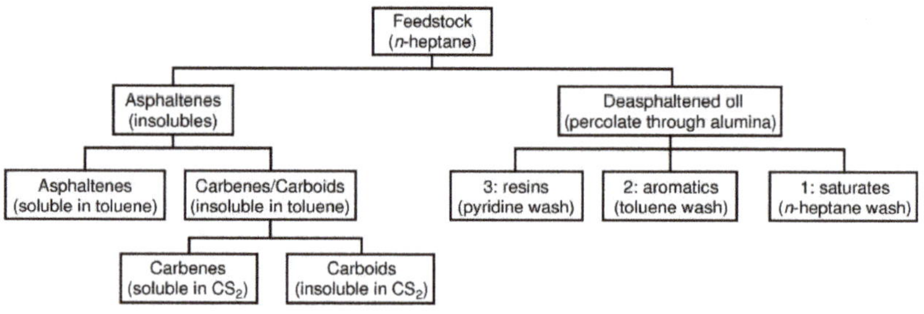

Figure 2. Bitumen composition (reprinted from Reference [39] with the permission of Elsevier).

Nowadays, the process defined by ASTM D-4124 [40] is divided into several steps. The first consists of the separation of asphaltenes with precipitation of n-heptane. Afterward, the maltenes in a solution of n-heptane are poured into a chromatographic column to separate the saturates; then, the aromatics are separated using 100% toluene and a 50/50 blend of toluene and methanol; finally, the resins are obtained using trichloroethylene.

Several methods are available to obtain the SARA fractions [41,42], but despite the proportions depending on the origins of the raw material, the use of different techniques gives slightly different results. Therefore, to compare different varieties of bitumen according to the SARA fractions, the use of the same fractionation method is always a good practice. A deeper description of the four distinct fractions is provided below.

Saturates represent about 5–15% by weight of bitumen and consist of an almost transparent liquid. They are mainly composed of aliphatic (branched, linear, and cyclic hydrocarbons) and have no polarity (rare aromatic rings or polar atoms). They function as a jellying agent for bitumen components, as they favor asphaltene flocculation and therefore the bitumen solid-elastic phase.

Aromatics represent about 30–45% by weight of bitumen and consist of a yellow-red oily liquid. They contain one or more aromatic rings and act as solvents for the peptized asphaltenes.

Asphaltenes represent about 5–20% by weight of bitumen and consist of a dark powder [43]. They have many aromatic rings and polar compounds. They are the main component associated with bitumen stiffness and viscosity [44].

Resins represent about 30–45% by weight of bitumen and consist of a black solid. They are similar to asphaltenes in terms of composition but they have a higher polarity [45]. They are the component that is associated with bitumen stability, as they behave as flocculent agents for the asphaltenes.

The bitumen behavior is determined by the relative amount of the components, but especially by the compatibility and the interactions among these homogeneous fractions.

2.3. Colloidal System

Several models have been proposed to understand the rheological properties of bitumen through its chemical composition. Although Rosinger had thought about a colloidal structure for bitumen in 1914 [46], today this intuition is attributed to Nellesteyn, who described the bituminous colloidal system in 1924 [47]. They described bitumen as the dispersion of asphaltene micelles (solid) in an oily phase (fluid) thanks to the presence of peptizing agents. The first description of the colloidal system is a structure determined by asphaltenes micelles immersed in a maltene solution. In particular, the asphaltenes are covered by the maltenic polar part (resins), which acts as a peptizing agent for the asphaltenes themselves, and everything is immersed in the so-called oils, flocculating agents for asphaltenes.

In the following years, Pfeiffer and Saal [48] introduced the difference between two colloidal systems (sol and gel), which represent the two colloidal limit systems for all bitumen.

In particular, if resins keep asphaltenes highly peptized (or dispersed) in the oily phase so that micelles are not interacting, the associated bitumen is characterized by a sol model. This model results in a very viscous (not elastic) behavior at low temperatures and a Newtonian liquid behavior at high temperatures. If resins are not very effective in peptizing asphaltenes, which become fully interconnected, a gel model is obtained. This model results in non-Newtonian fluid (viscoelastic) behavior at high temperatures and elastic solid behavior at low temperatures.

Most of the bitumen shows intermediate characteristics between these two structures, which represent the limit cases. The coexistence of the sol-type micelles and the gel structure as a function of temperature and aggregation state of micelles (ratio among asphaltenes, resins, aromatics, and saturates) is defined as a gel-sol model (Figure 3).

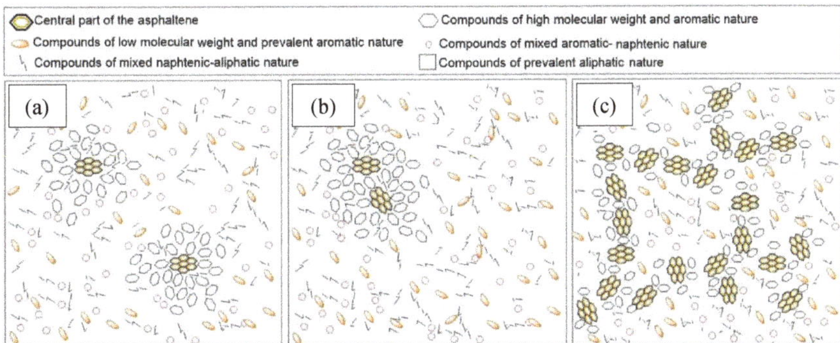

Figure 3. Schematic representation of the colloidal model of bitumen: (**a**) sol, (**b**) flocculated asphaltene micelles, (**c**) gel (reprinted from Reference [49] with the permission of Elsevier).

In 1971, Gaestel et al. [50] introduced the concept of the Colloidal Index (CI) or Instability Index, whose empirical expression is reported below.

$$CI = \frac{\text{Asphaltenes} + \text{Saturates}}{\text{Aromatics} + \text{Resins}}$$

Generally, this index has a value between 0.5 and 2.7 for the most used bitumen. The bitumen shows a clear gel behavior if the colloidal index is greater than 1.2, while the behavior is closer to a sol model if the colloidal index is lower than 0.7.

2.4. Test Methodologies to Investigate Bitumen Chemistry

Nowadays, a wide variety of methodologies are available to analyze the chemical properties of bitumen. Each technique has some issues, because the results are different as a function of the nature of the binder and the process conditions under which it is analyzed. Therefore, to deeply investigate the chemical properties of bitumen, it is a good practice to combine multiple chemical tests, together with a rheological and traditional characterization.

The most frequently used techniques for the chemical analysis of bitumen are summarized in Table 2 and described hereafter.

Table 2. Most frequently used techniques for the chemical analysis of a bitumen.

Technique	Type of Analysis	Parameters Used
AFM: Atomic Force Microscope	Microscopic	Microstructure and micro-mechanical properties of bitumen
FTIR: Fourier Transform Infrared spectroscopy	Chemical	Quantity of carbonyl and sulphoxide groups
TLC FID: Thin Film Chromatography with Flame Ionization Detection	Chemical	Saturated, aromatic, asphaltene and resin content
HP-GPC: Gel Permeation High-Pressure Chromatography	Chemical	Number of chemical groups and molecular weights

2.4.1. Atomic Force Microscope (AFM)

The AFM test is a non-destructive analysis that allows representing of the surface morphology of a bitumen sample, as well as information regarding stiffness, cohesion, and molecular interactions at a microscopic level. The fundamental principle on which this test is based is very easy to understand. The device is equipped with a flexible cantilever, which is linked to a piezoelectric component and has a tip at the extremity. During the test, the tip slides on the bitumen surface while its position is measured through a laser system and the resistance to the tip movement, which depends on the distance between atoms, is registered. By coupling this information, a detailed scansion of the bitumen sample surface at a microscopic (atomic) scale is collected [51].

The AFM technique allows the identification of three major phases of different rheology and composition (Figure 4):

1. Catana phase or bee phase, which are a sort of hills in the undulated pattern of the AFM image;
2. Peri phase (from Greek peri = around), surrounding the bees and characterized by a certain roughness;
3. Para phase (from Greek para = neighbor), next to the peri phase and typically flat [52–54].

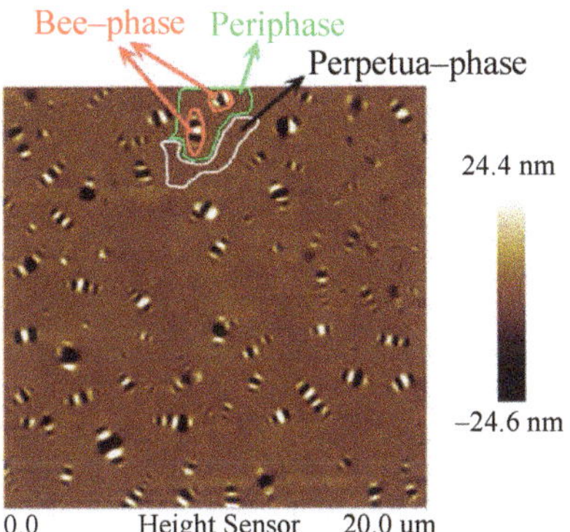

Figure 4. Three major phases of the superficial morphological representation of bitumen (reprinted from Reference [55]).

2.4.2. Fourier Transform Infrared (FTIR) Spectroscopy

FTIR spectroscopy is widespread technology for the identification and analysis of organic compounds. In particular, FTIR is a method for determining the molecular structure of a material by measuring the atom oscillations (rotations and vibrations). During the test, infrared radiations hit the sample, whose atomic functional groups absorb part of these radiations. In particular, the specific wavenumber of the absorbed radiation is a function of the vibration mode of the functional group. Through the application of the Fourier transform, the absorbance spectrum of the sample is obtained.

Each functional group has different vibration modes, whose number is related to the number of atoms and type of bond. For instance, the symmetrical molecules that include two atoms (e.g., diatomic nitrogen N_2) have no absorption in the IR spectrum, while asymmetrical diatomic molecules (e.g., carbon monoxide CO) do. For more complex functional groups, for example methylene (-CH_2), the vibration modes (Figure 5) include six types of oscillations [56].

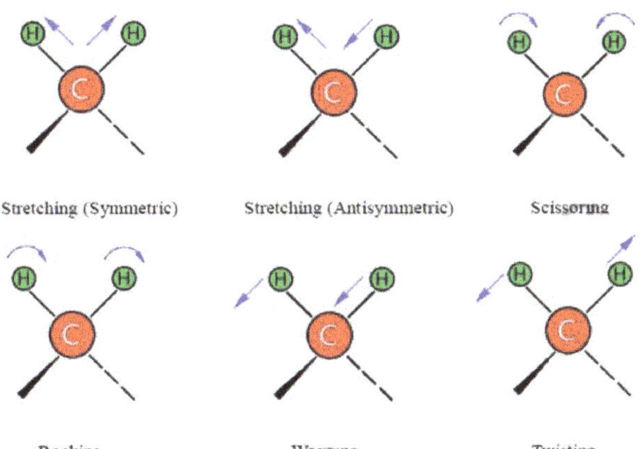

Figure 5. Vibration modes of the methylene group (-CH2) (reprinted from Reference [57] with the permission of Elsevier).

The bitumen characterization through FTIR spectroscopy provides different information. The analysis of the peak position in the spectrum allows the identification of the functional groups. Moreover, the height of the peaks allows the quantification of the functional group concentration [57].

2.4.3. TLC-FID: Thin Film Chromatography with Flame Ionization Detection

TLC–FID is used to quantify the content of each SARA fraction in a bitumen. The bitumen is initially blended with cyclohexane solvent and a small amount of solution is put on a quartz rod, known as a chromarod. Then, the procedure is repeated using three more solvents (n-hexane, toluene, dichloromethane). The separated fractions (i.e., asphaltenes, resins, aromatics, and saturates) are located in the chromarod series at, respectively, 0 cm, 2.5 cm, 5 cm, and 10 cm, as a function of the decreasing polarity. Finally, the chromarod is analyzed by Flame Ionization Detection (FID), which step-by-step ionizes the different zones corresponding to the four SARA fractions, and allows the estimation of their percentages in the bitumen [58,59].

2.4.4. HP-GPC: Gel Permeation High-Pressure Chromatography

HC-GPC is a test method that allows quantifying of the molecular size distribution of bitumen. The binder sample is dissolved in tetrahydrofuran (THF), and the solution is put in a column for chromatographic analysis. Gel (stationary phase) is used as a stabilizing agent that slows down the permeation of light components, while the high pressure allows increasing the test speed and the separation efficiency.

Typically, thirteen slices of the chromatographic pattern are assumed in order to discriminate between large (slices 1–5), medium (slices 6–9), and small (slices 10–13) molecular sizes [60].

The molecular weight can be represented in terms of average molecular weight by molecule weight (M_w) and average molecular weight by molecule number (M_n). In addition to M_w and M_n, their ratio (M_w/M_n), defined as "molecular weight dispersion index", is also often considered in the study of bitumen molecular size distribution [51]

3. Bitumen Aging

Bitumen aging is defined as the series of chemical transformations that the material undergoes and that results in the variation of its physical characteristics [61]. In general, two different aging processes, namely short-term aging and long-term aging, are identified.

Short-term aging is the phenomenon that bitumen suffers during HMA manufacturing (mixing, hauling, paving, and compacting) because of the high processing temperatures (>150 °C). Long-term aging is the phenomenon that affects bitumen during the entire service life of the mix, which is subjected to traffic and environmental stresses. The severity of this process is mainly related to the bitumen's exposure to air, which depends on the mix air voids and the position of the HMA layer within the pavement structure [39].

Despite some studies [62,63] that have identified several variables that affect bitumen aging, the most widely recognized mechanisms include:

- Physical and steric hardening (reversible mechanisms);
- Loss of low-weight components (volatiles) by evaporation;
- Oxidation, with the consequent changes at the molecular level that cause a change in the SARA fractions.

The oxidation and evaporation of volatiles, which are irreversible processes, are accelerated during the HMA production and paving when the bitumen is hot [64]. When the mix reaches the air temperature, the evaporation of volatiles becomes much less influential, while oxidation continues in the long-term aging. Furthermore, the greater importance of oxidation compared to that of physical hardening is given by the different nature of these processes: oxidation is irreversible, while physical hardening can be recovered. For these reasons, oxidation is considered the main process in the aging of bitumen.

These mechanisms are described in detail in the following sections.

3.1. *Physical and Steric Hardening*

Physical hardening deals with the changes in the bitumen viscoelastic properties due to the material cooling below the glass transition region. However, the process does not entail any change in the chemical structure and is reversed when the bitumen is re-heated to air temperatures [65]. Assuming that the material volume consists of the volume of the oscillating molecules and the free volume between the molecules [66,67], when the bitumen temperature decreases, both molecular mobility and free volume reduces, maintaining the same proportion between occupied and free volume. When the glass transition temperature is reached, the free volume decrease becomes slower than the decrease of molecule oscillation, entailing a kind of "over-hardening" for the bitumen [68].

Physical hardening should not be confused with steric hardening. Steric hardening is a chemical process in which the bitumen molecules rearrange and form wax compounds in the maltenes, due to the presence of linear alkanes in the asphaltenes [69]. The process happens at intermediate temperatures but takes three times longer than physical hardening [70]. Even steric hardening is reversible. While the former type of hardening occurs within 1–2 days at temperatures below the glass transition temperature of bitumens ($-35/-15$ °C), the latter is manifested at room temperature, and requires days or even weeks. The steric hardening is related to the inner reorganization of the binder molecules; it is associated with the formation of ordered structures by waxes in the maltenes phase that is influenced by the linear alkanes present in the asphaltenes fraction. It is a reversible process because it can be removed by heating or mechanical work [71].

3.2. Evaporation of the Volatile Components

The evaporation of saturated and aromatic components has been also reported as an aging mechanism of bitumen. In particular, this phenomenon is mainly related to short-term aging, as it depends on the temperature to which the bitumen is subjected during the mixing and installation phases [72]. It has been quantified that the loss of volatiles can be double for a temperature increase of 10 °C during HMA manufacturing at the plant [73]. The volatile evaporation causes the unbalancing of the SARA fractions, determining the predominance of resins and asphaltenes over saturates and aromatics. Consequently, the bitumen that results is harder, stiffer, more viscous, and more fragile. The evaporation of volatile compounds is an irreversible mechanism that significantly affects bitumen aging, even if to a lower extent than the oxidation process [74].

3.3. Oxidation

Thurston and Knowles, in 1941, demonstrated how bitumen components, in particular asphaltenes and resins, absorb oxygen [63]. It is widely accepted today to consider the oxidative process of bitumen as the most important mechanism that happens during aging.

Bitumen oxidation is an irreversible process that deals with the "capture" of oxygen atoms by the bitumen components (particularly the asphaltenes), which undergo an alteration of their chemical characteristics. Moreover, this aging process could be photo-catalyzed in the case of the bitumen in pavement surface layers, in particular for polymer-modified binders [23]. As oxidation is due and depends on the access to oxygen in the mixture, the voids content, the HMA layer depth, the bitumen content, and the presence of cracking are factors that can influence the quantity of bitumen exposed, and therefore the quantity of potentially aged bitumen.

Within bitumen morphology, oxidation includes dehydrogenation, the reaction of the alkyl sulfides into sulphoxides, and the reaction of the benzyl carbons into ketones, which in turn form carboxylic acids with dicarboxylic anhydride. These reactions can be quantitatively determined by functional group analysis through FTIR spectroscopy. A typical infrared spectrogram is shown in Figure 6. The absorbance bands around 1690 cm^{-1} are due to the increase in C=O bonds (carbonyl groups)—for example, ketones, carboxylic acids, and anhydrides—while those around 1030 cm^{-1} are due to the increase in S=O bonds (sulphoxide groups). Consequently, the peak areas of the two wavenumbers can be considered as concentration measurements of carbonyl compounds and sulphoxides, respectively [75].

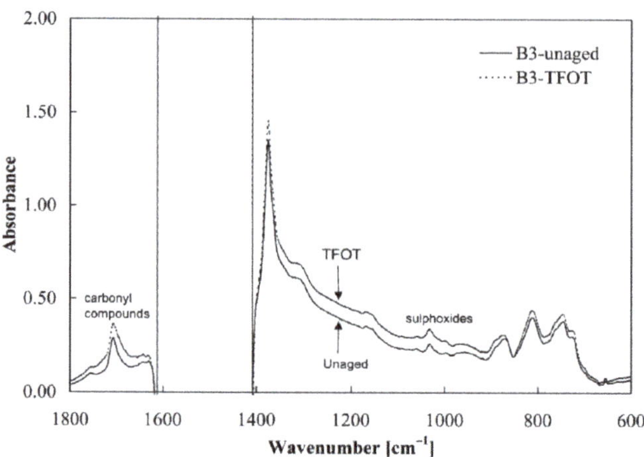

Figure 6. Effect of aging on bitumen FTIR spectrogram (reprinted from Reference [75] with the permission of Elsevier).

It is important to highlight that the carbonyls, ketones, and sulfoxides generated through oxidation are characterized by a marked polarity. Therefore, they associate with the polar groups in the bitumen forming agglomerates with a high molecular weight [76]. These large and "heavy" clusters, which typically involve the asphaltene fraction, determine the reduction of the molecular mobility within the bitumen colloidal system, resulting in increased viscosity, stiffness, and brittleness [77].

Temperature is a key factor in the oxidation phenomenon. In particular, the degree of oxidation is doubled every time the temperature increases by 10 °C (after 100 °C) [73]. The influence of temperature was also observed in the laboratory by Lu et al. [78]. They noted that it takes 4–8 longer times to obtain the same aging when the pressure aging vessel (PAV) temperature is decreased from 100 to 75 °C.

As previously said, the oxidation reaction can be promoted and accelerated when the bitumen functional groups are excited by UV radiations. This issue has been neglected for many years as it only involves the upper pavement layer, because of the low ability of the radiations to penetrate in depth. However, the amplifying effect on the aging of the bituminous binder due to ultraviolet radiation should be considered, particularly in the most exposed surfaces of geographic regions characterized by high levels of solar radiation and humidity [79]. Many authors associated the increase in bitumen viscosity, as a consequence of oxidation, with the number of radiations that invested the material [80–82]. In particular, Zeng et al. [81] observed the detrimental effects of ultraviolet radiation in association with high temperatures in promoting oxidation. Afanasieva et al. [82] specified that bitumen is highly oxidized when subjected to radiations with a wavelength coinciding with the UVB range (280–315 nm).

3.4. Laboratory Aging Methods

To date, the most frequently used methodologies to age bitumens are the thin film oven test (TFOT), rolling thin film oven test (RTFOT), pressure aging vessel (PAV), and ultraviolet test (UV). Most of them are often characterized by increases in temperature, oxygen pressure, or a combination of these two in order to generate the aging conditions, which are as close as possible to the conditions in which real bitumen can be found. While the first two methodologies are mainly adopted to reproduce short-term aging, which occurs during storage, mixing, hauling, and laying of an HMA, the last ones can simulate long-term aging that occurs during the service life of the pavement.

3.5. Laboratory Aging Assessment Methods: General Results

As regarding the different SARA fractions, aging can generally be summarized as follows:

- the saturates remain almost unchanged;
- the aromatics decrease;
- the resins see a small increase;
- the asphaltenes increase.

In particular, the general effect of oxidation is the shift of each SARA fraction towards the next component in the polarity scale. As explained before, the four bitumen fractions are ranked, as a function of the increasing polarity, as follows: saturates, aromatics, resins, and asphaltenes. Actually, the saturates only show little changes between the unaged and long-term-aged binder, so they can be considered an unreactive fraction [27]. The next fractions, aromatics and resins, oxidize and respectively shift into resins and asphaltenes. Since aromatics evolve into resins, but there is poor supply from the saturates (which are mainly inert), a global decrease of the aromatic content with aging is observed. Differently, the resin content only experiences a slight increase or decrease since there is the contemporary uptake of the oxidized resins into asphaltenes and of the oxidized aromatics into resins [83].

The relationship between resins and asphaltenes plays a crucial role in aging: the asphaltene fraction is the component that grows the most; at the same time, the resin content increases to a lesser extent, facilitating the mutual contacts between asphaltenes. When the ratio between aromatics and resins is not high enough to allow the peptization of the asphaltene micelles, or when the solvation capacity of the system is insufficient, the micelles tend to bond to each other [84]. This determines the formation of larger irregular structures in which voids are present (filled by the external liquid of the component micelles). Thus, in terms of colloidal models, aged bitumen tends to assume a gel structure, causing a stiffer and more brittle behavior [85]. The decrease of the maltenes affects the stability of the colloidal system and determines the flocculation of the asphaltenes. Moreover, the higher amount of asphaltenes increase the propensity of the asphaltenes themselves to micellization and agglomeration [86].

Li et al. [51] quantified the variation of SARA fractions during aging. In this study, five different bitumens were separated into the SARA fractions before and after aging. Figure 7 highlights how, during aging, the asphaltenes increase while the contents of aromatics and resins decrease. Moreover, the amount of saturates tends to remain stable. This phenomenon was investigated in terms of the colloidal index (CI), which proved to grow with aging, particularly in the long-term step.

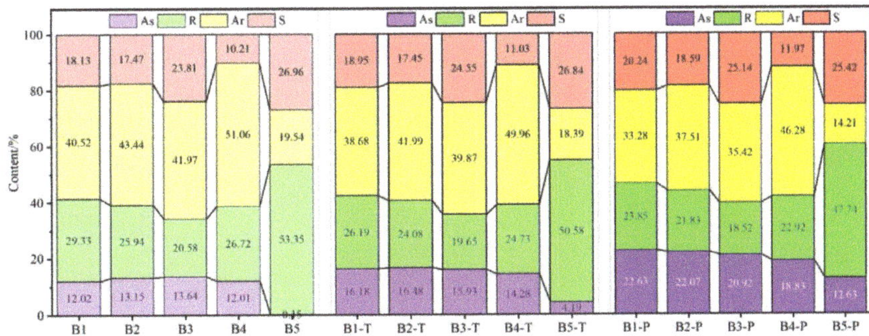

Figure 7. Proportion of bitumen SARA fractions with aging (reprinted from Reference [51] with the permission of Elsevier).

Mirwald et al. [83] tried to evaluate (using FTIR), not only the changes in quantities but also the chemical modifications within individual SARA fractions during aging. In particular, an unaged bitumen and three long-term-aged bitumens were separated into their SARA fractions, which were subsequently analyzed through FTIR spectroscopy. The absorbance spectra showed that:

- The saturates spectrum remains unchanged, confirming that this fraction is little affected by aging;
- The aromatic spectrum shows some changes (an increase of the peaks in the carbonyl band, main aromatic band and across the entire fingerprint area), but no significant increase in sulfoxides;
- The interpretation of the resin spectrum is complex, because of the position of this fraction in the polarity gradient. In general, aging results in uptake from 2-quinolones and carbonyls into ketones and in the growth of the sulfoxides content;
- The asphaltenes spectrum shows significant variations in the fingerprint area. In particular, the aging determines an increase in sulfoxides (1030 cm^{-1}), main aromatic (1600 cm^{-1}), and carbonyl (1700 cm^{-1}) peaks.

Thus, the increase of the carbonyls affects the aromatics and the resins, while the increase of the sulfoxides affects the resins and the asphaltenes.

Using GPC, Li et al. evaluated the molecular weight of the bitumen components [51]. Table 3 shows that the molecular weight increases when moving from saturates, to aromatics, to resins and asphaltenes. In particular, the saturates have the shortest molecular chains, denoting an approximate structure. Differently, the asphaltenes have the longest chains and the most complex structure. In addition, they include a higher number of polar groups that have a low sensitivity to temperature changes. Therefore, when the asphaltene content increases, the bitumen tends to maintain its mechanical properties (stiffness, viscosity), even when increasing temperature [51].

Table 3. Molecular weight of the bitumen SARA fractions (reprinted from Reference [51] with the permission of Elsevier).

Component	M_n	M_w	M_w/M_n	Features
Saturated (S)	506	673	1.329	Low molecular region
Aromatics (Ar)	648	1220	1.882	Transition region
Resins (R)	907	2761	3.045	Transition region
Asphaltenes (As)	1898	14,660	7.725	High molecular region

FTIR spectroscopy has been used for many years to analyze the bituminous binders [87], particularly to investigate the polymer modification mechanisms and the effects of aging. Table 4 summarizes the main changes noticed in the bitumen FTIR spectrum as a consequence of aging.

Table 4. Change of key FTIR bands during aging.

Chemical Group	Bond	Approximate Wavenumber (cm^{-1})	Change with Aging	References
Sulphoxide	S=O	1030	Increase	[57,88–92]
Carbonyl	C=O	1690	Increase	
Aliphatics (plan deformation)	CH_2, CH_3	1460, 1375	Small decrease	
Aromatics	C=C	1600	Small increase	[88,90,93]
Aliphatics (asymmetric or symmetric stretching	CH_2, CH_3	2923, 2853	Small decrease	[93,94]
Polarity	O–H	3450	Increase	[93,94]

The more significant changes in the FTIR spectrum associated with bitumen oxidation are the rise carbonyl C=O (1690 cm^{-1}) and sulfoxide S=O (at 1030 cm^{-1}) bands. This has been observed in both laboratory [90,95] and site [88,89] aged bitumen. In particular, the work by Mouillet et al. [90] specified that increases in the S=O and C=O bands are mainly related to short-term aging and long-term aging, respectively. Even the aromatic C=C band (1600 cm^{-1}) shows a slight increase that is associated with the increase of the resins and especially the asphaltenes, which include condensed aromatic rings. The polarity band (around 3450 cm^{-1}) is highly marked in the spectra of resin and asphaltene components, but this has not been exactly associated with the bitumen aging process [93]. Finally, the aliphatic CH_2 and CH_3 bands showed a slight decrease with increased aging [94].

In order to quantify the effects of bitumen oxidation, two indices have been introduced:

- Carbonyl index: $I_{C=O} = \frac{A_{1690}}{Aref}$
- Sulphoxide index: $I_{S=O} = \frac{A_{1030}}{Aref}$

where A_{1690} is the area of the C=O peak centered at 1690 cm^{-1}, A_{1030} is the area of the S=O peak centered at 1030 cm^{-1}, and A_{ref} is the area of the reference ethylene and methyl peaks, centered at 1460 and 1375 cm^{-1}, respectively [96].

With the aim to estimate the effects of bitumen aging on its properties, the variation of the indices in unaged, short-term-, and long-term-aged bitumen can be calculated. When increasing the aging, the heights and the areas of the peaks in correspondence of the wave numbers of 1690 cm^{-1} and 1030 cm^{-1} (respectively for the carbonyl C=O and sulphoxide S=O bands) increase, so the two indices also increase [97–100]. Moreover, a recent study [101] proposed the Chemical Aging Index (CAI), calculated as $I_{C=O}$ plus $I_{S=O}$, in order to better understand the variation of both indices during aging.

Regarding the AFM test, changes of the particular bee-shaped structure can be used to investigate the aging process. Nowadays, there is still some discussion about the nature and conformation of these structures. There is a certain agreement that the bee structures are associated with wax crystallization [102]. For this reason, it was hypothesized that the "bees" are correlated to the asphaltenes [103,104]. A study by dos Santos et al. [105] demonstrated that the valleys, due to their lower thickness and roughness, are less strong than the hills. Other studies proposed that the formation of the bee structure is associated not only with asphaltenes but also with resins and aromatics, and so it is strictly correlated to the bitumen performance [106,107]. One more point of discussion is related to the presence of these phases in the whole bitumen volume or only on the surface [108].

Li et al. [51] tried to identify the changes in the bee structures during short- and long-term aging in different bitumens. Before aging, these show the distinct elliptic bee structures, which are short and thick. After aging, different behaviors were observed for the various bitumens. In general, the bee structures became larger and irregular when increasing the aging level. In addition, peculiar phenomena happened in the AFM diagrams, such as the increase of the bee stripes number or the formation of sunk regions, columnar peaks, blocky structures, or cracks.

A study by Lu et al. [109] investigated wax-including bitumens in comparison with a non-waxy bitumen using AFM. They noted that aging led to a decrease in the number and an increase in the dimensions of the bee structures in the bitumen with wax, probably because of the highly increased stiffness of the aged binder and/or the reduced compatibility between the saturated crystalline fraction and the more polar bitumen matrix. On the other hand, for a non-waxy bitumen, no structure was observed either on the unaged or PAV-aged state. Moreover, after 60 h of PAV aging, the content of asphaltenes increased from 22% to almost 25%. This demonstrated that asphaltenes are not the fraction responsible for the structures, unless they contain n-heptane insoluble crystallizable materials.

Zhang et al. [110] also studied the aging effect on the bee structures. They confirmed that aging entails an increase in the bee structure number, dimension, and roughness (Figure 8), which is associated with a contemporary decrease of the penetration and increase of the ring and ball softening point and viscosity.

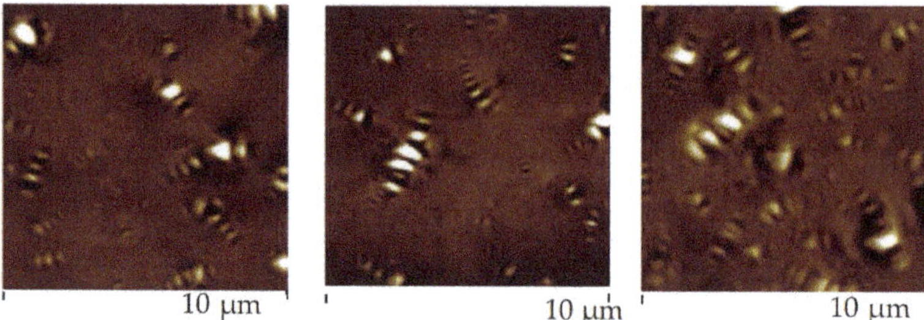

Figure 8. AFM images of unaged, short-term-aged, and long-term-aged bitumen (reprinted from Reference [110]).

4. Bitumen Rejuvenation

Including RAP in HMA requires a careful study of the mixture, because its mechanical properties are strongly conditioned by the presence of the aged bitumen contained in RAP. In particular, improper use of RAP can lead to premature cracking linked to the excessive stiffness of the bitumen (thermal and fatigue cracking) [111–114]. In order to achieve the proper mechanical properties for the HMA, additives are included in the mixture. In the following paragraph, the concept of rejuvenator is provided, clarifying which are the main distinctions within this category and explaining the benefits associated with their use.

A rejuvenator is an agent that allows renovating the properties that the bitumen loses with aging [115]. Thus, rejuvenators should reduce aged bitumen viscosity and stiffness and improve ductility [24]. However, the many products that can be used for this aim may act at a different level, for example, in the colloidal system or the chemical morphology. The literature does not seem very clear about this; the term "rejuvenator" is often used for any additive, without specifying what the effective mechanisms of action in bitumen are. According to different authors [24,116,117], rejuvenators can be classified, based on the effect, as:

- Softening agents (also called fluidifying agents or rheological rejuvenators), which include:
    - Incompatible softeners, which mainly have a viscosity lowering effect;
    - Soluble softeners, which restore the balance in the SARA composition by re-enriching the maltene fraction;
- Real rejuvenators or compatibilizers, which help to renovate the physical and chemical characteristics of the bitumen through the disruption of the intermolecular associations between the asphaltenes.

The softening agents are usually based on extracts of lubricating oils. Slurry oil, flux oil, and lube stock belong to this category. They include a proper content of maltenic, naphthenic, or aromatic components aiming to rebalance the SARA fractions of the aged binder, characterized by a lower concentration of maltenes. The softening agent allows an increase in bitumen ductility and a decrease in viscosity and brittleness by merely supplying oily components to maltene fraction, but does not achieve any change in the complex structure. The real rejuvenators are additives that allow restoring both the physical and chemical properties of bitumen. In particular, they are able to restore the agglomeration condition of asphaltenes to their original state by promoting their re-dispersion. Regarding their composition, real rejuvenators should have a high content of aromatics, which allows keeping the asphaltenes dispersed and maintains a low content of saturates, which have poor compatibility with the asphaltenes.

The efficacy of the rejuvenation process is strictly related to the adequate dispersion within the aged bitumen. Lee et al. first faced this aspect, claiming that a mechanical mixing could allow a uniform distribution [118]. A few years before, Carpenter and Wolosick divided the diffusion of a rejuvenator into an aged binder into four different steps. In the first phase, the rejuvenator forms a low-viscosity layer around the particles of aggregates covered with aged bitumen. Then, the additive starts to penetrate the binder by softening it. In the third phase, the rejuvenator penetrates the aged bitumen, and the viscosity of the internal and external surfaces gradually decreases. Finally, in the last phase, over time, the rejuvenator manages to reach all the aged bitumen [119]. Noureldin and Wood [120] and Huang et al. [121] confirmed this theory some years later. Karlsson and Isacsson [122], found out that the diffusion of the rejuvenator into the aged bitumen is mainly influenced by the viscosity of the maltene phase. Therefore, it can be facilitated by raising the temperature or adding oil [123].

4.1. Rejuvenators for Hot Recycled Mix Asphalt (HRMA)

During the last decades, many studies have tried to estimate the efficacy of different rejuvenators. According to the literature, there are many different sources from which we can obtain rejuvenators. In general, according to their nature, rejuvenators can be classified as additives derived from oil and biological additives. For instance, Zaumanis et al. [124] and Dony et al. [125] stated that although the two categories may show differences, both additives can be used successfully to soften the aged bitumen and allow the fulfillment of the requirements in terms of penetration, softening point, and rheological characteristic.

Besides fuels and bitumens, rejuvenators able to improve the properties of an aged binder are often produced from the processing of crude oil. These products include aromatic extracts, paraffinic oils, naphthenic oils, and spent motor oils [126]. These petroleum products have been the main agents used for many years to improve the characteristics of aged bitumen. In recent decades, research has moved towards finding materials with rejuvenating effects for aged bitumen that have lower costs and are environmentally sustainable. These are called bio-rejuvenators, and they are products such as tall oil, rapeseed oil, soybean oil, sunflower oil, corn oil, used cooking oil, castor oil residues, and organic oils. Several studies promote these rejuvenators, highlighting that they are able to restore the original properties of bitumen, while being eco-sustainable at the same time.

Among the numerous studies carried out during the last decades, the most relevant and suitable are reported hereafter, together with a brief description of the used rejuvenators and the main results.

Bocci et al. [127,128] and Grilli et al. [129] focused on the evaluation of the mechanical and performance characteristics of HMA made with a high amount of RAP and a bio-rejuvenator. They found out that, using appropriate dosages of this additive, acceptable performance, similar to that reached with virgin materials, can be achieved. Thus, the evaluation of the correct dosage of rejuvenators plays a fundamental role in obtaining a mixture that can easily meet the specifications.

Krol et al. [130] and Somé et al. [131] evaluated the effects of several vegetable oils (rapeseed oil, soybean oil, sunflower oil, flaxseed oil) on the mechanical characteristics of bitumen and mixes. Furthermore, they also developed chemical processes to make new additives. These studies aimed to improve the performance of bituminous binders using only very cheap raw materials.

Zargar et al. [132] studied the possibility of using waste cooking oil (WCO) as a rejuvenator in aged bitumen. They highlighted that by adding a specific dosage of this additive, the same values of penetration index, softening point, and viscosity of a virgin bitumen can be obtained. Moreover, increasing the amount of WCO, these properties can be further improved. Gökalp and Emre Uz [133] confirmed that using WCO leads to improvements in penetration index, viscosity, and fatigue resistance.

During the last years, many studies have compared the effect of WCO with respect to waste engine oils (WEO) in restoring the original properties of an aged bitumen. In particular, Joni et al. [134] stated that the effect of WCO is greater than that of WEO on the properties of an aged binder; therefore, to obtain the same results, the required amount of WEO to be added is greater than that of WCO. Li et al. [135] confirmed that both WCO and WEO have rejuvenating properties, but WCO has higher efficacy than WEO, which implies the use of higher dosages. Moreover, Al Mamun et al. [136] investigated different contents of RAP and WCO/WEO rejuvenators. Differently from the previous studies, they noted that WEO allows a higher reduction of indirect tensile strength and stiffness with respect to WCO under the same dosage, but WCO allows a higher content of RAP to be recycled.

The differences between bio-additives and aromatic additives were also studied from a molecular point of view by FTIR. Noor et al. [92] showed that using biological additives involves an increase of particular functional groups (C=O) in bitumen. Moreover, Cavalli et al. [137] stated that using rejuvenators such as seed oil or tall oil can increase the carbonyl and sulphoxide indices since they contain functional groups C=O and S=O. Investigating the performance of these indices is a good way to qualitatively understand the effects of a rejuvenating agent on an aged binder. Zhang et al. [138], through the analysis of the carbonyl, sulphoxide, and aromatic indices, assessed the degree of aging and rejuvenation of some mixtures including bio-additives.

Zeng et al. [139] tested castor oil as a rejuvenating agent, and they found out that using this additive, it is possible to obtain an improvement of the rheological properties of the aged binder.

The studies carried out by Elkashef et al. [140] and Nayak and Sahoo [141] focused on the use of soybean oil and Pongamia oil. They concluded that both products are good rejuvenators, capable of restoring the rheological properties that the binder lost with aging.

Kehzen et al. [142] tried to evaluate tung oil (also called "China wood oil") as an additive, showing how its addition improves the elasticity of an aged binder. In addition, with suitable amounts of tung oil, good performances of the mixture at high temperatures are also ensured.

Several studies regard the effects of rejuvenators engineered and "built" in the laboratory. In particular, Zhang et al. [143] tested a rejuvenating agent made of rubber oil, plasticizers, and surfactants, and demonstrated that the blend of these ingredients allows the binder to recover the malleability and ductility lost during the aging processes. Rzek et al. [144] obtained a rejuvenator by modifying pyrolytic condensate of scrap tires with tire crumb. The results confirmed that the addition of this product improved the properties of the binder. Moreover, mechanical and rheological tests showed that the amount of RAP can reach up to 60% when this engineered rejuvenator is added to the mixture.

Zaumanis et al. [124,145,146] tested the rejuvenating effect of many products of both biological and hydrocarbon nature, such as cotton seed oil, vegetable oil, used cooking oil, used motor oil, aromatic extract, distilled tall oil, distilled tall oil, exhausted motor oil, naphthenic oil, and aromatic extract.

Radenberg et al. [147] carried out a study testing 21 different types of rejuvenators commonly used in the road sector; the products were distinguished between "rheologically effective" and "chemically effective". While the former led to an increase in the maltenic phase, the latter reversed the effects of the oxidation process in the agglomerated compounds.

Table 5 presents recent publications concerning the use of additives to rejuvenate the aged bitumen in RAP. For a correct understanding, a brief description of the nature of the additives tested is also included.

Table 5. Most used additives, biological (green) and derived from oil (gold), used to rejuvenate aged bitumen.

Additive	Dosage *	Description	References
Tall oil	4–20%	Tall oil is an organic product deriving from the kraft process, a procedure for converting wood into wood pulp, the main component of the paper. It contains fatty acids, acid resins, and surfactants.	[124,127,145,146,148,149]
Exhausted vegetable cooking oil (mix of the main oils used for frying)	1–20%	The chemical composition of these additives mainly contains fatty acids and methyl esters, with both oleophilic and hydrophilic properties.	[92,132,133,135,136,145,146, 150–155]
Sunflower oil	5–9%	It is the oil extracted from sunflower seeds. Contains triglycerides, with a high content of linoleic acid. It has a high content of polyunsaturated fatty acids.	[130,131,137]
Linseed oil	6–9%	It is the oil obtained by squeezing previously dried or toasted flax seeds. It is mainly composed of triglycerides. It is one of the vegetable oils with the highest concentrations of acidolinolenic acid.	[130,131]
Soybean oil	6–9%	It is obtained by extraction from soybeans through a special process called "crushing" with the use of chemical solvents. It too is mainly composed of triglycerides.	[130,131,140]
Rapeseed oil	1.5–9%	It is a vegetable oil produced from rapeseed seeds. It occurs naturally in many varieties. The resulting oil, therefore, depends on the characteristics of the rapeseed from which it is extracted. The chemical composition includes fatty acids and methyl esters.	[130,131,156]
Castor oil	5–50%	It is a very valuable vegetable oil, which is extracted from the seeds of the castor plant. It is mainly composed of acylglycerides, and the main fatty acid present is ricinoleic acid.	[139,141]
Pongamia oil	5–15%	It is a fixed oil derived from the seeds of the Millettia pinnata tree. Typically, Pongamia oil is made up of glycerides, especially triglycerides. It is considered a fluxing agent rather than a rejuvenator.	[141]
Tung oil	2–8%	Also called China wood oil, it is the oil extracted from Aleurites fordii seeds. It is mainly composed of triglycerides and is considered a drying oil, with extremely short polymerization times.	[142]
Cashew oil	5%	It is an oil that derives from natural resins that fill the interstitial spaces of the honeycomb structure of the cashew shell. The resin is made up of 80–85% of anacardial acids (o-pentadeca dienylsalicylic acid) and the remaining fraction is cardol and methylcardol.	[137]
Corn oil	1.5–9%	It is an oil extracted from the germs of the seeds of Zea mays, a graminaceous plant native to North America. It has a composition similar to sunflower oil, very rich in linoleic acid. It is mainly composed of triglycerides	[157]
Cotton seed oil	12%	It is the vegetable oil extracted from the seeds of cotton plants. It is mainly composed of triglycerides.	[124]
Oleic acid	2.5–4.5%	It is an 18-carbon monounsaturated carboxylic acid of the omega-9 series. In the form of triglyceride, it is an important component of animal fats, and is the most abundant constituent of the majority of vegetable oils.	[158]
Organic oil from wood waste	2–12.4%	A very wide range of types of timber can be used, such as Red Maple, Magnolia, Balsam, Poplar, Linden, Beech and Pine.	[159,160]
Vegetable waste fat	12%	Material composed of waste grease produced by catering processes.	[145]

Table 5. *Cont.*

Additive	Dosage *	Description	References
Pig manure	2–10%	It is the product of the fermentation of pig manure mixed with solid material used as bedding.	[159–161]
Algae additive	10%	This is a bio-oil extracted from algae leaves or blooms through pyrolysis, and it is rich in phenolic compounds.	[161]
Waste engine oil	1–20%	It is the waste lubricating oil used by engines. It is mainly produced from paraffinic oil.	[124,125,134–136,146,162–164]
Rubber powder from pyrolysis of used tires	5–12%	Pyrolysis is a thermochemical decomposition process of organic materials, obtained by applying heat in the complete absence of an oxidizing agent. The pyrolytic product from tires pyrolysis contains high concentrations of polycyclic aromatic hydrocarbons.	[144,165]
Aromatic extract	5–9%	Aromatic extracts are refined products from crude oil and constitute one of the most traditional classes of rejuvenators. Their chemical structure includes aromatic polar rings.	[124,146,166,167]
Naphthenic oil	50–400%	Naphthenic oils are high-quality pure naphthenic mineral bases, obtained by hydrogen refining of selected crude oil.	[146]
"Soft" bitumen	5%	Bitumen with a high penetration value and low stiffness. It is typically classified as a fluxing agent since it does not restore the physical and chemical properties of the aged binder. However, this binder can lead to a decrease in bitumen blend viscosity.	[168,169]

* The dosages refer to the weight of the aged bitumen.

4.2. Rejuvenating Mechanisms

As previously explained, aging determines morphological changes in the bitumen: in particular, it causes more intense molecular interactions by introducing polar groups, which lead to an increase of the colloidal agglomerates into the bitumen volume. As the molecular structure strongly influences the rheological properties, any alteration of the balances and interactions of the polar and non-polar components in the bitumen can determine a modification of the thermal and mechanical properties in a mixture.

From a chemo-morphological point of view, rejuvenation is the inverse process of aging. True rejuvenation breaks the molecular aggregations and rebalances the SARA fractions, leading to an improvement in the rheological properties of the bitumen [170]. Remembering the classification made in the previous section, real rejuvenators are additives capable of both replenishing the lost volatile components and flaking the large agglomerates of asphaltenes into much smaller compounds. The studies by Pahlavan et al. [161,170,171], which explain in detail the effect of rejuvenation in the aged bitumen morphology, are summarized hereafter.

The approach adopted by Pahlavan et al. to investigate the rejuvenating process is based on molecular dynamics (MD) simulation and quantum mechanical studies through density functional theory (DFT). It starts from the analysis of the asphaltene monomer and the related oxidized dimer, which consists of a structure resulting from the interaction of two monomers of asphaltenes.

The electrons of the C=O groups located around the oxidized asphaltenes (red dots in Figure 9) change the distribution of π electrons on these flat-shaped molecules. This determines an insufficiency of electrons at the π bonds in the central part of the asphaltenes with a consequent decrease of the repulsive forces between the asphaltenes. This decrease leads to the formation of the dimer (two asphaltene monomers).

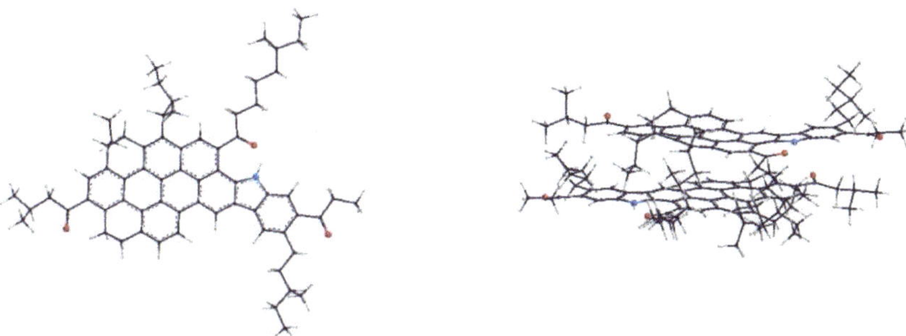

Figure 9. Asphaltene monomer (**left**) and dimer (**right**) molecular structures in the oxidized state (reprinted with permission from [161], Copyright 2019, American Chemical Society).

The forces that control the degree of asphaltenes stacking are concentrated in the aromatic motifs present in the asphaltenes. Figure 10 shows the most probable sites of interaction between a rejuvenator and an asphaltene dimer. The zones A, B, and C represent the areas where the rejuvenator likely finds a lower opposition by the hydrocarbon chains when it approaches the dimer. In these regions, the rejuvenation process happens in two steps that the author named the "lock and key" step and "intercalating" step. The first step ("lock and key") relates to zones A and B, where the adhesion between the asphaltenes in the dimer is stronger due to the polar groups. When the rejuvenator reaches the zones, it detaches the aromatic ring planes allowing access to more additive. In particular, the rejuvenator can intercalate in zone C, previously less accessible, and further deagglomerate the dimer by physically imposing a hindrance between the asphaltenes. The effectiveness of this second step depends on the chemistry of the rejuvenator. If the additive has polar properties, it creates interference on the Van der Waals forces that keep the polyaromatic cores stuck. However, the efficacy is higher if the additive has hydrocarbons with donor CH sites, which can interact with the π electrons in the asphaltenes and thus disrupt the bound of the dimer.

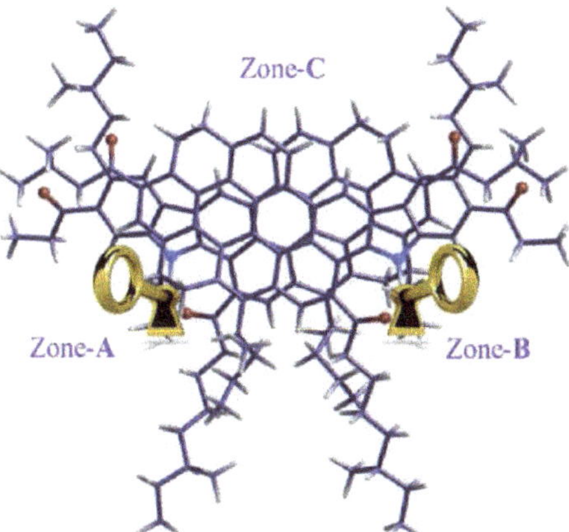

Figure 10. More likely interaction sites for biological rejuvenator with an asphaltene dimer (reprinted with permission from [161], Copyright 2019, American Chemical Society).

Of course, the rejuvenating effect of an additive is higher if it achieves both the steps, in particular, because the "intercalating" is noticeably facilitated when preceded by the "lock and key".

Figure 11 shows the side views of the dimer before and after rejuvenation. The simulation allowed observing an increase in the distance between the asphaltene sheets when the rejuvenator is used. In particular, in zones A and B (on the right in the side views), the increase of the gap between the polyaromatic cores was higher than in zone C (on the left in the side views).

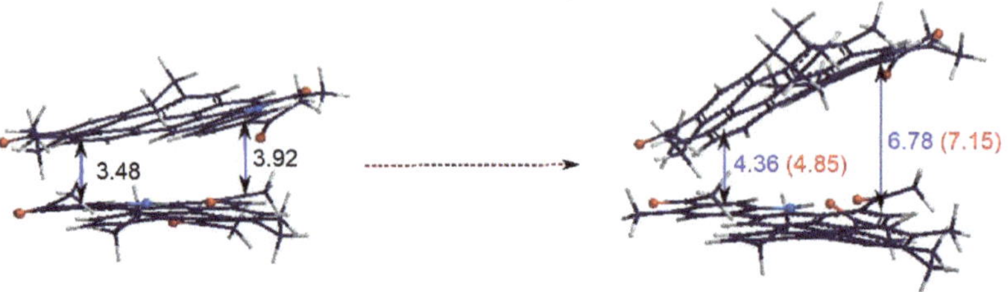

Figure 11. Intermolecular gaps between the aromatic rings in zones A/B and C for an oxidized dimer before (**left**) and after (**right**) rejuvenation. Blue and red numbers indicate the distances (in Å) achieved with two different rejuvenators (Reprinted with permission from [161], Copyright 2019, American Chemical Society).

4.3. Laboratory Rejuvenating Assessment Methods: General Results

As mentioned in the previous chapters, one way to evaluate the aging and rejuvenation of bitumen is to monitor the progress of the four SARA fractions. With oxidative aging, aromatics and resins are converted into asphaltenes leading to an increase in the contacts between the micelles in the colloidal system, increasing bitumen stiffness. The application of rejuvenators should increase the maltene content of the aged binder and consequently restore the proportions of the system.

In this regard, Zadshir et al. [172] tested an aged bitumen (RTFOT and 2 PAV) rejuvenated using three additives of different natures: the first additive was based on organic oil from pig manure (BB), the second derived from vegetable oils (VB), and the third was hydrocarbon-based (PB). The chemical composition of the biological additive (BB) shows that it is characterized by a high content of asphaltenes and resins and a low content of saturates and aromatics (Figure 12). The high content of resins stabilizes the asphaltenes and the relative micelles of the colloidal system, but the addition of further asphaltenes is typically discouraged in favor of a higher content of aromatics. Despite the structure of bio-based asphaltenes being different from that of hydrocarbon-based asphaltenes [173], the lack of aromatics and abundance of asphaltenes do not tend to rebalance the SARA composition. Conversely, the vegetable oil (VB) additive shows a great content of resins and aromatics, with low percentages of saturated and asphaltenes. The high amount of aromatic components can balance the maltenic fraction lost during the aging process, causing an increase in the colloidal stability of the aged binder. Lastly, the hydrocarbon-based additive (PB) is characterized by a large percentage of aromatics and a lower percentage of asphaltenes in comparison to the biological binder (BB).

The evolution of the percentages of the SARA fractions using these rejuvenators can also be analyzed by monitoring the Colloidal Index (CI). In particular:

- Adding the biological rejuvenator (BB) (10% by weight) causes a decrease in the CI from 0.61 of the aged binder to 0.51 (bringing it back to values similar to that of virgin bitumen). However, by increasing the percentage of the same additive up to 30%, the index increases to the unit value (exceeding even that of the aged binder without additives);
- Adding the vegetable-oil-based additive (VB), a progressive decrease in the index by increasing the percentage of rejuvenator occurs. This means that the stability of the binder is increased as the additive dosage increases. With an additive content of 30%, a CI lower than that of virgin bitumen is reached.

- Adding the hydrocarbon-based rejuvenator (PB), the percentage of asphaltenes decreases and this leads to a slight decrease in the CI. Increasing the percentage of additive from 10% to 30%, there is not a further decrease.

From the comparison between the three rejuvenators, it is clear that the content of resins and aromatics brought by the additives plays a fundamental role in the process. Resins are more effective than aromatics because of their high polarity and ability to disperse the asphaltenes micelles in the maltenic phase and consequently stabilize the colloidal system.

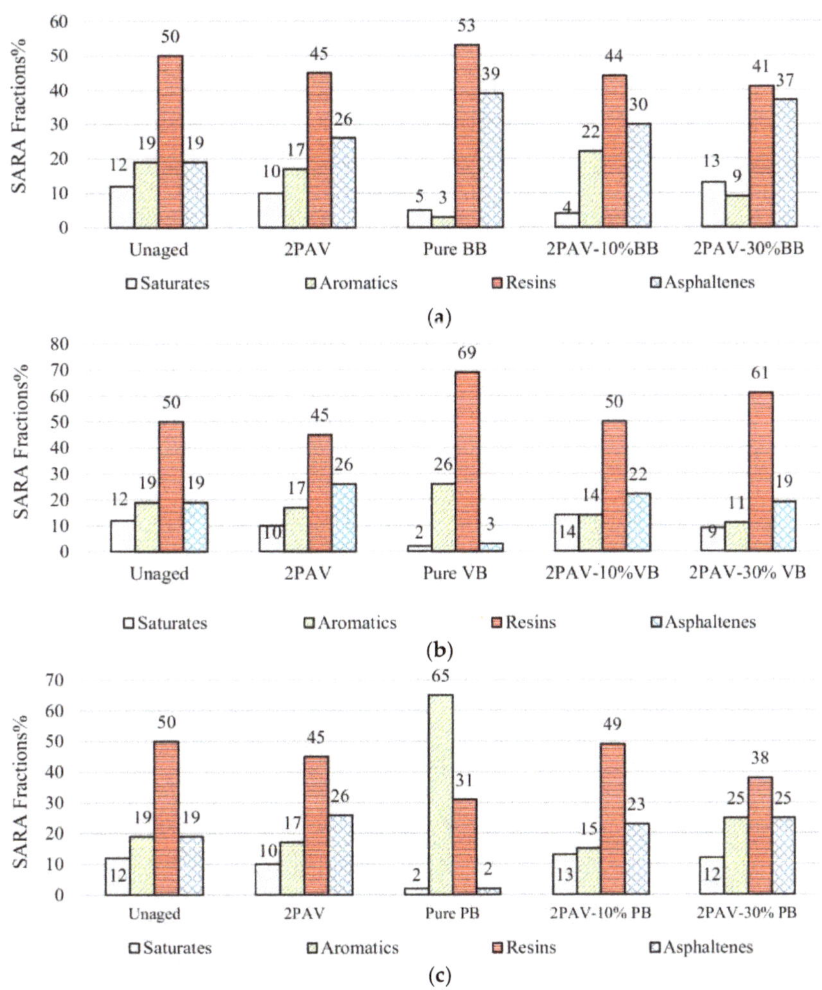

Figure 12. SARA fractions of virgin binder, aged binder, additives, and rejuvenated binder: (**a**) biological additive, (**b**) vegetable oil, (**c**) hydrocarbon-based additive (reprinted from Reference [172] with the permission of Elsevier).

A very important aspect to consider in the rejuvenating process is the amount of additive that has to be introduced into the aged bitumen to ensure the greatest benefit to the process. The results of CI help to find an optimal rejuvenator content for each type of additive. However, it has to be considered that for some products, as in the case of the biological additive, due to the different nature of the asphaltene molecules that compose it, the CI may increase with increasing rejuvenator dosage, but an improvement in physical and rheological behavior of the bitumen is still achieved.

The same study [172] also investigated the molecular size distribution of the rejuvenated bitumens through GPC. The results show that:

- the large molecules (LMS) increase from 83% for the non-aged binder to 87% for the aged binder, at the expense of the percentage of medium-sized molecules (MMS), which is reduced by 14% to 10%. The increase in LMS is a consequence of the increase in the number of asphaltenes in the system and their agglomeration.
- the addition of a rejuvenator tends to decrease the percentage of larger LMS molecules by increasing the presence of medium-sized molecules (MMS).

Further conclusions, related to the GPC test, were also obtained by Cao et al. [174], who tested the effects of waste cooking oil as a rejuvenator for aged bitumen. The main conclusions obtained in this study are the following:

- The aged binder gets higher Mw and Mn values than the virgin one, denoting the formation of larger molecules in the binder during the aging process. Compared to the virgin bitumen, the higher poly-dispersion of the aged binder indicates that there is a greater distribution of molecular weights.
- Adding waste cooking oil with different dosages, there is no chemical reaction between the additive and the aged binder. The decrease in Mw and the poly-dispersion is due to a physical dilution.

Figure 13 shows the curves obtained by the GPC of the virgin, aged, and subsequently rejuvenated bitumen with different dosages of additive.

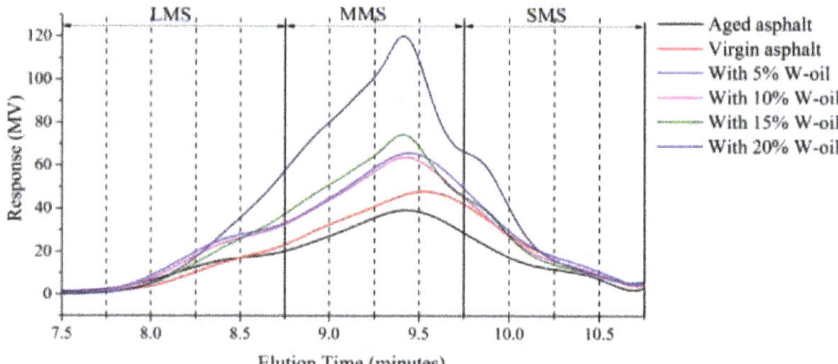

Figure 13. GPC curves of the regenerated binder with different dosages of waste cooking oil (reprinted from Reference [174] with the permission of Elsevier).

Compared to the virgin binder (red line), the aged one (black line) has a higher percentage of LMS compared to a lower small and medium-size molecules. Adding the waste cooking oil, the LMS tends to decrease; the typical LMS concentration of the virgin binder can be recovered by adding 20% additive. Increasing the percentage of waste cooking oil, MMS greatly increase and SMS decrease. However, the percentage of MMS and SMS cannot be brought back to the original values of the virgin binder. Therefore, it can be concluded that by adding waste cooking oil, the distribution of the molecular size cannot be totally restored.

FTIR analysis can be used to investigate the evolution of the functional groups present in an aged binder when it is rejuvenated with additives. Referring to the study made by Cao et al. [174], hereafter, a picture showing the FTIR spectra of the waste cooking oil, the virgin binder, and the rejuvenated one is reported (Figure 14). Figure 15 shows the FTIR spectra of the functional groups S=O, C=O, and C-C in more detail. Additive and bitumen do not reach together during the rejuvenating process. In fact, the rejuvenated binder spectra are a union of those of the aged binder and the rejuvenator, without any new groups [155,175]. In terms of carbonyl index ($I_{C=O}$) and sulphoxide index ($I_{S=O}$), both increased with aging and were decreased by adding rejuvenators. This effect is related to the physical dilution of the binder.

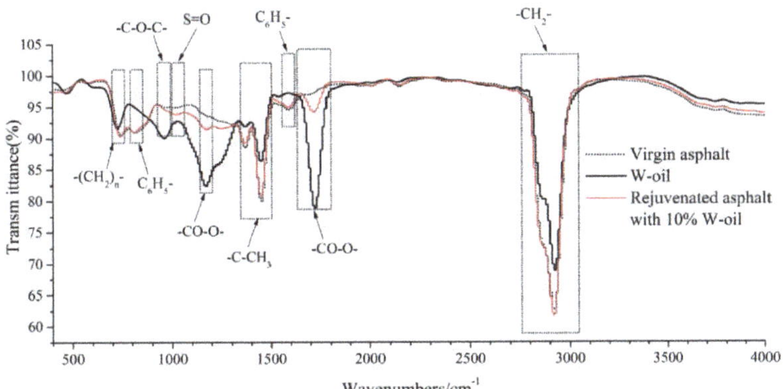

Figure 14. FTIR spectra of WCO (W-oil), virgin, and rejuvenated bitumen (reprinted from Reference [174] with the permission of Elsevier).

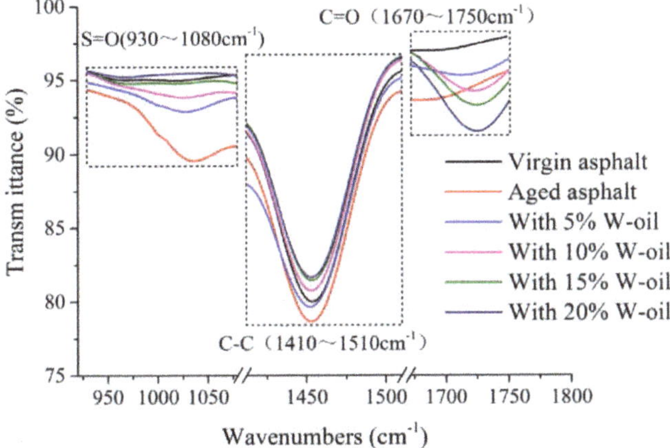

Figure 15. FTIR spectra of S=O, C=O, and C-C (reprinted from Reference [174] with the permission of Elsevier).

A study conducted by Noor et al. [92] focused on the analysis of the functional groups via FTIR when a biological or aromatic rejuvenator is added to a bituminous binder. Qualitative analyses were conducted by studying the spectra of the virgin bituminous binder, the biological additive, and the aromatic additive, in order to identify the representative functional groups. The functional groups identified in the bitumen are the same as those identified in the aromatic rejuvenator (both derived from petroleum). When the aromatic additive is added into the bitumen, there are no differences in the absorbance spectrum, as the functional groups are the same. On the other hand, when the biological rejuvenator is added, the resulting spectrum shows two distinct peaks in correspondence of the functional groups C=O (1744 cm^{-1}) and C-O (1162 cm^{-1}). Therefore, from the quantitative point of view, these bands can be a good reference for the identification of a specific biological rejuvenator in the bituminous binder.

Recently, Menapace et al. [176] investigated the AFM images when adding tall oil to two different types of aged bitumen:

- TOAS: blend of virgin and aged bitumen extracted from Recycled Asphalt Shingles (RAS) from re-roofing or roof removal projects;
- MWAS: blend of virgin and aged bitumen extracted from RAS from the excess material obtained during the shingles' production.

Regarding the TOAS-type bitumen, after long-term aging, there is a significant increase in roughness and a decrease in the phase contrast of the colloidal structure (Figure 16). Bee structures appear in greater numbers but less defined than they were in virgin bitumen. The reduced size of the bee-shaped structures that have formed with aging indicates a relatively lower molecular mobility than for the virgin binder. Adding tall oil, there is a reduction of the dispersed domains and an increase of the matrix area. At the same time, a slight decrease in roughness and phase contrast are noted, and the bee structures are no longer visible. After re-aging (second PAV) of the rejuvenated binder, the surface of the matrix decreases, the roughness increases, the phase-contrast decreases, and some bee-shaped structures reappear. It is interesting to understand how the bee-shaped structures come out again after the PAV re-aging, while they are not present in the rejuvenated blend. It is possible to hypothesize that the species used for the construction of the bee domains require a certain molecule agglomeration to create bee-shaped structures. The rejuvenator lowers the degree of association, thus breaks the bee-shaped structures originally present. Aging, on the other hand, helps the molecules to cluster again and reform the bee-shaped structures.

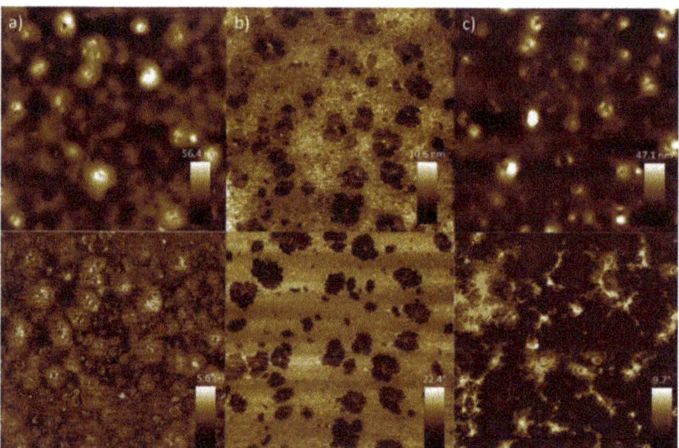

Figure 16. Height (**above**) and phase (**below**) images of aged TOAS binder (**a**), regenerated with tall oil (**b**) and re-aged (**c**) (reprinted from Reference [176] with the permission of Elsevier).

In the MWAS bitumen, the long-term aging seems to form rippled structures that are slightly different from the typical bees (Figure 17). When the rejuvenator is used, the dispersed domains degenerate on the binder surface as if they were liquid, to form a single interconnected phase. The rippled structures are still present, and the overall surface roughness slightly increases, while the phase-contrast shows a small decrease. After a second long-term aging process, numerous bee-shaped structures appear on the surface (c). The edges of these structures are not as smooth as in the original ones, but show a more irregular conformation. This indicates that the molecules can hardly move and reach colloidal stability.

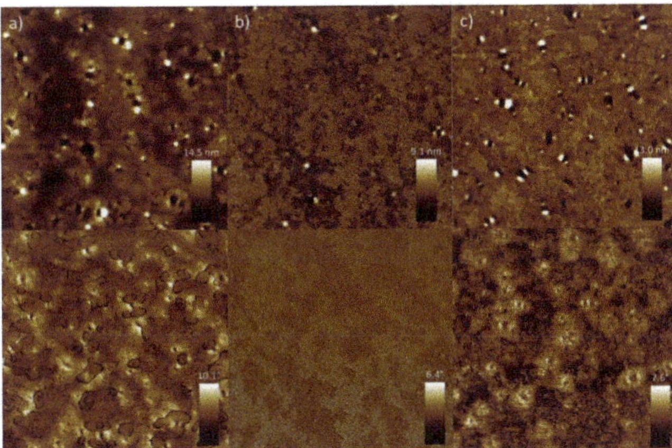

Figure 17. Height (**above**) and phase (**below**) images of the MWAS aged binder (**a**), regenerated with tall oil (**b**) and re-aged (**c**) (reprinted from Reference [176] with the permission of Elsevier).

In conclusion, the rejuvenator (tall oil), through the separation of the agglomerates formed by aging, allows the attenuation or even elimination of the bee-shaped structures, which, however, reappear when the binder ages again.

More recently, Ganter et al. [177] tried to evaluate the evolution of the bee structures during the aging and rejuvenating steps. For this purpose, three different rejuvenators, coming from bio (R1, R2) and oil (R3) sources were used. A PmB 25/55-55 bitumen was first short- and long-term aged (RTFOT + PAV) and then rejuvenated with the three additives listed before. They highlighted that the virgin binder displays clear bee-like structures and with increasing aging, the number of bees rises. The interesting aspect is that the addition of the three different rejuvenators causes a very different change in the surface morphology of the bitumen: size and quantities of the bee-like structure differ completely as a function of additive type and origin.

5. Aging of a Rejuvenated Bitumen (Re-Aging)

The growing use of RAP in the road sector is leading to new pavement mixtures composed not only of virgin materials, but also of old recycled bitumen and rejuvenators. Therefore, it is very important to understand the changes that occur in the properties of an HRMA when it is subjected to aging during its service life (which indeed is a "second aging" or "re-aging" for the RAP material). Hereafter, a literary review regarding the re-aging changes in the chemical and rheological properties of the recycled bitumen is provided.

Generally, the second aging in a rejuvenated binder is less harmful with respect to the aging of a virgin bitumen. Based on the studies carried out by Mazzoni et al. [178] and Bocci et al. [179], it is reasonable to expect that HMA including a high amount of RAP and a rejuvenator suffers fewer aging phenomena and can be less stiff than an alike mixture with no RAP.

From the chemical point of view, in a recent study, Ingrassia et al. [2] focused on evaluating the possibility of recycling binders that were already rejuvenated. Two virgin binders (one ordinary bitumen and a bio-binder) and two aged binders (one recovered from RAP and one "Bio-RAP" binder produced in the laboratory) were combined to reproduce the RAP hot recycling process. The chemical properties before and after aging were investigated using FTIR analysis through the evaluation of the $I_{C=O}$ and $I_{S=O}$ indices and an additional parameter, named chemical aging index AI_{FTIR}, and calculated with the following equation:

$$AI_{FTIR}\frac{(Ico + Iso)_{aged}}{(Ico + Iso)_{unaged}}$$

Basically, the analysis of AI_{FTIR} shows that the susceptibility to aging of recycled blends is significantly lower than that of the reference virgin binder. In fact, recycled bituminous blends (already containing a certain amount of oxidized binder) are less susceptible to further long-term aging.

Moreover, Sa-da-Costa et al. [180] tried to analyze, from the chemical point of view, the effects of aging on a rejuvenated bituminous binder. Several types of bitumen were analyzed in this study. As expected, an increase in the content of asphaltenes was found, confirming the oxidative transformations that contribute to increasing the polarity of the bitumen components in terms of SARA fraction. Furthermore, the oxidation of the bitumen involves a general increase in the carbonyl functional groups (C=O). Although precise correlations have not been found out, it can be stated that the chemical properties of an aged rejuvenated bitumen (second aging) are very close to that of an aged virgin binder (primary aging).

The second aging is of fundamental importance to understand the condition of the HRMA at the end of its second useful life. Continuous recycling of RAP would allow an increase in circularity related to the road construction field. For this reason, in the next future, more studies regarding this aspect need to be carried out.

6. Conclusions

In the sector of road pavement engineering, the easiest way to promote the circular economy is to encourage the use of RAP. One of the main issues regarding the use of this recycled material is the aged bitumen contained in it. Since the aging and rejuvenating process of bituminous pavements can significantly affect their durability and service life, it also is fundamental to achieve a complete knowledge of these aspects from a chemical point of view.

As explained in detail in the previous sections, bitumen aging involves different phenomena: loss of volatiles, oxidation, and physical and steric hardening. These result in the unbalancing of the SARA fractions and the formation of large molecular agglomerates, which in turn affect the mobility of the fractions in the colloidal system and determine the binder embrittlement. In order to restore the properties that bitumen loses with aging, rejuvenators can be used. A good rejuvenator can rebalance the SARA components and, despite oxidation being mainly irreversible, it can disrupt the asphaltene clusters and re-establish the proper molecular mobility.

In the present review, different chemical analyses were described as tools to investigate bitumen aging and rejuvenation processes. Table 6 provides an overview of the discussed approaches. All the techniques showed promising results but, due to the limitations of each method, the contemporary use of different approaches is highly recommended to have a precise and clear overview of how bitumen ages and rejuvenates. Moreover, the definition of a multi-testing protocol for the characterization of the effects of aging and rejuvenation at a chemical level can represent the basis for future application on more complex binders (polymer-modified bitumens, bitumens including extenders or nanoparticles).

Table 6. Advantages and limitations of the different techniques for the study of bitumen aging and rejuvenation.

Technique		Investigation of Aging	Investigation of Rejuvenation
AFM	Advantages	The bitumen chemo-morphological degradation with aging can be studied using AFM.	The rejuvenation process has been proven to influence the surface morphology of the aged bitumen, specifically the bee structures.
	Limitations	Since the approach is very recent, there is still a gap of knowledge on associating the AFM phase evolution with other chemical and mechanical properties.	The experimental approach is still at an early stage, and further research is necessary to deeply understand how to exploit this powerful tool.
FTIR	Advantages	It allows determining the severity of aging through the change of specific bands (particularly sulfoxide and carbonyl).	The presence of the rejuvenator in an HMA can be detected by comparing the spectra of the pure additive and the recovered bitumen.
	Limitations	It does not discriminate what happens to the bitumen colloidal system, but mainly focuses on the oxidation effects.	Since most additives have peculiar bands in correspondence to the sulfoxide and carbonyl bands, it is difficult to quantify the rejuvenator content in the recovered bitumen.
TLC-FID	Advantages	The evolution of the SARA fractions with aging allows estimating the severity of the phenomena.	It allows understanding the efficacy of the rejuvenation process in restoring the SARA proportioning.
	Limitations	There is not a clear correlation between the SARA proportions and the bitumen mechanical behavior.	Again, the main issue with this technique is related to the poor association between SARA proportions after rejuvenation and the effective improvement of the aged bitumen rheological properties.
HP-GPC	Advantages	It allows quantifying the effects of SARA fraction shifting and the agglomeration of the asphaltenes due to aging.	The analysis of the molecular weight distribution allows one to understand if a rejuvenator can really detach the asphaltene clusters or only has a dilution effect.
	Limitations	There is still uncertainty on how the different aging phenomena (oxidation, loss of volatiles, etc.) influence the molecular weight distribution.	It could be used to estimate the degree of blending between aged and virgin bitumen with/without rejuvenators, but no precise procedures have been defined yet.

7. Recommendations for Future Research

The state of the art provided in the present paper highlighted that, despite extensive research having been moved to innovative and improved binders, there are still important aspects about neat bitumen that deserve to be studied to fully understand the complex chemistry of this material and somehow predict the behavior of the bituminous mixtures when in service. In light of the themes discussed in the previous sections, future works should be addressed to:

1. Combine the results of the chemical tests at the binder scale with the results of the mechanical and rheological tests at both the binder and mixture scale. Understanding what happens to the bitumen from the chemo-morphological point of view is fundamental, but should be correlated to the corresponding effects on the material performance in order to have the research focused on the practical outcomes. To this goal, for instance, the IFSTTAR research team recently identified a relationship between the bitumen molecular weight distribution and the phase angle of the complex modulus, and proposed a tool, the δ-method, to determine the molecular weight distribution from rheological tests [181]. Within the RILEM TC 264-RAP, scientists are trying to find links between the mechanical characteristics of HRMA and the FTIR spectrum of the extracted bitumen, while also aiming to estimate the presence or even the content of a rejuvenator in a mix from the FTIR binder spectrum.
2. Evaluate new solutions to hinder, restrict, or slow down the bitumen's aging. Several investigations showed that the use of a straight-run bitumen, instead of a visbreaker

one, can reduce the aging susceptibility of an HMA in both the short and long term [29,30,182]. However, poor attention is paid by road authorities, boards for standardization, and HMA manufacturers on the bitumen's origin and production process. In a similar scope, further research should also be focused on additives with antioxidant effect, that is, with the ability to reduce the bitumen's propensity to oxidation. Some products are currently available on the market with the declared effect of hindering bitumen oxidation, but scientific studies are required to deeply understand their behavior at both the chemical and mechanical levels.

3. Understand the interaction between old and fresh bitumen in hot recycling. The topics of RAP bitumen degree of activation (DoA) and RAP/virgin bitumen degree of blending (DoB) are actually among the most studied worldwide [183–193]. However, because of the huge complexity of the problem (which is influenced by many factors such as RAP bitumen content, nature, and aging state; HMA production process; type, dosage, and method of addition of the rejuvenators; hauling, paving, and compaction procedure, etc.), univocal protocols to classify different RAP materials according to the DoA or to estimate the DoB during pavement construction have not been defined.
4. Identify a method to allow precise quality controls on HRMA. This objective, which is maybe utopian, is one of the most crucial. Technical specifications currently provide controls on the HMA mechanical performance to limit the amount of RAP in the mix and encourage the use of rejuvenators. However, a solution to estimate how much RAP and how much rejuvenator have been included in a mix should be found, possibly including a series of physical, chemical, microscopic, and rheological analyses on the raw materials (RAP and its components, rejuvenator, virgin bitumen), the laboratory, and the plant-produced mixtures preliminarily to full road construction.

Author Contributions: Writing—original draft preparation, E.P.; writing—review and editing, E.B. Both authors have read and agreed to the published version of the manuscript.

Funding: This research received no external funding.

Institutional Review Board Statement: Not applicable.

Informed Consent Statement: Not applicable.

Data Availability Statement: Not applicable.

Conflicts of Interest: The authors declare no conflict of interest.

References

1. Geissdoerfer, M.; Savaget, P.; Bocken, N.M.P.; Hultink, E.J. The Circular Economy—A new sustainability paradigm? *J. Clean. Prod.* **2017**, *143*, 757–768. [CrossRef]
2. Ingrassia, L.P.; Lu, X.; Ferrotti, G.; Conti, C.; Canestrari, F. Investigating the "circular propensity" of road bio-binders: Effectiveness in hot recycling of reclaimed asphalt and recyclability potential. *J. Clean. Prod.* **2020**, *255*, 120193. [CrossRef]
3. Zaumanis, M.; Mallick, R.B. Review of very high-content reclaimed asphalt use in plant-produced pavements: State of the art. *Int. J. Pavement Eng.* **2015**, *16*, 39–55. [CrossRef]
4. Willis, J.R.; Turner, P.; Julian, G.; Taylor, A.J.; Tran, N.; De Padula, G.F. *Effects of Changing Virgin Binder Grade and Content on Rap Mixture Properties*; Report No. 12-03; National Center for Asphalt Technology(NCAT): Auburn, AL, USA, 2012.
5. Bocci, E.; Prosperi, E. Analysis of different reclaimed asphalt pavements to assess the potentiality of RILEM cohesion test. *Mater. Struct.* **2020**, *53*, 1–14. [CrossRef]
6. Zaumanis, M.; Mallick, R.B.; Frank, R. 100% Hot Mix Asphalt Recycling: Challenges and Benefits. *Transp. Res. Procedia* **2016**, *14*, 3493–3502. [CrossRef]
7. Tarsi, G.; Tataranni, P.; Sangiorgi, C. The challenges of using reclaimed asphalt pavement for new asphalt mixtures: A review. *Materials* **2020**, *13*, 4052. [CrossRef]
8. Polacco, G.; Filippi, S.; Merusi, F.; Stastna, G. A review of the fundamentals of polymer-modified asphalts: Asphalt/polymer interactions and principles of compatibility. *Adv. Colloid Interface Sci.* **2015**, *224*, 72–112. [CrossRef]
9. Wang, H.; Lu, G.; Feng, S.; Wen, X.; Yang, J. Characterization of Bitumen Modified with Pyrolytic Carbon Black from Scrap Tires. *Sustainability* **2019**, *11*, 1631. [CrossRef]
10. Celauro, C.; Saroufim, E.; Mistretta, M.C.; La Mantia, F.P. Influence of short-term aging on mechanical properties and morphology of polymer-modified bitumen with recycled plastics from waste materials. *Polymers* **2020**, *12*, 1985. [CrossRef] [PubMed]

11. Feng, Z.; Cai, F.; Yao, D.; Li, X. Aging properties of ultraviolet absorber/SBS modified bitumen based on FTIR analysis. *Constr. Build. Mater.* **2021**, *273*, 121713. [CrossRef]
12. Sun, G.; Li, B.; Sun, D.; Yu, F.; Hu, M. Chemo-rheological and morphology evolution of polymer modified bitumens under thermal oxidative and all-weather aging. *Fuel* **2021**, *285*, 118989. [CrossRef]
13. Cuciniello, G.; Leandri, P.; Filippi, S.; Lo Presti, D.; Polacco, G.; Losa, M.; Airey, G. Microstructure and rheological response of laboratory-aged SBS-modified bitumens. *Road Mater. Pavement Des.* **2021**, *22*, 372–396. [CrossRef]
14. Rodriguez-Fernandez, I.; Cavalli, M.C.; Poulikakos, L.M.B. Recyclability of asphalt mixtures with crumb rubber incorporated by dry process: A laboratory investigation. *Materials* **2020**, *13*, 2870. [CrossRef] [PubMed]
15. Lin, P.; Liu, X.; Apostolidis, P.; Erkens, S.; Ren, S.; Xu, S.; Scarpas, T.; Huang, W. On the rejuvenator dosage optimization for aged SBS modified bitumen. *Constr. Build. Mater.* **2021**, *271*, 121913. [CrossRef]
16. Liu, H.; Fu, L.; Jiao, Y.; Tao, J.; Wang, X. Short-Term Aging Effect on Properties of Sustainable Pavement Asphalts Modified by Waste Rubber and Diatomite. *Sustainability* **2017**, *9*, 996. [CrossRef]
17. Cheraghian, G.; Wistuba, M.P. Ultraviolet aging study on bitumen modified by a composite of clay and fumed silica na-noparticles. *Sci. Rep.* **2020**, *10*, 1121. [CrossRef] [PubMed]
18. Cheraghian, G.; Wistuba, M. Effect of Fumed Silica Nanoparticles on Ultraviolet Aging Resistance of Bitumen. *Nanomaterials* **2021**, *11*, 454. [CrossRef]
19. Dell'Antonio Cadorin, N. Victor Staub de Melo, J.; Borba Broering, W.; Luiz Manfro, A.; Salgado Barra, B. Asphalt nanocomposite with titanium dioxide: Mechanical, rheological and photoactivity performance. *Constr. Build. Mater.* **2021**, *289*, 123178. [CrossRef]
20. Moretti, L.; Fabrizi, N.; Fiore, N.; D'andrea, A. Mechanical characteristics of graphene nanoplatelets-modified asphalt mixes: A comparison with polymer-and not-modified asphalt mixes. *Materials* **2021**, *14*, 2434. [CrossRef]
21. Han, M.; Muhammad, Y.; Wei, Y.; Zhu, Z.; Huang, J.; Li, J. A review on the development and application of graphene based materials for the fabrication of modified asphalt and cement. *Constr. Build. Mater.* **2021**, *285*, 122885. [CrossRef]
22. Anwar, W.; Ahmad, N.; Khitab, A.; Faizan, M.; Tayyab, S.; Saeed, M.; Imran, M. Performance augmentation of asphalt binder with multi-walled carbon nanotubes. *Proc. Inst. Civ. Eng. Transp.* **2021**, *174*, 130–141.
23. Petersen, J.C. *Review of the Fundamentals of Asphalt Oxidation: Chemical, Physicochemical, Physical Property, and Durability Relationships*; Transportation Research Circular Number E-C140; Transportation Research Board of the National Academies: Washington, DC, USA, 2009. [CrossRef]
24. Loise, V.; Caputo, P.; Porto, M.; Calandra, P.; Angelico, R.; Rossi, C.O. A review on Bitumen Rejuvenation: Mechanisms, materials, methods and perspectives. *Appl. Sci.* **2019**, *9*, 4316. [CrossRef]
25. Paliukaitė, M.; Vaitkus, A.; Zofka, A. Evaluation of bitumen fractional composition depending on the crude oil type and production technology. In Proceedings of the 9th International Conference Environmental Engineering, Vilnius, Lithuania, 22–23 May 2014.
26. de Klerk, A. *Unconventional Oil: Oilsands*; Elsevier Ltd.: Amsterdam, The Netherlands, 2020; ISBN 9780081028865.
27. Giavarini, C. Stability of bitumens produced by thermal processes. *Fuel* **1981**, *60*, 401–404. [CrossRef]
28. Speight, J.G. Visbreaking: A technology of the past and the future. *Sci. Iran.* **2012**, *19*, 569–573. [CrossRef]
29. Giavarini, C. Visbreaker and straight-run bitumens. *Fuel* **1984**, *63*, 1515–1517. [CrossRef]
30. Giavarini, C.; Saporito, S. Oxidation of visbreaker bitumens. *Fuel* **1989**, *68*, 943–946. [CrossRef]
31. Mortazavi, M.; Moulthrop, J.S. *The SHRP Materials Reference Library*; SHRP Report A-646; Strategic Highway Research Program, National Research Program: Washington, DC, USA, 1993; p. 228.
32. Petersen, J.C. Chemical Composition of Asphalt As Related to Asphalt Durability: State of the Art. *Transp. Res. Rec.* **1984**, *999*, 13–30. [CrossRef]
33. Redelius, P.; Soenen, H. Relation between bitumen chemistry and performance. *Fuel* **2015**, *140*, 34–43. [CrossRef]
34. Boussingault, J.B. Mémoire sur la composition des bitumes. *Ann. Chim. Phys.* **1837**, *64*, 141–151.
35. Richardson, C. *The Modern Asphalt Pavement*, 2nd ed.; Wiley: Hoboken, NJ, USA, 1910.
36. Krishnan, J.M.; Rajagopal, K.R. Review of the uses and modeling of bitumen from ancient to modern times. *Appl. Mech. Rev.* **2003**, *56*, 149–214. [CrossRef]
37. Rostler, F.S. Fractional Composition: Analytical and Functional Significance. In *Bituminous Materials: Asphalts, Tars, and Pitches*; Interscience: New York, NY, USA, 1965; Volume 2.
38. Corbett, L.W. Composition of Asphalt Based on Generic Fractionation, Using Solvent Deasphaltening, Elution-Adsorption Chromatography, and Densimetric Characterization. *Anal. Chem.* **1969**, *41*, 576–579. [CrossRef]
39. Lesueur, D. The colloidal structure of bitumen: Consequences on the rheology and on the mechanisms of bitumen modification. *Adv. Colloid Interface Sci.* **2009**, *145*, 42–82. [CrossRef] [PubMed]
40. American Society for Testing and Materials. *ASTM D4124-09: Standard Test Method for Separation af Asphalt into Four Fractions*; ASTM: West Conshohocken, PA, USA, 2018.
41. Leroy, G. Bitumen analysis by thin layer chromatography (IATROSCAN). In Proceedings of the 4th Eurobitume Congress, Madrid, Spain, 4–6 October 1989.
42. Ecker, A. The application of Iatroscan-technique for analysis of bitumen. *Pet. Coal* **2001**, *43*, 51–53.
43. Speight, J.G. Petroleum asphaltenes. Part 1. Asphaltenes, resins and the structure of petroleum. *Oil Gas Sci. Technol.* **2004**, *59*, 467–477. [CrossRef]

44. Branthaver, J.F.; Petersen, J.C.; Robertson, R.E.; Duvall, J.J.; Kim, S.S.; Harnsberger, P.M.; Mill, T.; Ensley, E.K.; Barbour, F.A.; Scharbron, J.F. *Binder Characterization and Evaluation—Volume 2: Chemistry*; SHRP Report A-368; National Research Council: Washington, DC, USA, 1994.
45. Koots, J.A.; Speight, J.G. Relation of petroleum resins to asphaltenes. *Fuel* **1975**, *54*, 179–184. [CrossRef]
46. Rosinger, A. Beiträge zur Kolloidchemie des Asphalts. *Kolloid-Z* **1914**, *15*, 177–179. [CrossRef]
47. Nellensteyn, F.J. The constitution of asphalt. *J. Inst. Pet. Technol.* **1924**, *10*, 311–325.
48. Pfeiffer, J.P.; Saal, R.N.J. Asphaltic bitumen as colloid systems. *J. Phys. Chem.* **1940**, *44*, 139–149. [CrossRef]
49. Behnood, A.; Modiri Gharehveran, M. Morphology, rheology, and physical properties of polymer-modified asphalt binders. *Eur. Polym. J.* **2019**, *112*, 766–791. [CrossRef]
50. Gaestel, C.; Smadja, R.; Lamminan, K.A. Contribution à la connaissance des propriétés des bitumes routiers. *Rev. Gen. Routes Aérodromes* **1971**, *466*, 85–95.
51. Li, Z.; Fa, C.; Zhao, H.; Zhang, Y.; Chen, H.; Xie, H. Investigation on evolution of bitumen composition and micro-structure during aging. *Constr. Build. Mater.* **2020**, *244*, 118322. [CrossRef]
52. Fischer, H.R.; Dillingh, E.C.; Hermse, C.G.M. On the microstructure of bituminous binders. *Road Mater. Pavement Des.* **2014**, *15*, 1–15. [CrossRef]
53. Masson, J.F.; Leblond, V.; Margeson, J. Bitumen morphologies by phase-detection atomic force microscopy. *J. Microsc.* **2006**, *221*, 17–29. [CrossRef]
54. Tarpoudi Baheri, F.; Rico Luengo, M.; Schutzius, T.M.; Poulikakos, D.; Poulikakos, L.D. The effect of additives on water condensation on bituminous surfaces. In Proceedings of the RILEM International Symposium on Bituminous Materials (ISBM), Lyon, France, 14–16 December 2020.
55. Yang, Z.; Zhang, X.; Zhang, Z.; Zou, B.; Zhu, Z.; Lu, G.; Xu, W.; Yu, J.; Yu, H. Effect of aging on chemical and rheological properties of bitumen. *Polymers* **2018**, *10*, 1345. [CrossRef]
56. Fahrenfort, J.; Visser, W.M. On the determination of optical constants in the infrared by attenuated total reflection. *Spectrochim. Acta* **1962**, *18*, 1103–1116. [CrossRef]
57. Hou, X.; Lv, S.; Chen, Z.; Xiao, F. Applications of Fourier transform infrared spectroscopy technologies on asphalt materials. *Meas. J. Int. Meas. Confed.* **2018**, *121*, 304–316. [CrossRef]
58. Sharma, M.K.; Yen, T.F. *Asphaltene Particles in Fossil Fuel Exploration, Recovery, Refining, and Production Processes*; Springer: Berlin/Heidelberg, Germany, 1994; ISBN 9781461360452.
59. Masson, J.F.; Price, T.; Collins, P. Dynamics of bitumen fractions by thin-layer chromatography/flame ionization detection. *Energy Fuels* **2001**, *15*, 955–960. [CrossRef]
60. Xiao, F.; Amirkhanian, S.N. HP-GPC Approach to Evaluating Laboratory Prepared Long-Term Aged Rubberized Asphalt Binders. In Proceedings of the GeoHunan International Conference, Changsha, China, 3–6 August 2009; Volume 41043, pp. 42–48. [CrossRef]
61. Peralta, J.; Hilliou, L.; Silva, H.; Machado, A.; Pais, J.; Oliveira, J.R.M. Rheological quantification of bitumen aging: Definition of a new sensitive parameter. *Appl. Rheol.* **2010**, *20*, 63293. [CrossRef]
62. Traxler, R.R. Durability of asphalts cements. In Proceedings of the Asphalt Paving Technology, San Francisco, CA, USA, 18–20 February 1963; Volume 32, pp. 44–58.
63. Thurston, R.R.; Knowles, E.C. Asphalt and Its Constituents. Oxidation at Service Temperatures. *Ind. Eng. Chem.* **1941**, *33*, 320–324. [CrossRef]
64. Miró, R.; Martínez, A.H.; Moreno-Navarro, F.; Rubio-Gámez, C. Effect of ageing and temperature on the fatigue behaviour of bitumens. *Mater. Des.* **2015**, *86*, 129–137. [CrossRef]
65. Hesp, S.A.M.; Iliuta, S.; Shirokoff, J.W. Reversible aging in asphalt binders. *Energy Fuels* **2007**, *21*, 1112–1121. [CrossRef]
66. Struik, L. *Physical Hardening in Amorphous Polymers and Other Materials*; Elsevier: Amsterdam, The Netherlands, 1978.
67. Ferry, J.D. *Viscoelastic Properties of Polymers*, 3rd ed.; Wiley: New York, NY, USA, 1980.
68. Santagata, E.; Baglieri, O.; Dalmazzo, D.; Tsantilis, L. Experimental investigation on the combined effects of physical hardening and chemical ageing on low temperature properties of bituminous binders. In *8th RILEM International Symposium on Testing and Characterization of Sustainable and Innovative Bituminous Materials*; RILEM Bookseries; Springer: Berlin/Heidelberg, Germany, 2016; Volume 11, pp. 631–641. [CrossRef]
69. Tauste, R.; Moreno-Navarro, F.; Sol-Sánchez, M.; Rubio-Gámez, M.C. Understanding the bitumen ageing phenomenon: A review. *Constr. Build. Mater.* **2018**, *192*, 593–609. [CrossRef]
70. Frolov, I.N.; Bashkirceva, N.Y.; Ziganshin, M.A.; Okhotnikova, E.S.; Firsin, A.A. The steric hardening and structuring of paraffinic hydrocarbons in bitumen. *Pet. Sci. Technol.* **2016**, *34*, 1675–1680. [CrossRef]
71. Fernández-Gómez, W.D.; Rondón Quintana, H.; Reyes, F. A review of asphalt and asphalt mixture aging. *Ing. Investig.* **2013**, *33*, 5–12.
72. Bocci, E.; Prosperi, E.; Mair, V.; Bocci, M. Ageing and cooling of hot-mix-asphalt during hauling and paving—A laboratory and site study. *Sustainability* **2020**, *12*, 8612. [CrossRef]
73. Hunter, R.; Self, A.; Read, J. *The Shell Bitumen Handbook*, 6th ed.; ICE Publishing: London, UK, 2015.
74. Zupanick, M.; Baselice, V. Characterizing asphalt volatility. *Transp. Res. Rec.* **1997**, *1586*, 1–9. [CrossRef]
75. Lu, X.; Isacsson, U. Effect of ageing on bitumen chemistry and rheology. *Constr. Build. Mater.* **2002**, *16*, 15–22. [CrossRef]

76. Cortés, C.; Pérez-Lepe, A.; Fermoso, J.; Costa, A.; Guisado, F.; Esquena, J.; Potti, J.J. Envejecimiento Foto-Oxidativo de Betunes Asfálticos. Comunicación 21; In Proceedings of the V Jornada Nacional ASEFMA. 2010, pp. 227–238. Available online: https://www.itafec.com/descargas/envejecimiento-foto-oxidativo-de-betunes-asf%C3%A1lticos (accessed on 3 May 2021).
77. Tarsi, G.; Varveri, A.; Lantieri, C.; Scarpas, A.; Sangiorgi, C. Effects of Different Aging Methods on Chemical and Rheological Properties of Bitumen. *J. Mater. Civ. Eng.* **2018**, *30*, 04018009. [CrossRef]
78. Lu, H.; Talon, Y.; Redelius, P. Aging of bituminous binders—Laboratory tests and field data. In Proceedings of the 4th Euroasphalt and Eurobitume Congress, Copenhagen, Denmark, 22–23 May 2008; European Asphalt Pavement Association: Brussels, Belgium, 2008.
79. Wu, S.; Pang, L.; Liu, G.; Zhu, J. Laboratory Study on Ultraviolet Radiation Aging of Bitumen. *J. Mater. Civ. Eng.* **2010**, *22*, 767–772. [CrossRef]
80. Martinez, G.; Caicedo, B. Asfalticas, Efecto de la radiacion ultravioleta en el envejecimiento de ligantes y mezclas. In Proceedings of the Congresso Ibero-Latinoamericano del Asfalto CILA XIII, San Josè de, Costa Rica, Costa Rica, 20–25 November 2005.
81. Zeng, W.; Wu, S.; Wen, J.; Chen, Z. The temperature effects in aging index of asphalt during UV aging process. *Constr. Build. Mater.* **2015**, *93*, 1125–1131. [CrossRef]
82. Afanasieva, N.; Alvarez, M.; Ortiz, M. Rheological characterization of aged asphalt. *Ciencia Tecnol. Futur.* **2002**, *2*, 121–134.
83. Mirwald, J.; Werkovits, S.; Camargo, I.; Maschauer, D.; Hofko, B.; Grothe, H. Understanding bitumen ageing by investigation of its polarity fractions. *Constr. Build. Mater.* **2020**, *250*, 118809. [CrossRef]
84. Mastrofini, D.; Scarsella, M. The application of rheology to the evaluation of bitumen ageing. *Fuel* **2000**, *79*, 1005–1015. [CrossRef]
85. National Academies of Sciences, Engineering, and Medicine. *Evaluating the Effects of Recycling Agents on Asphalt Mixtures with High RAS and RAP Binder Ratios*; NCHRP Report 927; The National Academies Press: Washington, DC, USA, 2015. [CrossRef]
86. Fini, E.H.; Buabeng, F.S.; Abu-Lebdeh, T.; Awadallah, F. Effect of introduction of furfural on asphalt binder ageing characteristics. *Road Mater. Pavement Des.* **2016**, *17*, 638–657. [CrossRef]
87. Hofko, B.; Porot, L.; Cannone Falchetto, A.; Poulikakos, L.; Huber, L.; Lu, X.; Mollenhauer, K.; Grothe, H. FTIR spectral analysis of bituminous binders: Reproducibility and impact of ageing temperature. *Mater. Struct.* **2018**, *51*, 45. [CrossRef]
88. Lamontagne, J.; Dumas, P.; Mouillet, V.; Kister, J. Comparison by Fourier transform infrared (FTIR) spectroscopy of different ageing techniques: Application to road bitumens. *Fuel* **2001**, *80*, 483–488. [CrossRef]
89. Chávez-Valencia, L.E.; Manzano-Ramírez, A.; Alonso-Guzmán, E.; Contreras-García, M.E. Modelling of the performance of asphalt pavement using response surface methodology-the kinetics of the aging. *Build. Environ.* **2007**, *42*, 933–939. [CrossRef]
90. Mouillet, V.; Lamontagne, J.; Durrieu, F.; Planche, J.P.; Lapalu, L. Infrared microscopy investigation of oxidation and phase evolution in bitumen modified with polymers. *Fuel* **2008**, *87*, 1270–1280. [CrossRef]
91. Karlsson, R.; Isacsson, U. Application of FTIR-ATR to Characterization of Bitumen Rejuvenator Diffusion. *J. Mater. Civ. Eng.* **2003**, *15*, 157–165. [CrossRef]
92. Noor, L.; Wasiuddin, N.M.; Mohammad, L.N.; Salomon, D. Use of Fourier Transform Infrared (FT-IR) Spectroscopy to Determine the Type and Quantity of Rejuvenator Used in Asphalt Binder. In *Recent Developments in Pavement Engineering: Proceedings of the 3rd GeoMEast International Congress and Exhibition, Egypt 2019 on Sustainable Civil Infrastructures—The Official International Congress of the Soil-Structure Interaction Group in Egypt (SSIGE)*; Springer: Berlin/Heidelberg, Germany, 2020; Volume 1, pp. 70–84. [CrossRef]
93. Tachon, N. Nouveaux types de liants routiers a hautes performances, a teneur en bitume reduite par addition de produits organiques issus des agroressources. Ph.D. Thesis, L'Institut National Polytechnique de Toulouse, Touluse, France, 2008.
94. Araújo, M.F.A.S.; Lins, V.C.F.; Pasa, V.M.D. Effect of Ageing on Porosity of Hot Mix Asphalt. *Braz. J. Pet. Gas* **2011**, *5*, 011–018. [CrossRef]
95. Siddiqui, M.N.; Ali, M.F. Studies on the aging behavior of the Arabian asphalts. *Fuel* **1999**, *78*, 1005–1015. [CrossRef]
96. Dony, A.; Ziyani, L.; Drouadaine, I.; Pouget, S.; Faucon-Dumont, S.; Simard, D.; Mouillet, V.; Poirier, J.E.; Gabet, T.; Boulange, L.; et al. MURE National Project: FTIR spectroscopy study to assess ageing of asphalt mixtures. In Proceedings of the 6th Eurasphalt & Eurobitume Congress, Prague, Czech Republic, 1–3 June 2016.
97. Mikhailenko, P.; Bertron, A.; Ringot, E. Methods for Analyzing the Chemical Mechanisms of Bitumen Aging and Rejuvenation with FTIR Spectrometry. In *8th RILEM International Symposium on Testing and Characterization of Sustainable and Innovative Bituminous Materials*; RILEM Bookseries; Springer: Berlin/Heidelberg, Germany, 2016; Volume 11. [CrossRef]
98. Gabrielle do Nascimiento Camargo, I.; Hofko, B.; Mirwald, J.; Grithe, I. Effect of Thermal and Oxidative Aging on Asphalt Binders Rheology and Chemical Composition. *Materials* **2020**, *13*, 4438. [CrossRef]
99. Yan, Y.; Yang, Y.; Ran, M.; Zhou, X.; Zou, L.; Guo, M. Application of Infrared Spectroscopy in Prediction of Asphalt Aging Time History and Fatigue Life. *Coatings* **2020**, *10*, 959. [CrossRef]
100. Nivitha, M.R.; Prasad, E.; Krishnan, J.M. Ageing in modified bitumen using FTIR spectroscopy Ageing in modified bitumen using FTIR spectroscopy. *Int. J. Pavement Eng.* **2015**, *17*, 565–577. [CrossRef]
101. Poulikakos, L.D.; Cannone Falchetto, A.; Wang, D.; Porot, L.; Hofko, B. Impact of asphalt aging temperature on chemo-mechanics. *RSC Adv.* **2019**, *9*, 11602–11613. [CrossRef]
102. Dos Santos, S.; Poulikakos, L.D.; Partl, M.N. Crystalline structures in tetracosane-asphaltene films. *RSC Adv.* **2016**, *6*, 41561–41567. [CrossRef]

103. Hofko, B.; Eberhardsteiner, L.; Füssl, J.; Grothe, H.; Handle, F.; Hospodka, M.; Grossegger, D.; Nahar, S.N.; Scarpas, A.; Schmets, S.J. Impact of maltene and asphaltene fraction on mechanical behavior and microstructure of bitumen. *Mater. Struct.* **2016**, *49*, 829–841. [CrossRef]
104. Lu, X.; Langton, M.; Olofsson, P.; Redelius, P. Wax morphology in bitumen. *J. Mater. Sci.* **2005**, *40*, 1893–1900. [CrossRef]
105. dos Santos, S.; Partl, M.N.; Poulikakos, L. Newly observed effects of water on the microstructures of bitumen surface. *Constr. Build. Mater.* **2014**, *71*, 618–627. [CrossRef]
106. Cavalli, M.C.; Mazza, E.; Zaumanis, M.; Poulikakos, L.D. Surface nanomechanical properties of bio-modified reclaimed asphalt binder. *Road Mater. Pavement Des.* **2019**, *22*, 1407–1423. [CrossRef]
107. Bearsley, S.; Forbes, A.; Haverkamp, R.G. Direct observation of the asphaltene structure in paving-grade bitumen using confocal laser-scanning microscopy. *J. Microsc.* **2004**, *215*, 149–155. [CrossRef]
108. Ganter, D.; Franzka, S.; Shvartsman, V.V.; Lupascu, D.C. The phenomenon of bitumen 'bee" structures–bulk or surface layer–a closer look. *Int. J. Pavement Eng.* **2020**, 1–9. [CrossRef]
109. Lu, X.; Sjövall, P.; Soenen, H.; Blom, J.; Makowska, M. Oxidative aging of bitumen: A structural and chemical investigation. *Road Mater. Pavement Des.* **2021**, *14*, 25–37. [CrossRef]
110. Zhang, M.; Wang, X.; Zhang, W.; Ding, L. Study on the Relationship between Nano-Morphology Parameters and Properties of Bitumen during the Ageing Process. *Materials* **2020**, *13*, 1472. [CrossRef] [PubMed]
111. Coleri, E.; Sreedhar, S. Strategies to Improve Performance of Reclaimed Asphalt Pavement- Recycled Asphalt Shingle Mixtures. *Int. J. Pavement Eng.* **2019**, *22*, 201–212. [CrossRef]
112. Mannan, U.A.; Islam, R.; Tarefder, R.A. Fatigue Behavior of Asphalt Containing Reclaimed Asphalt Pavements Effects of recycled asphalt pavements on the fatigue life of asphalt under different strain levels and loading frequencies. *Int. J. Fatigue* **2015**, *78*, 72–80. [CrossRef]
113. Wu, Z.; Zhang, C.; Xiao, P.; Li, B.; Kang, A. Performance Characterization of Hot Mix Asphalt with High RAP Content and Basalt Fiber. *Materials* **2020**, *13*, 3145. [CrossRef]
114. Solanki, P.; Zaman, M.; Adje, D.; Hossain, Z. Effect of Recycled Asphalt Pavement on Thermal Cracking Resistance of Hot-Mix Asphalt. *Int. J. Geomech* **2015**, *15*, A4014001. [CrossRef]
115. Roberts, F.L.; Kandhal, P.S.; Brown, E.R.; Lee, D.Y.; Kennedy, T.W. *Hot Mix Asphalt Materials, Mixture Design and Construction*, 2nd ed.; NAPA Research and Education Foundation: Lanham, MD, USA, 1996.
116. Tabatabaee, H.A.; Kurth, T.L. Analytical investigation of the impact of a novel bio-based recycling agent on the colloidal stability of aged bitumen. *Road Mater. Pavement Des.* **2017**, *18*, 131–140. [CrossRef]
117. De Bock, L.; Vansteenkiste, S.; Vanelstraete, A. *Categorisation and Analysis of Rejuvenators for Asphalt Recycling*; Report No. 21; Belgian Road Research Centre: Brussels, Belgium, 2020.
118. Lee, C.; Terrel, R.; Mahoney, J. Test for Efficiency of Mixing of Recycled Asphalt Paving Mixtures. In *Transportation Research Record 911*; TRB: Washington, DC, USA, 1983.
119. Carpenter, S.; Wolosick, J. Modifier Influence in the Characterization of Hot-Mix Recycled Material. In *Transportation Research Record 777*; TRB: Washington, DC, USA, 1980.
120. Noureldin, S.; Wood, L. Rejuvenator Diffusion in Binder Film for Hot-Mix Recycled Asphalt Pavement. In *Transportation Research Record 1115*; TRB: Washington, DC, USA, 1987.
121. Huang, B.; Li, G.; Vukosavljevic, D.; Shu, X.; Egan, B. Laboratory Investigation of Mixing Hot-Mix Asphalt with Reclaimed Asphalt Pavement. In *Transportation Research Record 1929*; TRB: Washington, DC, USA, 2005.
122. Karlsson, R.; Isacsson, U. Investigations on Bitumen Rejuvenator Diffusion and Structural Stability. *Asph. Paving Technol. Assoc. Asph. Paving Technol. Tech. Sess.* **2003**, *72*, 463–501.
123. Oliver, J. *Diffusion of Oils in Asphalts*; Report No. 9; Proceedings of Australian Road Research Board: Vermont South, Victoria, Australia, 1975.
124. Zaumanis, M.; Mallick, R.; Frank, R. Evaluation of different recycling agents for restoring aged asphalt binder and performance of 100% recycled asphalt. *Mater. Struct.* **2014**, *48*, 2478–2488. [CrossRef]
125. Dony, A.; Colin, J.; Bruneau, D.; Drouadaine, I.; Navarro, J. Reclaimed asphalt concretes with high recycling rates: Changes in reclaimed binder properties according to rejuvenating agent. *Constr. Build. Mater.* **2013**, *41*, 175–181. [CrossRef]
126. Behnood, A. Application of rejuvenators to improve the rheological and mechanical properties of asphalt binders and mixtures: A review. *J. Clean. Prod.* **2019**, *231*, 171–182. [CrossRef]
127. Bocci, E.; Grilli, A.; Bocci, M.; Gomes, V. Recycling of high percentages of reclaimed asphalt using a bio-rejuvenator—A case study. In Proceedings of the 6th Eurasphalt & Eurobitume Congress, Prague, Czech Republic, 1–3 June 2016.
128. Bocci, E.; Cardone, F.; Grilli, A. Mix design and volumetric analysis of hot recycled bituminous mixtures using a bio-additive. In Proceedings of the AIIT International Congress on Transport Infrastructure and Systems, Rome, Italy, 10–12 April 2017; pp. 267–274.
129. Grilli, A.; Bocci, E.; Bocci, M. Hot recycling of reclaimed asphalt using a bio-based additive. In Proceedings of the 8th International RILEM-SIB Symposium, Ancona, Italy, 7–9 October 2015.
130. Król, J.B.; Kowalski, K.J.; Niczke, L.; Radziszewski, P. Effect of bitumen fluxing using a bio-origin additive. *Constr. Build. Mater.* **2016**, *114*, 194–203. [CrossRef]

131. Somé, C.; Pavoine, A.; Chailleux, E.; Andrieux, L.; DeMArco, L.; Philippe Da, S.; Stephan, B. Rheological behaviour of vegetable oil-modified asphaltite binders and mixes. In Proceedings of the 6th Eurasphalt & Eurobitume Congress; Prague, Czech Republic, 1–3 June 2016.
132. Zargar, M.; Ahmadinia, E.; Asli, H.; Karim, M. Investigation of the possibility of using waste cooking oil as a rejuvenating agent for aged bitumen. *J. Hazard. Mater.* **2012**, *223*, 254–258. [CrossRef] [PubMed]
133. Gökalp, I.; Emre Uz, V. Utilizing of Waste Vegetable Cooking Oil in bitumen: Zero tolerance aging approach. *Constr. Build. Mater.* **2019**, *227*, 116465. [CrossRef]
134. Joni, H.H.; Al-Rubaee, R.H.A.; Al-zerkani, M.A. Rejuvenation of aged asphalt binder extracted from reclaimed asphalt pavement using waste vegetable and engine oils. *Case Stud. Constr. Mater.* **2019**, *11*, e00279. [CrossRef]
135. Li, H.; Dong, B.; Wang, W.; Zhao, G.; Guo, P.; Ma, Q. Effect of Waste Engine Oil and Waste Cooking Oil on Performance Improvement of Aged Asphalt. *Appl. Sci.* **2019**, *9*, 1767. [CrossRef]
136. Al Mamun, A.; Wahhab, A.H.I.; Dalhat, M.A. Comparative Evaluation of Waste Cooking Oil and Waste Engine Oil Rejuvenated Asphalt Concrete Mixtures. *Arab. J. Sci. Eng.* **2020**, *45*, 7987–7997. [CrossRef]
137. Cavalli, M.C.; Zaumanis, M.; Mazza, E.; Partl, M.M.; Poulikakos, L.D. Effect of ageing on the mechanical and chemical properties of binder from RAP treated with bio-based rejuvenators. *Compos. Part B Eng.* **2018**, *141*, 174–181. [CrossRef]
138. Zhang, R.; You, Z.; Wang, H.; Ye, M.; Yap, Y.K.; Si, C. The impact of bio-oil as rejuvenator for aged asphalt binder. *Constr. Build. Mater.* **2019**, *196*, 134–143. [CrossRef]
139. Zeng, M.; Li, J.; Zhu, W.; Xia, Y. Laboratory evaluation on residue in castor oil production as rejuvenator for aged paving asphalt binder. *Constr. Build. Mater.* **2018**, *189*, 568–575. [CrossRef]
140. Elkashef, M.; Williams, R.C.; Cochran, E.W. Thermal and cold flow properties of bio-derived rejuvenators and their impact on the properties of rejuvenated asphalt binders. *Thermochim. Acta* **2019**, *671*, 48–53. [CrossRef]
141. Nayak, P.; Sahoo, U.C. A rheological study on aged binder rejuvenated with Pongamia oil and Composite castor oil. *Int. J. Pavement Eng.* **2017**, *18*, 595–607. [CrossRef]
142. Kezhen, Y.; Yang, P.; Lingyun, Y. Use of tung oil as a rejuvenating agent in aged asphalt: Laboratory evaluations. *Construct. Build. Mater.* **2020**, *239*, 117783.
143. Zhang, J.; Sun, H.; Jiang, H.; Xu, X.; Liang, M.; Hou, Y.; Zhanyong, Y. Experimental assessment of reclaimed bitumen and RAP asphalt mixtures incorporating a developed rejuvenator. *Constr. Build. Mater.* **2019**, *215*, 660–669. [CrossRef]
144. Rzek, L.; Ravnikar Turk, M.; Marjan, T. Increasing the rate of reclaimed asphalt in asphalt mixture by using alternative rejuvenator produced by tire pyrolysis. *Constr. Build. Mater.* **2020**, *232*, 117177. [CrossRef]
145. Zaumanis, M.; Mallick, R.B.; Poulikakos, L.; Frank, R. Influence of six rejuvenators on the performance properties of reclaimed asphalt pavement (RAP) binder and 100% recycled asphalt mixtures. *Constr. Build. Mater.* **2014**, *71*, 538–550. [CrossRef]
146. Zaumanis, M.; Mallick, R.B.; Frank, R. Evaluation of rejuvenator's effectiveness with conventional mix testing for 100% reclaimed asphalt pavement mixtures. *Transp. Res. Rec.* **2013**, *2370*, 17–25. [CrossRef]
147. Radenberg, M.; Boetcher, S.; Sedaghat, N. Effect and efficiency of rejuvenators on aged asphalt binder German experiences. In Proceedings of the 6th Eurasphalts and Eurobitume Congress, Prague, Czech Republic, 1–3 June 2016.
148. Bearsley, S.R.; Haverkamp, R.G. Age hardening potential of tall oil pitch modified bitumen. *Road Mater. Pavement Des.* **2007**, *8*, 467–481. [CrossRef]
149. Mokhtari, A.; David Lee, H.; Williams, R.C.; Guymon, C.A.; Sholte, J.P.; Schram, S. A novel approach to evaluate fracture surfaces of aged and rejuvenator-restored asphalt using cryo-SEM and image analysis techniques. *Constr. Build. Mater.* **2017**, *133*, 301–313. [CrossRef]
150. Hugener, M.; Partl, M.N.; Morant, M. Cold asphalt recycling with 100% reclaimed asphalt pavement and vegetable oil-based rejuvenators. *Road Mater. Pavement Des.* **2014**, *15*, 239–258. [CrossRef]
151. Maharaj, R.; Harry, V.; Mohamed, N. The rheological properties of Trinidad asphaltic materials blended with waste cooking oil. *Prog. Rubber Plast. Recycl. Technol.* **2015**, *31*, 265–279. [CrossRef]
152. Asli, H.; Ahmadinia, E.; Zargar, M.; Karim, M.R. Investigation on physical properties of waste cooking oil—Rejuvenated bitumen binder. *Constr. Build. Mater.* **2012**, *37*, 398–405. [CrossRef]
153. Osmari, P.H.; Aragao, F.T.S.; Leite, L.F.M.; Simao, R.A.; da Motta, L.M.G.; Kim, Y.-R. Chemical, microstructural, and rheological characterization of binders to evaluate aging and rejuvenation. *Transp. Res. Rec.* **2017**, *2632*, 14–24. [CrossRef]
154. Azahar, W.N.A.W.; Jaya, R.P.; Hainin, M.R.; Bujang, M.; Ngadi, N. Chemical modification of waste cooking oil to improve the physical and rheological properties of asphalt binder. *Constr. Build. Mater.* **2016**, *126*, 218–226. [CrossRef]
155. Gong, M.; Yang, J.; Zhang, J.; Zhu, H.; Tong, T. Physical-chemical properties of aged asphalt rejuvenated by bio-oil derived from biodiesel residue. *Constr. Build. Mater.* **2016**, *105*, 35–45. [CrossRef]
156. Kowalski, J.K.; Krol, B.J.; Bankowski, W.; Radziszewski, P.; Sarnowski, M. Thermal and fatigue evaluation of asphalt mixtures containing RAP treated with a bio-agent. *Appl. Sci.* **2017**, *7*, 216. [CrossRef]
157. Zhao, K.; Wang, Y.; Chen, L.; Li, F. Diluting or dissolving? The use of relaxation spectrum to assess rejuvenation effects in asphalt recycling. *Constr. Build. Mater.* **2018**, *188*, 143–152. [CrossRef]
158. Ali, A.W.; Mehta, Y.A.; Nolan, A.; Purdy, C.; Bennert, T. Investigation of the impacts of aging and RAP percentages on effectiveness of asphalt binder rejuvenators. *Constr. Build. Mater.* **2016**, *110*, 211–217. [CrossRef]

159. Borghi, A.; Jimenez del Barco Carrion, A.; Lo Presti, D.; Giustozzi, F. Effects of laboratory aging on properties of biorejuvenated asphalt binders. *J. Mater. Civ. Eng.* **2017**, *29*, 1–3. [CrossRef]
160. Yang, X.; Mills-Beale, J.; You, Z. Chemical characterization and oxidative aging of bio-asphalt and its compatibility with petroleum asphalt. *J. Clean. Prod.* **2017**, *142*, 1837–1847. [CrossRef]
161. Pahlavan, F.; Samieadel, A.; Deng, S.; Fini, E. Exploiting Synergistic Effects of Intermolecular Interactions to Synthesize Hybrid Rejuvenators to Revitalize Aged Asphalt. *ACS Sustain. Chem. Eng.* **2019**, *7*, 15514–15525. [CrossRef]
162. Farooq, M.A.; Mir, M.S.; Sharma, A. Laboratory study on use of RAP in WMA pavements using rejuvenator. *Constr. Build. Mater.* **2018**, *168*, 61–72. [CrossRef]
163. Fernandes, S.R.M.; Silva, H.M.R.D.; Oliveira, J.R.M. Recycled stone mastic asphalt mixtures incorporating high rates of waste materials. *Constr. Build. Mater.* **2018**, *187*, 1–13. [CrossRef]
164. Romera, R.; Santamaria, A.; Pena, J.; Munoz, M.; Barral, M.; Garcia, E.; Janez, V. Rheological aspects of the rejuvenation of aged bitumen. *Rheol. Acta* **2006**, *45*, 474–478. [CrossRef]
165. Avsenik, L.; Klinar, D.; Tušar, M.; Perše, L.S. Use of modified slow tire pyrolysis product as a rejuvenator for aged bitumen. *Constr. Build. Mater.* **2016**, *120*, 605–616. [CrossRef]
166. Mogawer, W.S.; Austerman, A.; Roque, R.; Underwood, S.; Mohammad, L.; Zou, J. Ageing and rejuvenators: Evaluating their impact on high RAP mixtures fatigue cracking characteristics using advanced mechanistic models and testing methods. *Road Mater. Pavement Des.* **2015**, *16* (Suppl. 2), 1–28. [CrossRef]
167. Garcia, A.; Schlangen, E.; van de Ven, M.; Sierra-Beltran, G. Preparation of capsules containing rejuvenators for their use in asphalt concrete. *J. Hazard. Mater.* **2010**, *184*, 603–611. [CrossRef]
168. Ameri, M.; Mansourkhaki, A.; Daryaee, D. Evaluation of fatigue behavior of high reclaimed asphalt binder mixes modified with rejuvenator and softer bitumen. *Constr. Build. Mater.* **2018**, *191*, 702–712. [CrossRef]
169. Chen, J.-S.; Chen, S.-F.; Liao, M.-C.; Huang, S.W. Laboratory evaluation of asphalt blends of recycling agents mixed with aged binders. *J. Mater. Civ. Eng.* **2015**, *27*, 04014143. [CrossRef]
170. Pahlavan, F.; Hung, A.M.; Zadshir, M.; Hosseinnezhad, S.; Fini, E.H. Alteration of π-Electron Distribution to Induce Deagglomeration in Oxidized Polar Aromatics and Asphaltenes in an Aged Asphalt Binder. *ACS Sustain. Chem. Eng.* **2018**, *6*, 6554–6569. [CrossRef]
171. Pahlavan, F.; Mousavi, M.; Hung, A.M.; Fini, E.H. Characterization of oxidized asphaltenes and the restorative effect of a bio-modifier. *Fuel* **2018**, *212*, 593–604. [CrossRef]
172. Zadshir, M.; Oldham, D.J.; Hosseinnezhad, S.; Fini, E.H. Investigating bio-rejuvenation mechanisms in asphalt binder via laboratory experiments and molecular dynamics simulation. *Constr. Build. Mater.* **2018**, *190*, 392–402. [CrossRef]
173. Xu, Q.; Zhang, Z.; Zhang, S.; Wang, F.; Yan, Y. Molecular structure models of asphaltene in crude and upgraded bio-oil. *Chem. Eng. Technol.* **2014**, *37*, 5. [CrossRef]
174. Cao, X.; Wang, H.; Cao, X.; Sun, W.; Zhu, H.; Tang, B. Investigation of rheological and chemical properties asphalt binder rejuvenated with waste vegetable oil. *Constr. Build. Mater.* **2018**, *180*, 455–463. [CrossRef]
175. Chen, A.; Liu, G.; Zhao, Y.; Li, J.; Pan, Y.; Zhou, J. Research on the aging and rejuvenation mechanisms of asphalt using atomic force microscopy. *Constr. Build. Mater.* **2018**, *167*, 177–184. [CrossRef]
176. Menapace, I.; Garcia Cucalon, L.; Kaseer, F.; Arámbula-Mercado, E.; Epps Martin, A.; Masad, E.; King, G. Effect of recycling agents in recycled asphalt binders observed with microstructural and rheological tests. *Constr. Build. Mater.* **2018**, *158*, 61–74. [CrossRef]
177. Ganter, D.; Mielke, T.; Maier, M.; Lupascu, D.C. Bitumen rheology and the impact of rejuvenators. *Constr. Build. Mater.* **2019**, *222*, 414–423. [CrossRef]
178. Mazzoni, G.; Bocci, E.; Canestrari, F. Influence of rejuvenators on bitumen ageing in hot recycled asphalt mixtures. *J. Traffic Transp. Eng. Engl. Ed.* **2018**, *5*, 157–168. [CrossRef]
179. Bocci, E.; Mazzoni, G.; Canestrari, F. Ageing of rejuvenated bitumen in hot recycled bituminous mixtures: Influence of bitumen origin and additive type. *Road Mater. Pavement Des.* **2019**, *20*, S127–S148. [CrossRef]
180. Sá-da-Costa, M.; António Correia, D.; Farcas, F. Life cycle of bitumen: Ageing-regeneration-ageing. In Proceedings of the 6th Eurasphalts and Eurobitume Congress, Prague, Czech Republic, 1–3 June 2016.
181. Themeli, A.; Chailleux, E.; Farcas, F.; Chazallon, C.; Migault, B. Molecular weight distribution of asphaltic paving binders from phase- angle measurements. *Road Mater. Pavement Des.* **2015**, *16*, 228–244. [CrossRef]
182. Camargo, I.; Ben Dhia, T.; Loulizi, A.; Hofko, B.; Mirwald, J. Anti-aging additives: Proposed evaluation process based on literature review. *Road Mater. Pavement Des.* **2021**, *22*, S134–S153. [CrossRef]
183. Lo Presti, D.; Vasconcelos, K.; Orešković, M.; Pires, G.M.; Bressi, S. On the degree of binder activity of reclaimed asphalt and degree of blending with recycling agents. *Road Mater. Pavement Des.* **2020**, *21*, 2071–2090. [CrossRef]
184. Ashtiani, M.Z.; Mogawer, W.S.; Austerman, A.J. A Mechanical Approach to Quantify Blending of Aged Binder from Recycled Materials in New Hot Mix Asphalt Mixtures. *J. Transp. Res. Board* **2018**, *2672*, 107–118. [CrossRef]
185. Coffey, S.; Dubois, E.; Mehta, Y.; Nolan, A.; Purdy, C. Determining the impact of degree of blending and quality of reclaimed asphalt pavement on predicted pavement performance using pavement ME design. *Constr. Build. Mater.* **2013**, *48*, 473–478. [CrossRef]

186. Ding, Y.; Huang, B.; Shu, X. Blending efficiency evaluation of plant asphalt mixtures using fluorescence microscopy. *Constr. Build. Mater.* **2018**, *161*, 461–467. [CrossRef]
187. Gundla, A.; Underwood, S. Evaluation of in situ RAP binder interaction in asphalt mastics using micromechanical models. *Int. J. Pavement Eng.* **2015**, *18*, 798–810. [CrossRef]
188. Kaseer, F.; Arámbula-Mercado, E.; Martin, A.E. A Method to Quantify Reclaimed Asphalt Pavement Binder Availability (Effective RAP Binder) in Recycled Asphalt Mixes. *Transp. Res. Rec. J. Transp. Res. Board* **2019**, *2673*, 1366. [CrossRef]
189. Shirodkar, P.; Mehta, Y.; Nolan, A.; Dubois, E.; Reger, D.; McCarthy, L. Development of blending chart for different degrees of blending of RAP binder and virgin binder. *Resour. Conserv. Recycl.* **2013**, *73*, 156–161. [CrossRef]
190. Shirodkar, P.; Mehta, Y.; Nolan, A.; Sonpal, K.; Norton, A.; Tomlinson, C.; Sauber, R. A study to determine the degree of partial blending of reclaimed asphalt pavement (RAP) binder for high RAP hot mix asphalt. *Constr. Build. Mater.* **2011**, *25*, 150–155. [CrossRef]
191. Stimilli, A.; Virgili, A.; Canestrari, F. New method to estimate the "re-activated" binder amount in recycled hot-mix asphalt. *Road Mater. Pavement Des.* **2015**, *16* (Suppl. 1), 442–459. [CrossRef]
192. Vassaux, S.; Gaudefroy, V.; Boulangé, L.; Pevere, A.; Michelet, A.; Barragan-Montero, V.; Mouillet, V. Assessment of the binder blending in bituminous mixtures based on the development of an innovative sustainable infrared imaging methodology. *J. Clean. Prod.* **2019**, *215*, 821–828. [CrossRef]
193. Vassaux, S.; Gaudefroy, V.; Boulangé, L.; Soro, L.J.; Pévère, A.; Michelet, A.; Mouillet, V. Study of remobilization phenomena at reclaimed asphalt binder/virgin binder interphases for recycled asphalt mixtures using novel microscopic methodologies. *Constr. Build. Mater.* **2018**, *165*, 846–858. [CrossRef]

MDPI

Review

A Review of the Utilization of Coal Bottom Ash (CBA) in the Construction Industry

Syakirah Afiza Mohammed [1,2], Suhana Koting [1,*], Herda Yati Binti Katman [3,*], Ali Mohammed Babalghaith [1], Muhamad Fazly Abdul Patah [4], Mohd Rasdan Ibrahim [1] and Mohamed Rehan Karim [5]

1 Center for Transportation Research, Department of Civil Engineering, Faculty of Engineering, Universiti Malaya, Kuala Lumpur 50603, Malaysia; syakirahafiza@unimap.edu.my (S.A.M.); bablgeath@hotmail.com (A.M.B.); rasdan@um.edu.my (M.R.I.)
2 Faculty of Civil Engineering Technology, Universiti Malaysia Perlis, Perlis 02600, Malaysia
3 Department of Civil Engineering, Putrajaya Campus, Universiti Tenaga Nasional (UNITEN), Jalan IKRAM-UNITEN, Kajang 43000, Malaysia
4 Department of Chemical Engineering, Faculty of Engineering, Universiti Malaya, Kuala Lumpur 50603, Malaysia; fazly.abdulpatah@um.edu.my
5 Transportation Science Society of Malaysia, Department of Civil Engineering, Faculty of Engineering, Universiti Malaya, Kuala Lumpur 50603, Malaysia; mrehan57@gmail.com
* Correspondence: suhana_koting@um.edu.my (S.K.); Herda@uniten.edu.my (H.Y.B.K.); Tel.: +60-3-7967-7648 (S.K.); +60-3-8921-2020 (ext. 2252) (H.Y.B.K.)

Abstract: One effective method to minimize the increasing cost in the construction industry is by using coal bottom ash waste as a substitute material. The high volume of coal bottom ash waste generated each year and the improper disposal methods have raised a grave pollution concern because of the harmful impact of the waste on the environment and human health. Recycling coal bottom ash is an effective way to reduce the problems associated with its disposal. This paper reviews the current physical and chemical and utilization of coal bottom ash as a substitute material in the construction industry. The main objective of this review is to highlight the potential of recycling bottom ash in the field of civil construction. This review encourages and promotes effective recycling of coal bottom ash and identifies the vast range of coal bottom ash applications in the construction industry.

Keywords: coal bottom ash; waste material; recycle; construction industry; civil engineering

Citation: Mohammed, S.A.; Koting, S.; Katman, H.Y.B.; Babalghaith, A.M.; Abdul Patah, M.F.; Ibrahim, M.R.; Karim, M.R. A Review of the Utilization of Coal Bottom Ash (CBA) in the Construction Industry. *Sustainability* **2021**, *13*, 8031. https://doi.org/10.3390/su13148031

Academic Editor: Edoardo Bocci

Received: 5 June 2021
Accepted: 12 July 2021
Published: 19 July 2021

Publisher's Note: MDPI stays neutral with regard to jurisdictional claims in published maps and institutional affiliations.

1. Introduction

The increasing price of oil and natural gas has made coal-fired power generation more economical, especially in countries with vast coal resources such as India, the United States of America (US), and China [1]. China consumes 50.2% of the coal in 2012, followed by US and India (11.7%), Japan (8.0%), Russia (3.3%), South Africa (2.5%), South Korea (2.4%), Germany (2.2%), Poland (2.1%), and Indonesia (1.4%) [2]. India, China, and Australia are projected to contribute 64% of the world coal production in 2040, a growth of about 4% compared to the 2012 coal production [3].

The increasing trend in coal consumption will continue mainly due to the high demand for electricity. Coal is rapidly gaining favor as an energy source for generating electricity, after gas [4]. The 0.9% increase in world coal consumption in 2019 was driven by Asia (1.8%). The utilization of coal as a global source of electricity generation is expected to increase to 47% by 2030 [2,5].

The high demand for coal production has resulted in the generation of a higher amount of industrial waste. Fly ash makes up 70–80% of the total coal ash wastes, and the remaining 10–20% is bottom ash [2,3,6]. Of the millions of tons of coal ash waste generated annually, 100 million metric tons (Mt) is bottom ash, and the remainder is fly ash [7]. The World of Coal Ash (WOCA) estimated that coal thermal power plants generate 780 million metric tons of coal bottom ash (CBA), of which 66% is by Asian countries, followed by Europe and the United States [8]. China produces the highest amount of coal ash of 395 million metric tons (Mt), followed by the US (118 Mt), India (105 Mt), Europe (52.6 Mt), and Africa (31.1 Mt). The Middle East and other countries contributed a small amount to the global coal ash generation [9]. Of the 105 million metric tons of coal produced in India [9], about 35 million metric tons is coal bottom ash produced by the power plants that generate electricity [10].

The wastes produced in electricity generation are boiler slag, fly ash, clinker, and bottom ash [11,12]. The physical properties and the chemical composition of bottom ash and fly ash differ because fly ash is lighter than the bottom ash collected in a hopper after falling through the bottom furnace. The bottom ash could be wet or dry bottom ash, depending on the type of boiler.

Bottom ash has a porous texture and angular particles; the size of the bottom ash generally ranges between sand and gravel particles and a small amount of slit–clay particles [11,12]. A large proportion of the bottom ash is fine particles that comprise 50–90% of bottom ash. The specific gravity of bottom ash is dependent on its chemical properties and ranges between 1.39 and 2.41 [13–15]. Coal bottom ash falls into the A-1-a class and well-graded sand groups of the AASHTO and USCS classification systems [16].

The disposal of bottom ash landfills has raised a grave environmental concern [16,17]. The high composition of heavy metal in bottom ash, relative to fly ash, increases the risk of groundwater pollution [18,19]. One way to deal with the increasing amount of CBA generated and the scarcity of land is by recycling and reusing CBA [20].

The large amount of coal bottom ash produced by the thermal power plants is one of the primary industrial wastes. Therefore, using CBA in the construction industry will save time and reduce landfill use, cost, and energy. The two key benefits of recycling CBA in civil construction are a considerable reduction in greenhouse gas emissions and solid waste generation by coal-fired thermal power plants [15]. Moreover, this approach will protect the environment from the harmful impact of this waste material. This paper presents a review of the utilization of coal bottom ash in civil construction.

2. Properties of Coal Bottom Ash

The specific properties of coal bottom ash are dependent on factors such as the coal source and type of coal. There are four types of coal: anthracite, bituminous, sub-bituminous, and lignite [21]. The type of coal is dependent on the types and amounts of carbon, the amount of heat energy the coal can produce, the level of carbon moisture, and other chemical elements [22]. Anthracite has the highest carbon content, followed by bituminous, sub-bituminous, and lignite. Generally, the types of coal used in energy generation are bituminous, sub-bituminous, and lignite. The geological formation of the coal determines its chemical composition; the CBA from the different types of coal have varying silica oxide (SiO_2), alumina oxide (Al_2O_3), and ferric oxide (Fe_2O_3) contents and characteristics that influence the research finding [22,23].

Bottom ash is washed before use to remove unnecessary materials such as pyrite, which could degrade the bottom ash in the presence of water. The bottom ash should also be free of dust to ensure the correct grain size distribution. Finally, the bottom ash is dried to remove humidity and excessive moisture, which affect the mixture's reliability [24].

2.1. Physical Properties

Table 1 shows the physical properties of CBA from various sources. The specific gravity of CBA ranges between 1.39 and 2.41 and its water absorption is between 6.8 and 32%. CBA has a Los Angeles abrasion of 55% and a moisture content of 0.43%. Its fineness modulus ranges between 1.5 and 3.44, and its specific surface area ranges from 3835.7 to 10,500 (cm^2/g).

Table 1. Physical properties of CBA.

References	Specific Gravity (No Unit)	Water Absorption (%)	Los Angeles Abrasion, (%)	Moisture Content (%)	Fineness Modulus (No Unit)	Surface Area (cm^2/g)
[25]	2.21	-	-	-	2.79	-
[26]	2.41	32	-	-	-	3835.75
[27]	1.8	-	-	-	-	10,500
[28]	2.08	6.8	-	-	1.5	-
[14]	2.22	20.15	-	-	2.71	-
[3]	1.88	11.61	-	-	3.44	-
[17,29]	1.39	31.58	-	-	1.37	
[30]	2.00	-	-	-	-	-
[31]	2.21	11.17	-	-	-	-
[32]	1.87	5.4	-	-	2.36	-
[33]	1.39	12.10	-	-	-	-
[24]	-	-	55	-	-	-
[34]	2.10	6.18	-	0.43	2.10	-
[25]	2.21	-	-	-	-	-

2.2. Chemical Properties

Table 2 presents the chemical composition of CBA from various power plants, where the percentages of the chemical composition are expressed by mass. Silicon dioxide (SiO_2), aluminum oxide (Al_2O_3), and calcium oxide (CaO) are the primary mineral compounds in coal rock [35]. The chemical composition of CBA comprises the major and minor components. The total percentages of the chemical composition for the CBA from different sources varies and is less 100%. The minor components cannot be traced due to their small amounts. Table 2 shows that coal bottom ash contains high amount of SiO_2, Al_2O_3, and Fe_2O_3 that improve pozzolanic effects, mixture interlock, and properties such as the strength [28].

Table 2. Chemical composition of CBA.

References	Chemical Composition (%)									
	SiO_2	Al_2O_3	Fe_2O_3	CaO	MgO	Na_2O	K_2O	TiO_2	P_2O_5	SO_3
[36]	62.33	25.52	4.16	1.00	0.94	0.08	3.25	0.84	0.12	-
[6]	65.02	19.18	6.86	1.76	2.00	0.85	-	0.93	0.04	-
[26]	52.5	17.65	8.30	4.72	0.58	-	-	2.17	-	0.84
[27]	50.49	27.56	10.93	4.19	1.24	0.57	0.82	2.23	0.24	0.10
[37]	47.1	23.1	5.7	7.8	1.5	0.7	5.3	1.2	-	1.5
[38]	59.82	27.76	3.77	1.86	0.70	1.61	0.33	-	-	1.39
[39]	52.2	27.5	6.0	5.9	1.7	-	0.6	1.53	0.74	0.13
[40]	58.7	20.1	6.2	9.5	1.6	0.1	1.0	-	1.0	0.4
[14]	62.32	27.21	3.57	0.50	0.95	0.70	2.58	2.15	-	-
[3]	45.30	18.10	19.84	8.70	0.97	-	2.48	3.27	0.351	0.352
[17]	56.44	29.24	8.44	0.75	0.4	0.09	1.24	-	-	-
[30]	64.45	15.89	7.77	3.92	2.45	0.89	1.6	-	<0.01	<0.01
[41]	68.9	18.67	6.5	1.61	0.53	0.24	1.52	1.33	-	-
[31]	52.1	18.34	11.99	6.61	4.85	2.43	1.57	0.87	-	-
[29]	47.53	20.69	5.99	4.17	0.82	0.33	0.76	-	-	1.00
[42]	57.76	21.58	8.56	1.58	1.19	0.14	1.08	-	-	0.02
[43]	54.8	28.5	8.49	4.2	0.35	0.08	0.45	2.71	0.28	-

2.3. Comparison between CBA and Different Waste Used as Aggregate Replacement

Researchers have investigated steel slag, coconut waste, recycled asphalt, recycled concrete, mining waste, glass, crumb rubber, palm oil shell, and palm oil clinker as natural aggregate replacement. Table 3 summarizes the physical properties of the wastes used as aggregate replacement in asphalt pavements and concrete production.

Except for steel slag, the specific gravities of the wastes listed in Table 3 are similar to that of CBA. Steel slag has a higher specific gravity of between 3.01 and 3.67 [44–47], although coconut waste and palm oil shells have high absorption values of between 13.8 and 25% because of their porous structure [48–52]. The Los Angeles abrasion for the wastes ranges between 11.29 and 25.3% [44,45,53–60], which is lower than the 55% for CBA [24]. The lower Los Angeles abrasion values indicate that the materials are tougher and more resistant to abrasion than CBA. The moisture content of the waste is between 0.11 and 9.1% [46,61–63]. The fineness modulus values of 6.53–6.78% for the coconut waste, recycled concrete, and palm oil shell are similar to CBA and are comparable to the natural aggregates, making them suitable for aggregate replacement.

Table 3. The physical properties of the wastes used as aggregates in asphalt pavement and concrete production.

Waste	Physical Property Parameters					Used As		Reference
	Specific Gravity (No Unit)	Water Absorption (%)	Los Angeles Abrasion (%)	Moisture Content (%)	Fineness Modulus (%)	Fine	Coarse	
Steel slag	3.41	1.49	11.29				√	[44]
	3.01	-	14.2				√	[45]
	3.42 * 3.58 **	3.31 * 4.23 **	-	1.56 * 2.8 **	-	√	√	[46]
	3.67	1.4	-	-	-		√	[47]
Coconut waste	1.15	21	-		6.78		√	[48,49]
	1.16	13.8	-	-	-		√	[50]
Recycled asphalt	2.68	0.20	22.2			√	√	[53]
	2.55	0.23	20.25			√	√	[54]
Recycled concrete	2.41 * 2.42 **	4.80 * 7.40 **	18.7			√	√	[55]
	2.18 * 2.42 **	2.69 * 4.28 **	24			√	√	[56–58]
	2.35	8.01	-	9.1	-	√	√	[61]
	2.42–2.44 * 2.415 **	6.5–6.8 * 9 **	-	-	-	√	√	[64]
	2.53	3.04	-	-	-		√	[65]
	2.44	5.65	-	-	6.92		√	[28]
Mining waste	2.34	0.86	20.5				√	[59]
	2.87	0.23	25.3			√	√	[60]
Glass	2.3	20–25	-	-	-		√	[66]
	2.45	0.36	-	-	-	√		[65]
Crumb rubber	1.15	-	-			√		[67]
	1.25	-	-			√		[68]
Palm oil shell	1.37	12.47	-	-	6.53		√	[51]
	1.3	25	-	-	-		√	[52]
Palm oil clinker	2.08	-	-			√		[69,70]
	1.51	5.5	-	0.31	-		√	[71]
	1.78	5.7	-	0.38	-		√	[62]
	1.18 * 2.15 **	4.35 * 5.75 **	-	0.28 * 0.11 **		√	√	[63]

* Coarse aggregate, ** fine aggregate.

2.4. Sustainability

Researchers are looking for alternatives to the conventional CBA disposal method that has given rise to environmental problems. Recycling coal bottom ash is the best solution for the high cost of disposing of the coal bottom ash waste [72], the decreasing disposal area, and the harmful environmental impact [73]. CBA is physically similar to natural aggregates and resembles Portland cement (PC) when pulverized into finer particles [74]. Previous studies have shown that its pozzolanic reactivity makes CBA a promising alternative for cement replacement in concrete and can reduce up to 90% of the concrete carbon footprints [75]. Replacing PC with CBA in concrete production can reduce CO_2 emissions [74,76]. Previous studies have also proven the beneficial impacts of using CBA as PC replacement on the environment and the problems associated with conventional concrete [18,77,78]. CBA is a green material that could reduce harmful environmental impact and promote sustainability in concrete production [74].

Asphalt construction requires a large amount of natural aggregates, namely 100% aggregates for the base and subbase courses, 95% for bituminous, and 87% for concrete pavements. The natural aggregates used to construct one kilometer of a surface course using a bituminous mixture could exceed 15,000 tons [7]. In recent years, natural aggregate replacement with CBA has reduced construction costs and minimized the need to harvest aggregates from natural resources.

3. Applications of Coal Bottom Ash

3.1. Pavement Construction

Prior studies on the utilization of coal ash waste in the construction industry focused more on fly ash than bottom ash. However, recent studies reported that bottom ash has some desired engineering properties that make it a feasible construction material. The minimum strength, stability, durability, and other specifications of the products incorporated with CBA must be complied with [79]. CBA has been used as an aggregate replacement, cement replacement, additive in bitumen, and filler in asphalt pavement. Table 4 summarizes the effect of CBA in pavement construction.

Yoo et al. [80] investigated the performance of HMA mixture incorporated with bottom ash as a partial replacement for fine aggregates at the ratio of 10, 20, and 30%. The incorporation of bottom ash using the Marshall Mix Design increased the optimum asphalt content by 10%. Increasing the bottom ash content from 10 to 30% did not affect the optimum asphalt content [80]. The presence of bottom ash in asphalt mixtures did not affect the moisture susceptibility of the asphalt mixtures relative to the control mixture. The asphalt mixtures containing bottom ash exhibited higher resistance towards fatigue cracking when subjected to repeated indirect tensile stiffness modulus (IDT) testing under dynamic loading. The research performed the Synthetic Precipitation Leaching Procedure (SPLP) test on the raw bottom ash to determine its toxicity and found that the toxicity concentration in bottom ash is within the permissible range.

Colonna et al. [24] investigated the impact of utilizing bottom ash as a partial replacement for fine aggregates in asphalt mixtures. The percentage bottom ash as a partial replacement is 15, 20, and 25%, and the optimum binder content (OBC) is 4.5% of the 60/70 asphalt binder. However, the OBC percentage for 20% of the CBA replacement is 5% higher. The researchers observed enhanced stability and reduced wearing resistance with higher CBA contents. However, the mixtures have good wearing resistance with a Cantabro index of less than 30%. The researchers concluded that the optimum CBA content is 15%. The leaching test showed that the hazardous substances leached by all samples are below the limit of detection and the trace substances are below the allowable limit.

Ksaibati [81] examined the feasibility of using CBA as a complete aggregate replacement. The coarse CBA and fine CBA were used in HMA and tested in field and laboratory assessments. The three tested samples were prepared using bottom ash from three different resources. Results showed that the optimum asphalt content was higher in the mixture containing bottom ash with no difference in the performance of the bottom ash mixture and control mixture after being in service. The laboratory analysis showed that all tested HMA mixtures have different low- and high-temperature cracking characteristics, indicating that the varying properties of the bottom ash from the various power plants may affect the strength of the asphalt mixture.

Hesami et al. [82] assessed the potential of using bottom ash in rigid pavements. The coal waste ash, coal waste powder, and limestone powder were used as cement replacement in the roller-compacted concrete pavement (RCCP) in varying percentages of 5 to 20%. The researchers determined the elasticity modulus, splitting tensile strength, flexural strength, and compressive strength of the pavements between day 7 to day 90. The water/cement ratio increased when using the coal waste ash and coal waste powder. The mixtures containing 5% CWS and coal waste ash had similar strength to the control mixture. The strength of RCCP mixtures decreased with higher coal waste ash contents. In summary, coal waste ash enhanced the mechanical properties when combined with limestone.

Ameli et al. [83] examined the performance of asphalt mixture incorporated with varying percentages of coal waste ash of 0, 25, 50, 75, and 100%. The coal ash waste was used as a replacement for the conventional filler. The research measured the rutting resistance and fatigue resistance of the mastics, stability, resilient modulus, dynamic creep, and moisture susceptibility of the asphalt mixtures. The results indicate that the addition of coal waste ash improved the fatigue behavior of the mastics, and the fatigue behavior was further enhanced when the mastics were modified with SBS. The replacement with coal waste ash reduced the Marshall stability, resilient modulus, rutting properties, and tensile strength of the asphalt mixtures, but improved the moisture resistance.

Xu, Chen [84] investigated the effect of coal waste ash that has been retreated from coal waste by reheating, on the asphalt mastic and asphalt mixtures. The coal waste was used as a replacement in varying percentages of 20, 40, 60, and 80%. Compared to fine limestone powder, coal waste ash had a lower density, higher alkalinity, smaller particle size, and larger interior air voids. The bitumen incorporated with coal waste ash had a lower penetration value, higher softening point, and better temperature stability. The incorporation of coal waste ash reduced the Marshall stability and rutting resistance of the asphalt mixture. However, the moisture susceptibility of the asphalt mixture improved significantly with the addition of CBA.

Other studies assessed the potential of using CBA as a filler in bituminous mixtures [85,86]. The high porosity and high specific surface properties of hydrated lime and/or bottom ash increased the value for the optimum binder content. The higher absorption value is due to the high porosity of the ashes, indicating that the high specific density of the mixture relative to the control mixture is due to the low specific densities of the bottom ash and lime. A higher value indicates a longer lifespan and better performance of the pavement. The pavement containing 30% hydrated lime and 70% bottom ash had the highest density, which is the recommended density for the road surface and intermediate layers with small infrastructures and light traffic [86]. However, the results of this research are contrary to the findings of an earlier study [85], which showed that using bottom ash as a filler substitute resulted in a higher rutting potential and a lower dynamic modulus than the control mixture.

Modarres and Ayar [87] examined the effect of using coal waste ash and coal waste powder as additives in the cold recycled mixture, which contains 100% reclaimed asphalt pavement (RAP) materials, using the emulsified cold recycling technology, where the additives dimensions are below 0.075 mm. The addition of coal waste ash and coal waste powder in varying percentages of 3, 5, and 7% enhanced the mechanical properties of the pavement. The higher pozzolanic content in the coal waste powder enhanced the resilient modulus, tensile strength, and Marshall stability. Relative to coal waste powder, coal waste ash had a better impact on moisture sensitivity and enhanced moisture damage resistance.

Table 4. Summary of the utilization of CBA in pavements.

References	Function	Effect on Pavement Performance
[83]	Filler in SMA mixture	• Reduced Marshall stability, resilient modulus, tensile strength, and fatigue properties of the pavements. • Improved moisture resistance.
[84]	Filler replacement	• Reduced pavement stability and rutting resistance. • Improved moisture resistance.
[80]	Fine aggregate in HMA	• Higher OBC with 10% CBA replacement. However, there was no significant difference with a higher percentage of CBA replacement. • No significant change in moisture susceptibility. • Improved fatigue resistance.
[82]	Cement replacement in roller-compacted concrete pavement	• Increased the water/cement ratio. • The RCCP mixtures containing higher amounts of coal waste ash had a lower strength. • Coal waste ash produced better mechanical properties when used in combination with limestone.
[88]	Filler replacement in asphalt mixture	• The high specific surface area of the bottom ash particles resulted in a higher percentage of asphalt binder and better mastic quality. • Lower CO_2 emissions in the processing of bottom ash compared to a commercial filler.
[89]	Filler replacement in HMA mixture	• Improved stability, resilient modulus, moisture resistance, and tensile strength.
[87]	An additive in cold recycled mixture	• Enhanced stability, resilient modulus, and tensile strength.
[24]	Fine aggregate in HMA	• Enhanced stability and Cantabro index.
[81]	Fine and coarse aggregate replacement	• The difference in performance was not significant after a particular period of service. The varying properties of CBA from various coal sources influence the strength of the asphalt mixture.

3.2. *Aggregate Replacement in Concrete Production*

There has been an increase in literature on using bottom ash as an aggregate substitute in concrete production due to its porous texture and low particle densities. The literature reported the promising potential of bottom ash as an aggregate and cement substitute in concrete, particularly to enhance the concrete's strength and microstructural properties. During the past several decades, there has been extensive research on using alternative materials in concrete manufacturing. The benefits of lightweight concrete are reduced weight, good thermal and sound insulation, durability, strength, low expansibility, ease of use in construction, and low cost [90].

Rafieizonooz et al. [3] investigated the performance of concrete incorporated with CBA. Varying percentages of 0, 20, 50, 75, and 100% of coal bottom ash were used as fine aggregate replacement and 20% of coal fly ash (CFA) as cement replacement in concrete. After curing at 91 and 180 days, the compressive strength of CBA and control concrete increased significantly. Despite the enhanced strength after a long curing period, the compressive strength between CBA and control concrete had no significant difference; however, the flexural strength and splitting tensile strength of the 75% CBA was much higher than the control concrete. The concrete with 50, 75, and 100% CBA had a lower drying shrinkage than the control concrete. The late effect is due to the delayed hydration and slow pozzolanic activity of the CFA and CBA.

According to Singh and Siddique [17,29], adding CBA to concrete mix reduced the workability and bleeding of the concrete. The cement in 38 MPa and 34 MPa concrete grade was replaced with varying percentages of 20, 30, 40, 50, 75, and 100% CBA, and superplasticizer was used as an admixture in the 34 MPa concrete grade. The compressive strength and splitting tensile strength of CBA concrete was similar to the control concrete after 90 days of ageing. The modulus of elasticity and abrasion resistance decreased with higher CBA contents. However, the abrasion resistance improved significantly with ageing. Based on the workability and strength properties results, the researcher recommended the optimum use of CBA in concrete as up to 30% for concrete without superplasticizer and up to 50% with superplasticizer.

Zhang and Poon [31] studied the properties of lightweight concrete incorporated with 0, 25, 50, 75, and 100% of bottom ash as a fine aggregate replacement with a 0.39 w/c ratio. The results showed that 100% bottom ash replacement resulted in comparable workability and compressive strength relative to the regular concrete. Bottom ash concrete has a low density, and a replacement with less than 50% bottom ash resulted in a high fc/D ratio, making the lightweight concrete suitable for structural purposes. The durability test showed that the lightweight aggregate concrete containing bottom ash had a high chloride penetration. The heat insulation property test showed that the thermal conductivity decreased with higher bottom ash contents without a significant loss of strength, making the lightweight aggregate concrete suitable for use energy-saving building envelope materials.

Jang et al. [19] investigated the ecofriendly porous concrete fabricated using coal bottom ash. The coal bottom ash was used as a coarse aggregate replacement, and geopolymer was used as the binder. The combination of coal bottom ash and geopolymer produced porous concrete with a higher compressive strength relative to the porous concrete fabricated from recycled aggregate and cement paste. However, the concrete had a lower compressive strength than the regular porous concrete fabricated using gravel and cement paste. Concerning environmental impact, the concentration of heavy metals leached from the porous concrete did not exceed the maximum permissible concentration. The researchers concluded that the porous concrete could effectively immobilize the heavy metals as solidified/stabilized products.

Researchers have also investigated the effect of using coal bottom ash as fine aggregate in self-compacting concrete on the split tensile strength [41]. The fine aggregates were replaced with varying percentages of 0, 10, 20, and 30% coal bottom ash using different water–cement ratios of 0.35, 0.40, and 0.45. The split tensile strength and density of the self-compacting concrete decreased with higher CBA contents. The highest tensile strength of the concrete containing CBA was 3.28 MPa at 10% CBA replacement and 0.35 water–cement ratio. However, this value was lower than the control sample, which had the highest tensile strength of 4.25 MPa.

Kim and Lee [32] investigated the chemical composition and physical properties of bottom ash particles to determine the feasibility of using the bottom ash as fine and coarse aggregates in high-strength concrete with a compressive strength of 60–80 MPa. The fine and coarse bottom ashes were replaced in varying percentages of 25, 50, 75, and 100%. Unlike coarse bottom ash, fine bottom ash had no impact on the fresh concrete flow characteristics. The low slump value of the fresh concrete incorporated with coarse aggregate is due to the complex shape and rougher texture of the bottom ash relative to the normal aggregates. The density of the high strength concrete containing 100% fine bottom ash and 100% coarse bottom ash was less than 2000 kg/m^3. The incorporation of 100% fine and coarse bottom ash reduced the modulus of elasticity by 49% relative to the control concrete. The compressive strength was not affected by the incorporation of bottom ash in the concrete mix. However, the flexural strength and modulus of rupture decreased linearly with higher percentages of bottom ash in the concrete mix.

3.3. *Cement Replacement in Concrete Production*

Besides using CBA as an aggregate replacement, researchers and technocrats investigated using CBA as cement replacement. The chemical properties of CBA are similar to cement as both are class F materials. The SiO_2 content of CBA is greater than 25%, and the content of SiO_2 + Al_2O_3 + Fe_2O_3 is higher than 70%, which meets the requirement for the recycling bottom ash as cement [25,91]. Moreover, replacing cement with CBA can reduce CO_2 emissions and improved energy conservation.

Cement is the most widely used material in civil constructions. However, the high amount of carbon emitted in cement production has a harmful impact on the environment. An estimated 50% of the total CO_2 emissions are from cement production. The production of each ton of cement releases 0.55 tons of CO_2, and an additional 0.39 tons of CO_2 is emitted during the baking and grinding processes, which are the key contributors to global warming [73].

According to Singh and Bhardwaj [6], replacing a certain percentage of Portland cement with ground coal bottom ash enhanced the resistance towards carbonation, chloride penetration, acid attack, and sulphate attack. The sound absorption coefficient of the concrete improved with the incorporation of both fine and medium CBA. The concrete containing up to 30% of CBA had a lower drying shrinkage after short-term and long-term curing periods. The physical and chemical properties of ground CBA are similar to FA. The concrete containing CBA had a comparable compressive strength to the control concrete, and enhanced flexural strength.

Another study evaluated the performance of the concrete incorporating CBA when exposed to seawater [26]. The original CBA was ground in a ball mill for 20 and 30 h, and 10% of the ground CBA was used as supplementary cementitious material. The concrete containing CBA with a fineness of 3836 and 3895 cm^2/g had a higher strength than the control mix after 180 days of curing in water and seawater. The concrete containing finer CBA was lighter and had reduced salt permeability. Moreover, the concrete containing CBA had a lower chloride penetration than the control mix.

A researcher investigated the short-term impact of replacing cement with 10% coal bottom ash on sulphate and chloride attack [26]. The coal bottom ash was ground for two h in a Los Angeles machine and 20 h in a ball mill grinder to obtain a particle size similar to the ordinary Portland cement. After demolding, the concrete samples were immersed in water for 28 days, followed by immersion in 5% sodium sulphate (Na_2SO_4) and 5% sodium chloride (NaCl) solution for additional curing of 28, 56, and 90 days. The coal bottom ash concrete cured with 5% Na_2SO_4 solution was comparable to the control concrete for up to 90 days. The concrete containing CBA exposed to NaCl solution for a short period had a lower strength than the control concrete, but its strength increased with a longer curing period. This result indicates that the incorporation of CBA enhanced the concrete's resistance towards an aggressive environment.

Aydin [90] investigated the utilization of bottom ash as cement replacement to produce ecofriendly building material. The suitability of using bottom ash as cement replacement was assessed in term of its physical and mechanical properties. The CBA was used as cement replacement in varying percentages of 0, 70, 80, 85, and 100% with 5% hydrated calcium lime. The resulting lightweight composite was suitable for civil engineering applications. The slump, flow, and dry unit weight decreased with higher CBA contents The concrete containing 70% CBA and 5% lime had a higher unconfined compressive strength and flexural strength than the composite containing only 70% CBA. The flexural strength of the final composite indicated that it is suitable for low- to medium-strength applications such as pavement, shotcrete lining, and base and subbase application.

3.4. Noise Barrier, Geotechnical Fill, Zeolite Composite, and Low-Cost Absorbent

Over the past four decades, extensive research has been conducted on noise barriers with different characteristics to protect the areas near roads, especially those with a high traffic volume [92]. Noise barrier wall, also known as the concrete wall, is one of the economic structures built to reduce noise pollution from transportation. Hannan et al. [25] investigated the production of concrete walls with varying CBA percentages that ranged from 0–100% fine aggregate replacement. The researchers reported that the values of fineness modulus of CBA were between 2.3 to 3.0, which is within the range as the fineness modulus of CBA was lower than the conventional fine aggregate specified in the BS 882:1992. The specific gravity of CBA was lower than the conventional fine aggregate due to the porous texture of CBA. The compressive strength of the concrete wall barriers did not increase linearly with higher CBA percentages, but increased with the concrete porosity, which is a good indicator for sound absorption structures. The sound absorption test to determine the acoustic performance showed that the walls containing 80–100% CBA were similar to the conventional wall and were class D absorbers; the remaining walls were class E absorbers. According to BS EN ISO 11,654:1997, the absorbers in class D absorb more than 30% of the sound, while class E absorbs between 15–25% of the sound [25].

Arenas et al. [30] studied the performance of road traffic noise reducing device prototypes from multilayer products composed of 80% bottom ash on a semi-industrial scale. The coarse bottom ash in this research had a Dp of >2.5 mm, and the Dp of the fine bottom ash was <2.5 mm. The measured acoustic performance parameters were the sound absorption coefficient and the airborne sound insulation in the reverberation room. The measured nonacoustic performance parameters were open void ratio, unit weight, compressive strength, Young's modulus, flexural strength, fracture energy, indirect tensile strength, characteristics length, impact strength, and fire resistance. The results showed that the bottom ash-based multilayer products were in the categories A2 and B3 of the sound absorption and assessment index, similar to other commercial products. The mechanical strength of the device containing bottom ash was lower than standard and porous concretes. The fire resistance of the coal bottom ash products remained unchanged after exposure to 47.8 min of fire, and only exhibited a slight discoloration.

Arenas et al. [93] evaluated the performance of highway noise barriers incorporated with bottom ash. Portland cement mixed with different sizes of coal bottom ash (coarse, medium, and fine) was used to produce mortar composite, and the results of the acoustic properties were compared with typical porous concrete. The CBA was separated into three particle size fractions, and a multilayer composite was fabricated using the fractions in three different layers. The wall incorporated with bottom ash had a density and compressive strength of 1470 kg/m^3 and 3.1 MPa. The acoustic properties of the composite containing CBA were comparable or better than the porous concrete. The wall fabricated using coarse CBA had the best sound absorption coefficient because of its porosity, while the wall fabricated with the finest CBA exhibited superior mechanical properties.

Researchers have assessed the potential of utilizing CBA in highway embankments. Theoretically, highway embankment materials must have high strength, density, and stability, good drainage properties, and low plasticity. Coal bottom ash has higher specific gravity than fly ash [94] because its high iron oxide content inhibits the transmission of dead load to the soil and supports the embankment. The application of high compaction resulted in low optimum water content and high maximum density. It also crushed the bottom ash particles and increased the density. Previous studies have shown that precautionary measures should be implemented when using bottom ash as an embankment material, especially upon including structural members and pipes in the ash [95], since mixtures of compacted ash can be corrosive.

A high bottom ash content could reduce hydraulic conductivity because of the significant effect of fine particles in bottom ash on permeability. The high permeability of kaolin mixed with CBA suits drainage application if used as backfill materials in embankments, particularly in areas with a high amount of annual rainfall [96]. The shear strength of the material or soil must be determined before beginning the construction because the ability to resist loading is a critical factor for an embankment. The friction angle of the bottom ash offers higher resistance to the rearrangement of the particles for sustained shearing due to the angular texture of bottom ash.

Researchers have investigated the conversion of coal bottom ash into zeolite X-carbon [97]. The Si and Al in CBA were the raw materials for zeolite, and the unburnt carbon was a source of activated carbon. NaOH and hydrothermal treatment at various times were used to alkali-fuse the CBA in the fabrication of zeolite X-carbon composite for hydrogen storage. The results showed that the optimum synthesis of zeolite X-carbon composite from CBA is by fusion through hydrothermal treatment at 90 °C for 15 h. The zeolite composite with the best crystallinity had a surface area of 185.824 m^2/gram, micropore diameter of 0.34 nm, mesopore diameter of 3 nm, and hydrogen uptake of up to 1.66% wt at 30 °C/20 psi. These results indicated that the composite is suitable for hydrogen storage.

Besides adsorption [98], reverse osmosis, flocculation, electroflotation, and precipitation techniques are also effective methods for eliminating pollutants from water [99]. Various low-cost adsorbents, such as clay materials [100,101], and agricultural wastes, such as apricot waste [102], soy meal hull [103], rice husk [104], sugar beet pulp [105], and wheat bran [106], have been reported as effective materials in eliminating hazardous chemicals from water. Jarusiripot [13] investigated the adsorption of dye using bottom ash as an absorbent. The International Union of Pure and Applied Chemistry (IUPAC) classifies bottom ash as mesopores with average pore diameter ranging between 3 and 7 nm. Before using it as an absorbent, the raw bottom ash was pretreated with chemical solutions, such as hydrochloric acid (HCl), nitric acid (HNO_3), and hydrogen peroxide (H_2O_2).

The results showed that bottom ash was an effective adsorbent for removing dye from wastewater even though it was not as efficient as the expensive commercially activated carbon. Because of the small surface area of the coal bottom ash, the dye absorption is dependent on the interaction between the charges of the dye molecules and the adsorbent surface. The superior absorption of the bottom ash with smaller particle sizes was due to the higher surface area of the absorbent. The adsorption capacity was also influenced by the chemical solutions used in the pretreatment process.

4. Critical Discussion

This review has shown that, at a higher percentage of CBA content, most of the CBA from various sources did not enhance the performance of the asphalt and concrete mixtures. Therefore, the CBA used as a replacement material in civil construction should be pretreated chemically or mechanically. The findings of previous research vary significantly, even at the same substitution ratio, because there was no control over the quality and constituents of the CBA. It is essential to perform a comprehensive assessment of the optimization of the materials used with the CBA because using CBA in combination with other materials, such as fly ash and superplasticizer, could enhance the mixture's performance. Despite its promising potential, there is a need to perform more comprehensive research to determine the cost–benefit and environmental impact of utilizing CBA and developing CBA as a green construction material. There is also a need to formulate a general guideline for using CBA as an alternative material in the construction industry.

5. Conclusions

The large volume of coal bottom offers a promising alternative for the aggregates used in the construction industry to prevent the depletion of natural aggregate resources. Previous studies have shown that coal bottom ash has a good potential as a substitute material, especially in the construction industry. However, further studies need to investigate turning bottom ash waste into applicable material. In summary, the use of coal bottom ash in the fields of construction, especially in civil engineering, can be commercialized through suitable design and appropriate construction procedures. A summary of this review is as follows.

1. CBA is a class F pozzolan that contains more than 70% of SiO_2, Al_2O_3, and Fe_2O_3.
2. The porous texture of CBA reduces the density of the pavement and concrete mixtures, making it a suitable lightweight aggregate in pavement and concrete production. Its porosity contributes to the ability of the CBA noise barrier wall to absorb more sound than the conventional wall.
3. The high-water absorption of the CBA particles increases the optimum binder content of the pavement mixture and the cement/water ratio in concrete production.
4. The pavement mixture incorporated with CBA exhibits enhanced moisture susceptibility. However, most research showed reduced Marshall stability, tensile strength, and resilient modulus. The low resilient modulus indicates that the mixture has a high elastic deformation, which increases its rutting resistance.
5. The optimum asphalt mixture performance is achieved with 10–30% CBA replacement. However, the performance can be enhanced by using CBA with other materials, such as fly ash, lime, and superplasticizer.
6. The concretes with higher CBA contents have reduced fresh concrete properties such as slump, bleeding, and flow because of the CBA's interlocking characteristics, rough texture, and irregular shape.
7. The properties of the concrete improved with the curing period, and after a certain curing period, the properties are superior to conventional concrete.
8. The utilization of CBA as cement replacement is beneficial to the environment because it reduces the CO_2 emissions in cement production. The heavy metals present in the raw CBA are below the permissible range.

Author Contributions: Conceptualization, S.A.M.; formal analysis, S.A.M. and S.K.; investigation, S.A.M. and A.M.B.; resources, S.K. and M.R.I.; writing—original draft preparation, S.A.M. and A.M.B.; writing—review and editing, S.K., H.Y.B.K., M.F.A.P., M.R.I. and M.R.K.; supervision, S.K. and M.R.K.; funding acquisition, S.K. and H.Y.B.K. All authors have read and agreed to the published version of the manuscript.

Funding: This study was funded by the Universiti Malaysia Perlis, grant name "RESMATE 9001- 00623" and the Universiti Tenaga Nasional (J5100D4103-BOLDREFRESH2025-CENTRE OF EXCELLENCE).

Institutional Review Board Statement: Not applicable.

Informed Consent Statement: Not applicable.

Data Availability Statement: All data used in this research can be provided upon request.

Acknowledgments: The authors would like to acknowledge the Universiti Malaya, Universiti Malaysia Perlis and Universiti Tenaga Nasional Malaysia.

Conflicts of Interest: The authors declare no conflict of interest.

References

1. Lior, N. Sustainable energy development: The present (2009) situation and possible paths to the future q. *Energy* **2010**, *35*, 3976–3994. [CrossRef]
2. Yao, Z.T.; Ji, X.S.; Sarker, P.K.; Tang, J.H.; Ge, L.Q.; Xia, M.S.; Xi, Y.Q. Earth-Science Reviews A comprehensive review on the applications of coal fl y ash. *Earth-Sci. Rev.* **2015**, *141*, 105–121. [CrossRef]
3. Rafieizonooz, M.; Mirza, J.; Salim, M.R.; Hussin, M.W.; Khankhaje, E. Investigation of coal bottom ash and fly ash in concrete as replacement for sand and cement. *Constr. Build. Mater.* **2016**, *116*, 15–24. [CrossRef]
4. Abubakar, A.U.; Baharudin, K.S. Tanjung Bin Coal Bottom Ash: From Waste to Concrete Material Tanjung Bin Coal Bottom Ash: From Waste to Concrete Material. *Adv. Mater. Res.* **2013**. [CrossRef]
5. Muthusamy, K.; Rasid, M.H.; Jokhio, G.A.; Mokhtar Albshir Budiea, A.; Hussin, M.W.; Mirza, J. Coal bottom ash as sand replacement in concrete: A review. *Constr. Build. Mater.* **2020**, *236*, 117507. [CrossRef]

6. Singh, N.; Shehnazdeep; Bhardwaj, A. Reviewing the role of coal bottom ash as an alternative of cement. *Constr. Build. Mater.* **2020**, *233*, 117276. [CrossRef]
7. Ahmaruzzaman, M. A review on the utilization of fly ash. *Prog. Energy Combust. Sci.* **2010**, *36*, 327–363. [CrossRef]
8. Heidrich, C.; Feuerborn, H.; Weir, A. Coal Combustion Products: A Global Perspective. In Proceedings of the World of Coal Ash (WOCA) Conference, Lexington, KY, USA, 22–25 April 2013; Available online: http://www.mcilvainecompany.com/Decision_Tree/subscriber/Tree/DescriptionTextLinks/International%20flyash%20perspective.pdf (accessed on 1 June 2021).
9. Sutcu, M.; Erdogmus, E.; Gencel, O.; Gholampour, A.; Atan, E.; Ozbakkaloglu, T. Recycling of bottom ash and fly ash wastes in eco-friendly clay brick production. *J. Clean. Prod.* **2019**, *233*, 753–764. [CrossRef]
10. Kumar, D.; Kumar, R.; Abbass, M. Study The Effect of Coal Bottom Ash on Partial Replacement of Fine Aggregate in Concrete With Sugarcane Molasses as an Admixture. *Int. J. Sci. Res. Educ.* **2016**. [CrossRef]
11. Souad, E.M.E.A.; Moussaoui, R.; Monkade, M.; Lahlou, K.; Hasheminejad, N.; Margaritis, A.; Bergh, W. Van den Lime Treatment of Coal Bottom Ash for Use in Road Pavements: Application to EL Jadida Zone in Morocco. *Materials* **2019**, *12*, 1–15.
12. Lokeshappa, B.; Kumar, A. Behaviour of Metals in Coal Fly Ash Ponds. *Apcbee Procedia* **2012**, *1*, 34–39. [CrossRef]
13. Jarusiripot, C. Removal of Reactive Dye by Adsorption over Chemical Pretreatment Coal based Bottom Ash. *Procedia Chem.* **2014**, *9*, 121–130. [CrossRef]
14. Baite, E.; Messan, A.; Hannawi, K.; Tsobnang, F.; Prince, W. Physical and transfer properties of mortar containing coal bottom ash aggregates from Tefereyre (Niger). *Constr. Build. Mater.* **2016**, *125*, 919–926. [CrossRef]
15. Mangi, S.A.; Wan Ibrahim, M.H.; Jamaluddin, N.; Arshad, M.F.; Memon, F.A.; Putra Jaya, R.; Shahidan, S. A Review on Potential Use of Coal Bottom Ash as a Supplementary Cementing Material in Sustainable Concrete Construction. *Int. J. Integr. Eng.* **2018**, *10*, 28–36. [CrossRef]
16. Rathnayake, M.; Julnipitawong, P.; Tangtermsirikul, S.; Toochinda, P. Utilization of coal fly ash and bottom ash as solid sorbents for sulfur dioxide reduction from coal fired power plant: Life cycle assessment and applications. *J. Clean. Prod.* **2018**, *202*, 934–945. [CrossRef]
17. Singh, M.; Siddique, R. Effect of coal bottom ash as partial replacement of sand on workability and strength properties of concrete. *J. Clean. Prod.* **2016**, *112*, 620–630. [CrossRef]
18. Menéndez, E.; Álvaro, A.M.; Hernández, M.T.; Parra, J.L. New methodology for assessing the environmental burden of cement mortars with partial replacement of coal bottom ash and fl y ash. *J. Environ. Manag.* **2014**, *133*, 275–283. [CrossRef]
19. Jang, J.G.; Ahn, Y.B.; Souri, H.; Lee, H.K. A novel eco-friendly porous concrete fabricated with coal ash and geopolymeric binder: Heavy metal leaching characteristics and compressive strength. *Constr. Build. Mater.* **2015**, *79*, 173–181. [CrossRef]
20. Singh, M.; Siddique, R. Compressive strength, drying shrinkage and chemical resistance of concrete incorporating coal bottom ash as partial or total replacement of sand. *Constr. Build. Mater.* **2014**, *68*, 39–48. [CrossRef]
21. Kim, R.G.; Li, D.; Jeon, C.H. Experimental investigation of ignition behavior for coal rank using a flat flame burner at a high heating rate. *Exp. Therm. Fluid Sci.* **2014**, *54*, 212–218. [CrossRef]
22. Gooi, S.; Mousa, A.A.; Kong, D. A critical review and gap analysis on the use of coal bottom ash as a substitute constituent in concrete. *J. Clean. Prod.* **2020**, *268*, 121752. [CrossRef]
23. Antoni; Klarens, K.; Indranata, M.; Al Jamali, L.; Hardjito, D. The use of bottom ash for replacing fine aggregate in concrete paving blocks. *MATEC Web Conf.* **2017**, *138*. [CrossRef]
24. Colonna, P.; Berloco, N.; Ranieri, V.; Shuler, S.T. Application of Bottom Ash for Pavement Binder Course. *Procedia Soc. Behav. Sci.* **2012**, *53*, 961–971. [CrossRef]
25. Hannan, N.I.R.R.; Shaidan, S.; Ali, N.; Bunnori, N.M.; Mohd Zuki, S.S.; Wan Ibrahim, M.H. Acoustic and non-acoustic performance of coal bottom ash concrete as sound absorber for wall concrete. *Case Stud. Constr. Mater.* **2020**, e00399. [CrossRef]
26. Mangi, S.A.; Wan Ibrahim, M.H.; Jamaluddin, N.; Arshad, M.F.; Putra Jaya, R. Short-term effects of sulphate and chloride on the concrete containing coal bottom ash as supplementary cementitious material. *Eng. Sci. Technol. Int. J.* **2019**, *22*, 515–522. [CrossRef]
27. Hashemi, S.S.G.; Mahmud, H.B.; Ghuan, T.C.; Chin, A.B.; Kuenzel, C.; Ranjbar, N. Safe disposal of coal bottom ash by solidification and stabilization techniques. *Constr. Build. Mater.* **2019**, *197*, 705–715. [CrossRef]
28. Singh, N.; Mithulraj, M.; Arya, S. Utilization of coal bottom ash in recycled concrete aggregates based self compacting concrete blended with metakaolin. *Resour. Conserv. Recycl.* **2019**, *144*, 240–251. [CrossRef]
29. Singh, M.; Siddique, R. Strength properties and micro-structural properties of concrete containing coal bottom ash as partial replacement of fine aggregate. *Constr. Build. Mater.* **2014**, *50*, 246–256. [CrossRef]
30. Arenas, C.; Leiva, C.; Vilches, L.F.; Cifuentes, H.; Rodríguez-Galán, M. Technical specifications for highway noise barriers made of coal bottom ash-based sound absorbing concrete. *Constr. Build. Mater.* **2015**, *95*, 585–591. [CrossRef]
31. Zhang, B.; Poon, C.S. Use of Furnace Bottom Ash for producing lightweight aggregate concrete with thermal insulation properties. *J. Clean. Prod.* **2015**, *99*, 94–100. [CrossRef]
32. Kim, H.K.; Lee, H.K. Use of power plant bottom ash as fine and coarse aggregates in high-strength concrete. *Constr. Build. Mater.* **2011**, *25*, 1115–1122. [CrossRef]
33. Topçu, I.B.; Bilir, T. Effect of bottom ash as fine aggregate on shrinkage cracking of mortars. *ACI Mater. J.* **2010**, *107*, 48–56. [CrossRef]

34. Onprom, P.; Chaimoon, K.; Cheerarot, R. Influence of Bottom Ash Replacements as Fine Aggregate on the Property of Cellular Concrete with Various Foam Contents. *Adv. Mater. Sci. Eng.* **2015**, *2015*. [CrossRef]
35. Milad, A.; Ali, A.S.B.; Babalghaith, A.M.; Memon, Z.A.; Mashaan, N.S.; Arafa, S.; Nur, N.I. Utilisation of waste-based geopolymer in asphalt pavement modification and construction; a review. *Sustainability* **2021**, *13*, 3330. [CrossRef]
36. Wie, Y.M.; Lee, K. Composition design of the optimum bloating activation condition for artificial lightweight aggregate using coal ash. *J. Korean Ceram. Soc.* **2020**, *57*, 220–230. [CrossRef]
37. Sigvardsen, N.M.; Ottosen, L.M. Characterization of coal bio ash from wood pellets and low-alkali coal fly ash and use as partial cement replacement in mortar. *Cem. Concr. Compos.* **2019**, *95*, 25–32. [CrossRef]
38. Ge, X.; Zhou, M.; Wang, H.; Liu, Z.; Wu, H.; Chen, X. Preparation and characterization of ceramic foams from chromium slag and coal bottom ash. *Ceram. Int.* **2018**, *44*, 11888–11891. [CrossRef]
39. Argiz, C.; Sanjuán, M.Á.; Menéndez, E. Coal Bottom Ash for Portland Cement Production. *Adv. Mater. Sci. Eng.* **2017**, *2017*. [CrossRef]
40. Oruji, S.; Brake, N.A.; Nalluri, L.; Guduru, R.K. Strength activity and microstructure of blended ultra-fine coal bottom ash-cement mortar. *Constr. Build. Mater.* **2017**, *153*, 317–326. [CrossRef]
41. Ibrahim, M.H.W.; Hamzah, A.F.; Jamaluddin, N.; Ramadhansyah, P.J.; Fadzil, A.M. Split Tensile Strength on Self-compacting Concrete Containing Coal Bottom Ash. *Procedia Soc. Behav. Sci.* **2015**, *195*, 2280–2289. [CrossRef]
42. Siddique, R. Compressive strength, water absorption, sorptivity, abrasion resistance and permeability of self-compacting concrete containing coal bottom ash. *Constr. Build. Mater.* **2013**, *47*, 1444–1450. [CrossRef]
43. Syahrul, M.; Sani, M.; Muftah, F.; Muda, Z. The Properties of Special Concrete Using Washed Bottom Ash (WBA) as Partial Sand Replacement. *Int. J. Sustain. Constr. Eng. Technol.* **2010**, *1*, 65–76.
44. Alinezhad, M.; Sahaf, A. Investigation of the fatigue characteristics of warm stone matrix asphalt (WSMA) containing electric arc furnace (EAF) steel slag as coarse aggregate and Sasobit as warm mix additive. *Case Stud. Constr. Mater.* **2019**, *11*, e00265. [CrossRef]
45. Ameli, A.; Hossein Pakshir, A.; Babagoli, R.; Norouzi, N.; Nasr, D.; Davoudinezhad, S. Experimental investigation of the influence of Nano TiO2 on rheological properties of binders and performance of stone matrix asphalt mixtures containing steel slag aggregate. *Constr. Build. Mater.* **2020**, *265*, 120750. [CrossRef]
46. Pang, B.; Zhou, Z.; Xu, H. Utilization of carbonated and granulated steel slag aggregate in concrete. *Constr. Build. Mater.* **2015**, *84*, 454–467. [CrossRef]
47. Wang, S.; Zhang, G.; Wang, B.; Wu, M. Mechanical strengths and durability properties of pervious concretes with blended steel slag and natural aggregate. *J. Clean. Prod.* **2020**, *271*, 122590. [CrossRef]
48. Mathew, S.P.; Nadir, Y.; Arif, M.M. Experimental study of thermal properties of concrete with partial replacement of coarse aggregate by coconut shell. *Mater. Today Proc.* **2020**, *27*, 415–420. [CrossRef]
49. Nadir, Y.; Sujatha, A. Durability Properties of Coconut Shell Aggregate Concrete. *KSCE J. Civ. Eng.* **2018**, *22*, 1920–1926. [CrossRef]
50. Kanojia, A.; Jain, S.K. Performance of coconut shell as coarse aggregate in concrete. *Constr. Build. Mater.* **2017**, *140*, 150–156. [CrossRef]
51. Ahmad Zawawi, M.N.A.; Muthusamy, K.; Abdul Majeed, A.P.P.; Muazu Musa, R.; Mokhtar Albshir Budiea, A. Mechanical properties of oil palm waste lightweight aggregate concrete with fly ash as fine aggregate replacement. *J. Build. Eng.* **2020**, *27*, 100924. [CrossRef]
52. Mo, K.H.; Alengaram, U.J.; Jumaat, M.Z.; Liu, M.Y.J.; Lim, J. Assessing some durability properties of sustainable lightweight oil palm shell concrete incorporating slag and manufactured sand. *J. Clean. Prod.* **2016**, *112*, 763–770. [CrossRef]
53. Devulapalli, L.; Kothandaraman, S.; Sarang, G. Evaluation of rejuvenator's effectiveness on the reclaimed asphalt pavement incorporated stone matrix asphalt mixtures. *Constr. Build. Mater.* **2019**, *224*, 909–919. [CrossRef]
54. Devulapalli, L.; Kothandaraman, S.; Sarang, G. Effect of rejuvenating agents on stone matrix asphalt mixtures incorporating RAP. *Constr. Build. Mater.* **2020**, *254*, 119298. [CrossRef]
55. Nwakaire, C.M.; Yap, S.P.; Yuen, C.W.; Onn, C.C.; Koting, S.; Babalghaith, A.M. Laboratory study on recycled concrete aggregate based asphalt mixtures for sustainable flexible pavement surfacing. *J. Clean. Prod.* **2020**, *262*, 121462. [CrossRef]
56. Pourtahmasb, M.S.; Karim, M.R. Performance Evaluation of Stone Mastic Asphalt and Hot Mix. *Adv. Mater. Sci. Eng.* **2014**, *2014*, 1–12.
57. Pourtahmasb, M.S.; Karim, M.R.; Shamshirband, S. Resilient modulus prediction of asphalt mixtures containing Recycled Concrete Aggregate using an adaptive neuro-fuzzy methodology. *Constr. Build. Mater.* **2015**, *82*, 257–263. [CrossRef]
58. Pourtahmasb, M.S.; Karim, M.R. Utilization of Recycled Concrete Aggregates in Stone Mastic Asphalt Mixtures. *Adv. Mater. Sci. Eng.* **2014**, *2014*. [CrossRef]
59. Huang, Q.; Qian, Z.; Hu, J.; Zheng, D. Evaluation of Stone Mastic Asphalt Containing Ceramic Waste Aggregate for Cooling Asphalt Pavement. *Materials* **2020**, *13*, 2964. [CrossRef] [PubMed]
60. Gautam, P.K.; Kalla, P.; Nagar, R.; Agrawal, R.; Jethoo, A.S. Laboratory investigations on hot mix asphalt containing mining waste as aggregates. *Constr. Build. Mater.* **2018**, *168*, 143–152. [CrossRef]
61. Bui, N.K.; Satomi, T.; Takahashi, H. Improvement of mechanical properties of recycled aggregate concrete basing on a new combination method between recycled aggregate and natural aggregate. *Constr. Build. Mater.* **2017**, *148*, 376–385. [CrossRef]

62. Hamada, H.M.; Yahaya, F.M.; Muthusamy, K.; Jokhio, G.A.; Humada, A.M. Fresh and hardened properties of palm oil clinker lightweight aggregate concrete incorporating Nano-palm oil fuel ash. *Constr. Build. Mater.* **2019**, *214*, 344–354. [CrossRef]
63. Abutaha, F.; Razak, H.A.; Ibrahim, H.A.; Ghayeb, H.H. Adopting particle-packing method to develop high strength palm oil clinker concrete. *Resour. Conserv. Recycl.* **2018**, *131*, 247–258. [CrossRef]
64. Omrane, M.; Kenai, S.; Kadri, E.H.; Aït-Mokhtar, A. Performance and durability of self compacting concrete using recycled concrete aggregates and natural pozzolan. *J. Clean. Prod.* **2017**, *165*, 415–430. [CrossRef]
65. Lu, J.X.; Yan, X.; He, P.; Poon, C.S. Sustainable design of pervious concrete using waste glass and recycled concrete aggregate. *J. Clean. Prod.* **2019**, *234*, 1102–1112. [CrossRef]
66. Adhikary, S.K.; Rudzionis, Z. Influence of expanded glass aggregate size, aerogel and binding materials volume on the properties of lightweight concrete. *Mater. Today Proc.* **2020**, *32*, 712–718. [CrossRef]
67. Wang, X.; Fan, Z.; Li, L.; Wang, H.; Huang, M. Durability evaluation study for crumb rubber-asphalt pavement. *Appl. Sci.* **2019**, *9*, 3434. [CrossRef]
68. Malarvizhi, G.; Senthil, N.; Kamaraj, C. A Study on Recycling Of Crumb Rubber and Low Density Polyethylene Blend on Stone Matrix Asphalt. *Int. J. Sci. Res. Publ.* **2012**, *2*, 1–16. [CrossRef]
69. Mohammed Babalghaith, A.; Koting, S.; Ramli Sulong, N.H.; Karim, M.R.; Mohammed AlMashjary, B. Performance evaluation of stone mastic asphalt (SMA) mixtures with palm oil clinker (POC) as fine aggregate replacement. *Constr. Build. Mater.* **2020**, *262*, 120546. [CrossRef]
70. Babalghaith, A.M.; Koting, S.; Ramli Sulong, N.H.; Karim, M.R.; Mohammed, S.A.; Ibrahim, M.R. Effect of palm oil clinker (POC) aggregate on the mechanical properties of stone mastic asphalt (SMA) mixtures. *Sustainability* **2020**, *12*, 2716. [CrossRef]
71. Nayaka, R.R.; Alengaram, U.J.; Jumaat, M.Z.; Yusoff, S.B.; Ganasan, R. Performance evaluation of masonry grout containing high volume of palm oil industry by-products. *J. Clean. Prod.* **2019**, *220*, 1202–1214. [CrossRef]
72. Inayat, A.; Inayat, M.; Shahbaz, M.; Sulaiman, S.A.; Raza, M.; Yusup, S. Parametric analysis and optimization for the catalytic air gasification of palm kernel shell using coal bottom ash as catalyst. *Renew. Energy* **2020**, *145*, 671–681. [CrossRef]
73. Singh, N.; Mithulraj, M.; Arya, S. Influence of coal bottom ash as fine aggregates replacement on various properties of concretes: A review. *Resour. Conserv. Recycl.* **2018**, *138*, 257–271. [CrossRef]
74. Ankur, N.; Singh, N. Performance of cement mortars and concretes containing coal bottom ash: A comprehensive review. *Renew. Sustain. Energy Rev.* **2021**, *149*, 111361. [CrossRef]
75. Kim, H.K.; Lee, H.K. Coal bottom ash in field of civil engineering: A review of advanced applications and environmental considerations. *KSCE J. Civ. Eng.* **2015**, *19*, 1802–1818. [CrossRef]
76. Atiyeh, M.; Aydin, E. Data for bottom ash and marble powder utilization as an alternative binder for sustainable concrete construction. *Data Br.* **2020**, *29*, 105160. [CrossRef]
77. Jamora, J.B.; Gudia, S.E.L.; Go, A.W.; Giduquio, M.B.; Loretero, M.E. Potential CO_2 reduction and cost evaluation in use and transport of coal ash as cement replacement: A case in the Philippines. *Waste Manag.* **2020**, *103*, 137–145. [CrossRef]
78. Bajare, D.; Bumanis, G.; Upeniece, L. Coal combustion bottom ash as microfiller with pozzolanic properties for traditional concrete. *Procedia Eng.* **2013**, *57*, 149–158. [CrossRef]
79. Gautam, P.K.; Kalla, P.; Jethoo, A.S.; Agrawal, R.; Singh, H. Sustainable use of waste in flexible pavement: A review. *Constr. Build. Mater.* **2018**, *180*, 239–253. [CrossRef]
80. Yoo, B.S.; Park, D.W.; Vo, H.V. Evaluation of Asphalt Mixture Containing Coal Ash. *Transp. Res. Procedia* **2016**, *14*, 797–803. [CrossRef]
81. Ksaibati, K.; Stephen, J. *Utilization of Bottom Ash in Asphalt Mixes*; No. MCP Report No. 99–104A; Department of Civil and Architectural Engineering, University of Wyoming: Laramie, WY, USA, 1999. Available online: https://citeseerx.ist.psu.edu/viewdoc/download?doi=10.1.1.593.9395&rep=rep1&type=pdf (accessed on 1 June 2021).
82. Hesami, S.; Modarres, A.; Soltaninejad, M.; Madani, H. Mechanical properties of roller compacted concrete pavement containing coal waste and limestone powder as partial replacements of cement. *Constr. Build. Mater.* **2016**, *111*, 625–636. [CrossRef]
83. Ameli, A.; Babagoli, R.; Norouzi, N.; Jalali, F.; Poorheydari Mamaghani, F. Laboratory evaluation of the effect of coal waste ash (CWA) and rice husk ash (RHA) on performance of asphalt mastics and Stone matrix asphalt (SMA) mixture. *Constr. Build. Mater.* **2020**, *236*, 117557. [CrossRef]
84. Xu, P.; Chen, Z.; Cai, J.; Pei, J.; Gao, J.; Zhang, J.; Zhang, J. The effect of retreated coal wastes as filler on the performance of asphalt mastics and mixtures. *Constr. Build. Mater.* **2019**, *203*, 9–17. [CrossRef]
85. Goh, S.W.; You, Z. A preliminary study of the mechanical properties of asphalt mixture containing bottom ash. *Can. J. Civ. Eng.* **2008**, *35*, 1114–1119. [CrossRef]
86. López-López, E.; Vega Zamanillo, Á.; Calzada-Pérez, M.; Taborga-Sedano, M.A. Use of bottom ash from thermal power plant and lime as filler in bituminous mixtures. *Mater. Constr.* **2015**, *65*, 1–7. [CrossRef]
87. Modarres, A.; Ayar, P. Coal waste application in recycled asphalt mixtures with bitumen emulsion. *J. Clean. Prod.* **2014**, *83*, 263–272. [CrossRef]
88. Suárez-Macías, J.; Terrones-Saeta, J.M.; Iglesias-Godino, F.J.; Corpas-Iglesias, F.A. Evaluation of physical, chemical, and environmental properties of biomass bottom ash for use as a filler in bituminous mixtures. *Sustainability* **2021**, *13*, 4119. [CrossRef]
89. Modarres, A.; Rahmanzadeh, M. Application of coal waste powder as filler in hot mix asphalt. *Constr. Build. Mater.* **2014**, *66*, 476–483. [CrossRef]

90. Aydin, E. Novel coal bottom ash waste composites for sustainable construction. *Constr. Build. Mater.* **2016**, *124*, 582–588. [CrossRef]
91. García Arenas, C.; Marrero, M.; Leiva, C.; Solís-Guzmán, J.; Vilches Arenas, L.F. High fire resistance in blocks containing coal combustion fly ashes and bottom ash. *Waste Manag.* **2011**, *31*, 1783–1789. [CrossRef]
92. Lacasta, A.M.; Penaranda, A.; Cantalapiedra, I.R.; Auguet, C.; Bures, S.; Urrestarazu, M. Acoustic evaluation of modular greenery noise barriers. *Urban For. Urban Green.* **2016**, *20*, 172–179. [CrossRef]
93. Arenas, C.; Leiva, C.; Vilches, L.F.; Cifuentes, H. Use of co-combustion bottom ash to design an acoustic absorbing material for highway noise barriers. *Waste Manag.* **2013**, *33*, 2316–2321. [CrossRef]
94. Awang, A.R.; Marto, A.; Makhtar, A.M. Geotechnical properties of tanjung bin coal ash mixtures for backfill materials in embankment construction. *Electron. J. Geotech. Eng.* **2011**, *16 L*, 1515–1531.
95. Kim, B.; Prezzi, M.; Rodrigo, S. Geotechnical Properties of Fly and Bottom Ash Mixtures for Use in Highway Embankments. *J. Geotech. Geoenviron. Eng.* **2007**, *1*, 1–7. [CrossRef]
96. Jorat, M.E.; Marto, A.; Namazi, E.; Amin, M.F.M. Engineering characteristics of kaolin mixed with various percentages of bottom ash. *Electron. J. Geotech. Eng.* **2011**, *16 H*, 841–850.
97. Widiastuti, N.; Hidayah, M.Z.N.; Praseytoko, D.; Fansuri, H. Synthesis of zeolite X-carbon from coal bottom ash for hydrogen storage material. *Adv. Mater. Lett.* **2014**, *5*, 453–458. [CrossRef]
98. Wang, S.; Boyjoo, Y.; Choueib, A.; Zhu, Z.H. Removal of dyes from aqueous solution using fly ash and red mud. *Water Res.* **2005**, *39*, 129–138. [CrossRef] [PubMed]
99. Daneshvar, N.; Ayazloo, M.; Khataee, A.R.; Pourhassan, M. Biological decolorization of dye solution containing Malachite Green by microalgae Cosmarium sp. *Bioresour. Technol.* **2007**, *98*, 1176–1182. [CrossRef] [PubMed]
100. Hajjaji, M.; Alami, A.; Bouadili, A. El Removal of methylene blue from aqueous solution by fibrous clay minerals. *J. Hazard. Mater.* **2006**, *135*, 188–192. [CrossRef] [PubMed]
101. Catalfamo, P.; Arrigo, I.; Primerano, P.; Corigliano, F. Efficiency of a zeolitized pumice waste as a low-cost heavy metals adsorbent. *J. Hazard. Mater.* **2006**, *134*, 140–143. [CrossRef]
102. Önal, Y. Kinetics of adsorption of dyes from aqueous solution using activated carbon prepared from waste apricot. *J. Hazard. Mater.* **2006**, *137*, 1719–1728. [CrossRef]
103. Arami, M.; Limaee, N.Y.; Mahmoodi, N.M.; Tabrizi, N.S. Equilibrium and kinetics studies for the adsorption of direct and acid dyes from aqueous solution by soy meal hull. *J. Hazard. Mater.* **2006**, *135*, 171–179. [CrossRef] [PubMed]
104. Srivastava, V.C.; Mall, I.D.; Mishra, I.M. Characterization of mesoporous rice husk ash (RHA) and adsorption kinetics of metal ions from aqueous solution onto RHA. *J. Hazard. Mater.* **2006**, *134*, 257–267. [CrossRef] [PubMed]
105. Aksu, Z.; Isoglu, I.A. Use of agricultural waste sugar beet pulp for the removal of Gemazol turquoise blue-G reactive dye from aqueous solution. *J. Hazard. Mater.* **2006**, *137*, 418–430. [CrossRef] [PubMed]
106. Papinutti, L.; Mouso, N.; Forchiassin, F. Removal and degradation of the fungicide dye malachite green from aqueous solution using the system wheat bran-Fomes sclerodermeus. *Enzyme. Microb. Technol.* **2006**, *39*, 848–853. [CrossRef]

 *sustainability*

Review

African Case Studies: Developing Pavement Temperature Maps for Performance-Graded Asphalt Bitumen Selection

Refiloe Mokoena *, Georges Mturi, Johan Maritz, Mohau Mateyisi and Peter Klein

Council for Scientific and Industrial Research (CSIR), Pretoria 0001, South Africa; gmturi@csir.co.za (G.M.); jmaritz@csir.co.za (J.M.); mmateyisi@csir.co.za (M.M.); pklein@csir.co.za (P.K.)
* Correspondence: rmokoena@csir.co.za

Abstract: The reliable performance of roads is crucial for service delivery, and it is a catalyst for domestic and cross-border spatial development. Paved national roads are expected to carry higher traffic volumes over time as a result of urbanization and to support the economic development in the continent. Increased traffic levels combined with expected increases in air temperatures as a result of global warming highlight the need to appropriately select bituminous road materials for a reliable performance of asphalt roads. The objective of the paper is to present African case studies on the development of temperature maps necessary for performance-graded bitumen selection for road design and construction. A consistent approach, that caters for the variability of geographical, environmental and climatic conditions, does not currently exist within the continent. Therefore, this paper discusses a series of critical components in the development of temperature maps for performance-graded bitumen including (i) pavement temperature models and climatic zones in Africa; (ii) the effect of urban heat islands on pavement temperature; (iii) sources of weather data and (iv) the mapping procedure to produce temperature maps. Characterizing the thermal properties of the pavement was found to be an important factor for reliably calculating expected road temperatures as well as the consideration of the ambient climate for a given location. During this study, the urban heat island effect was found to have little influence on the maximum pavement temperatures but a significant effect on the minimum pavement temperatures. Some areas of the urban district assessed in this investigation were found to increase by two performance grades according to the minimum temperature criteria. The recent observed weather data from weather stations are the most accurate means of measurement of the ambient environmental conditions necessary for performance-based specifications, but they are not always easily accessible, and therefore other sources of data, such as satellite data, may need to be used instead. With the expected temperature increases expected as a result of climate change, the use of Global Climate Models also opens new avenues for performance-based material selection in the African continent for expected climates as an alternative to traditional approaches based on historically observed weather.

Keywords: bitumen selection; performance-graded bitumen; asphalt pavement temperatures; temperature maps

Citation: Mokoena, R.; Mturi, G.; Maritz, J.; Mateyisi, M.; Klein, P. African Case Studies: Developing Pavement Temperature Maps for Performance-Graded Asphalt Bitumen Selection. *Sustainability* **2022**, *14*, 1048. https://doi.org/10.3390/su14031048

Academic Editor: Edoardo Bocci

Received: 1 October 2021
Accepted: 25 October 2021
Published: 18 January 2022

1. Introduction

Roads play a crucial role in the economic growth, development and accessibility of communities. In Africa, road infrastructure networks are the most heavily used form of transport and have a significant bearing on the inclusive and sustainable development of the region [1].

The reliable performance of roads is crucial for service delivery, and it is a catalyst for domestic and cross-border spatial development [2], especially considering that paved national roads are expected to carry increasing traffic volumes over time. Selecting the most suitable and durable road construction materials is therefore necessary to ensure a reliable serviceability. Taking the value of paved roads into consideration, a system for quantifying and measuring the in-field performance of the material is imperative. Most countries in Africa still rely on empirical methods for bituminous binder classification in asphalt road construction. However, South Africa has recently moved towards the implementation of a performance-graded (PG) specification founded on principles of pavement performance [3].

Establishing the climatic conditions that affect pavement temperatures is therefore necessary for characterizing the material's response to traffic loading and environmental conditions, given the visco-elastic nature of asphalt. The three failure mechanisms of asphalt layers addressed by the PG specification are (i) permanent deformation (rutting at high service temperatures); (ii) fatigue cracking at intermediate service temperatures; and (iii) brittle fracture at low service temperatures [4]. The in-service temperature and traffic conditions, in conjunction with the ageing effects of bituminous binders, play a pivotal role in selecting bitumen to achieve a given in situ performance within the asphalt layer. Pavement failures due to a poor selection of bituminous binders can be reduced by correctly specifying the appropriate binder for a given climate [4].

As per the Superpave method, historic air temperature data are required to calculate the maximum and minimum pavement temperatures for PG asphalt bitumen selection [5]. Using different sources of historical data, pavement temperature maps have been developed for South Africa [4] and Tanzania [6].

An appropriate binder selection results in asphalt that will withstand the imposed environmental and traffic loading conditions on roads. Temperature maps are an integral part of the climatic factors of a pavement's operating environment. These maps assist road engineers to select the most suitable bituminous binders for a given geographical location based on regional climatic conditions. Many influencing factors determine the usage of temperature maps such as the type of pavement temperature model used, prevalent climatic conditions, the effect of urban heat islands for localized planning in metropolitan municipalities, the geospatial mapping procedure and considerations as well as the source of required climate data and weather parameters, as illustrated in Figure 1.

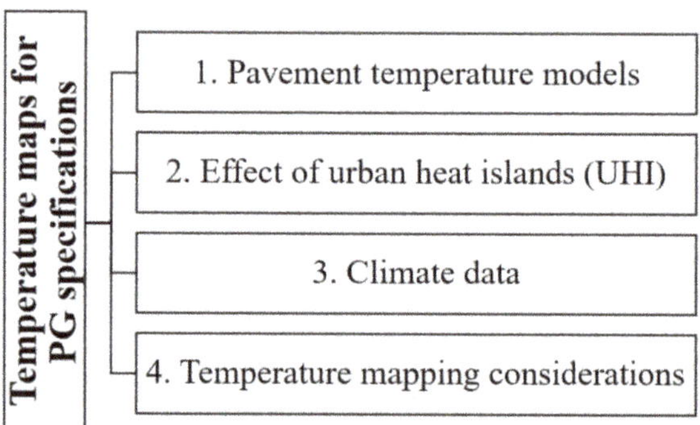

Figure 1. Influencing factors for the development of temperature maps for PG specifications.

A consistent approach, that caters for the variability of geographical, environmental and climatic conditions, does not currently exist within the continent. This paper will therefore explore these factors for the development of temperature maps and present an approach based on the previous experience and findings for the African continent.

2. Pavement Temperature Models

Defining the relationship between air and pavement temperatures is the foundation of developing temperature maps. It is also useful for other applications, such as maintenance and early failure detection [7–9]. Reliable models for determining the pavement temperatures used for pavement design are therefore important, as well as the methodology adopted for measuring air and road temperatures [10].

Most pavement temperature models are fundamentally based on the theory of heat transfer, which mathematically accounts for the transfer mechanisms of heat through conduction, convection and radiation. Some models also account for the energy lost through evapotranspiration [11,12]. The selection of the most appropriate model will be based on the required application, where the selected model will fall under one of the following categories [13]:

1. Analytical models
2. Numerical models or
3. Empirical models

Analytical pavement temperature models are concerned with obtaining an exact solution. Ref. [14] derived a mathematical solution to the partial differential equation governing one-dimensional heat conduction in a semi-infinite medium. This formed the basis of many later developments concerning pavement temperature models. Even though analytical models are focused on obtaining precise results, it should be noted that empirical assumptions are still often used within analytical solutions, and their applicability should be considered before use.

Numerical models use reasonable estimates to arrive at a solution. For South Africa, Williamson and Kirby [15] used a finite difference approach that relied on the thermal properties of pavement materials to determine pavement temperatures. This was followed by the introduction of a corresponding methodology in the mid-seventies [16]. The methodology required extensive solar radiation data which was not available in the country at the time, thus limiting its further implementation.

Empirical models are usually based on a regression analysis of the measured results. Although empirical models do not produce precise estimates of pavement temperatures, they are usually preferred because they require less computation and provide a simplified relationship between pavement temperatures and key weather parameters. However, the calibration against other areas within the country is critical in order to obtain representative results, especially considering how climates can vary quite substantially across a national scale, not to mention a continental scale. For this reason, and depending on the required accuracy, caution should be exercised when using empirical models that have been developed for a climate that is different from the one intended without calibration. Local knowledge of the area is also an added advantage in this regard.

The Strategic Research Highway Program (SHRP) developed the first Superpave™ method for asphalt mix design between 1987 and 1993. Considering the differences in climate, amongst other motivators, the method was investigated for use in South Africa by Everitt et al. [17]. This investigation led to the development of the Viljoen [18] algorithms in 2001, which were based on the empirical relationship between selected sky parameters and pavement temperatures. The linear regression model was able to predict 80% of the measured results within a 3 °C margin for Everitt's 1999 dataset. Other researchers in the country using additional data points further verified this model for certain regions in the country [4,19].

The mean error for the Viljoen [18] equations showed only a slight improvement when compared to the Superpave predictions. With the availability of a solar radiation dataset, the energy balance concept was reconsidered, and a numerical approach based on the method of Orthogonal Collocation on Finite Elements was employed. The numerical model also incorporated ambient air temperature, wind speed and relative humidity. The model was validated with a selected site in Pretoria, South Africa, and a comparison of predicted to actual road temperatures is shown in Figures 2 and 3.

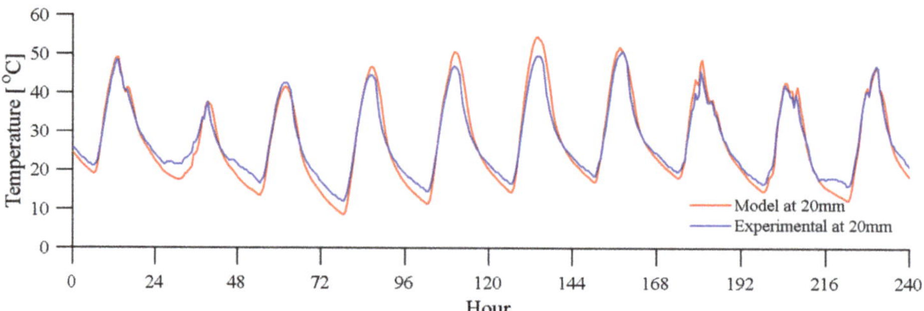

Figure 2. Validation of heat transfer model for a 40 mm asphalt pavement.

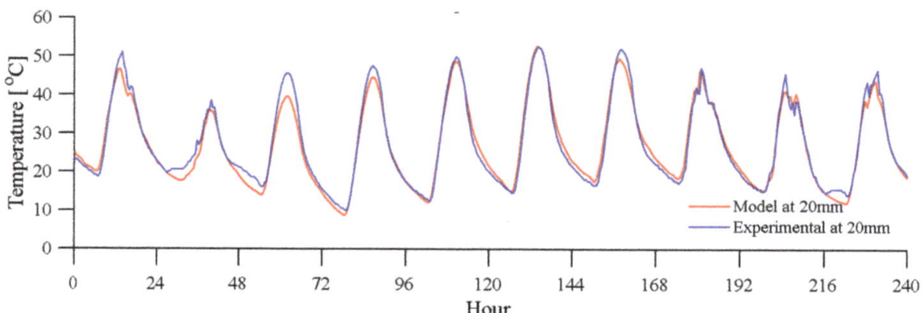

Figure 3. Validation of heat transfer model for a 90 mm asphalt pavement.

The dataset included both clear days and days with transient cloud cover. The recorded air temperature, solar irradiance and wind speed measurements were incorporated in the energy balance model. The daily recorded values are shown in Figure 2. As illustrated above, the pavement temperature relies on more than the air temperature. It is determined based on various weather, climatic factors, as well as pavement material properties. Accurate working models for determining pavement temperatures are therefore required to be verified for the different regions in the continent as the basis for road material selection.

A field experiment was conducted comparing the measured pavement temperature results and the predicted pavement temperatures using (i) a statistical model based on a regression analysis of the historical recorded asphalt temperatures in various locations in South Africa [18] and (ii) an energy balance numerical model with assumed thermal properties of the pavement materials. The numerical model also takes into consideration ambient weather conditions as well the overall pavement structure. The experiment was conducted in Pretoria, South Africa, over a period of 6 months, between September 2019 and February 2020, for two asphalt sections with depths of 40 mm and 90 mm, respectively. Comparisons of the predicted temperature results against the recorded pavement temperatures revealed that, in general, both models show a general agreement in the diurnal temperature patterns within the asphalt layer with varying daily weather conditions, as seen in Figures 1 and 2. However, this was not the case when predicting the daily extreme pavement temperatures necessary for the PG binder selection, as shown in Figures 4 and 5.

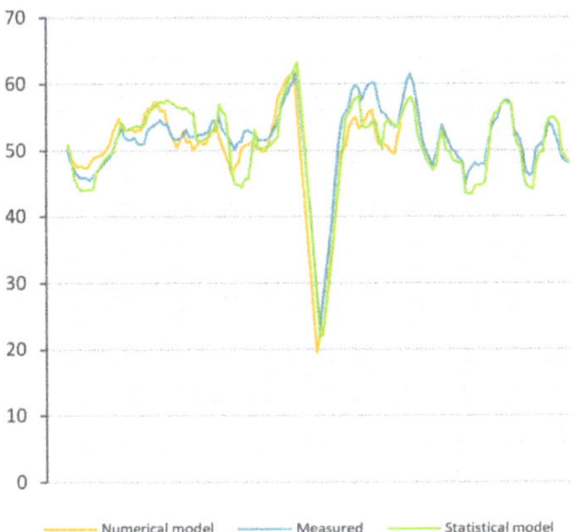

Figure 4. 7-Day maximum temperatures for a 40 mm asphalt pavement at a depth of 20 mm in Pretoria during the summer period.

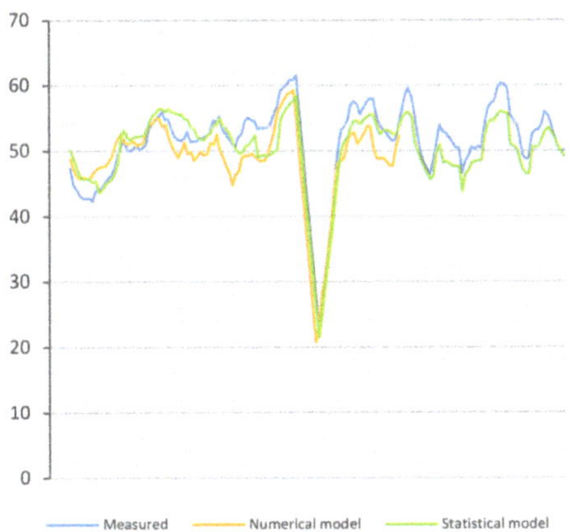

Figure 5. 7-Day maximum temperatures for a 90 mm asphalt pavement at a depth of 20 mm in Pretoria during the summer period.

On average, both the statistical model and the numerical model tend to underestimate the daily and 7-day maximum asphalt temperatures for the asphalt layers with both 40 mm and 90 mm (See Figures 6–9), where the majority of the predicted 7-day average pavement temperatures were observed to be less than the recorded pavement temperatures, particularly after a prolonged event of rain and high cloudy cover. However, this trend was observed specifically for the duration of the experiment after the week-long event of rain and high cloud cover, while the opposite was observed before the event. Larger deviations for maximum temperatures were observed using the numerical model, with the assumed thermal properties of the asphalt layer, compared to the statistical model. During the investigation, the maximum error for calculating the 7-day average maximum temperature using the numerical model was 9.5 °C, with a standard deviation of 3.2 °C for the 40 mm asphalt section. The maximum error obtained for the 90 mm asphalt layer was 8.1 °C and a standard deviation of 3.5 °C. The statistical model exhibited a slightly lower discrepancy, with a maximum error of 7.5 °C and a standard deviation of 2.6 °C for the 40 mm asphalt section. A maximum error of 6.0 °C was found for the 90 mm section and a corresponding 2.9 °C standard deviation. However, further research on the material thermal properties is warranted for conclusive results, given that assumed values from the literature were used during the above investigation.

Pavement temperature relies on more than air temperature; it is determined based on various weather, climatic factors, pavement material properties and the overall pavement structure; therefore, algorithms for determining pavement temperatures may need to be verified for the different regions in the continent as the basis for the road material selection.

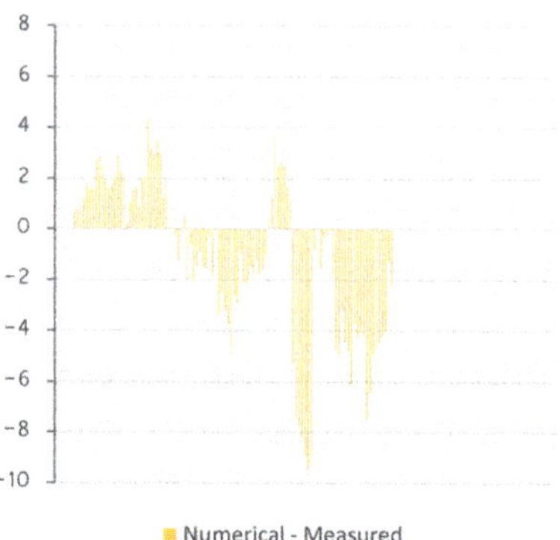

Figure 6. Difference between numerical model predictions and measured temperatures for a 40 mm asphalt pavement in Pretoria during the summer period.

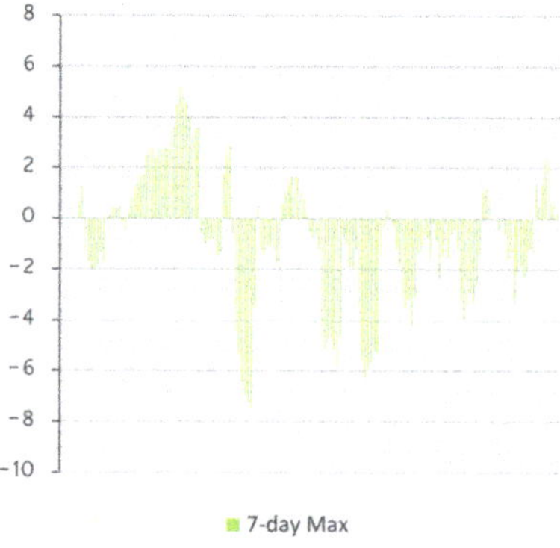

Figure 7. Difference between statistical model predictions and measured temperatures for a 40 mm asphalt pavement in Pretoria during the summer period.

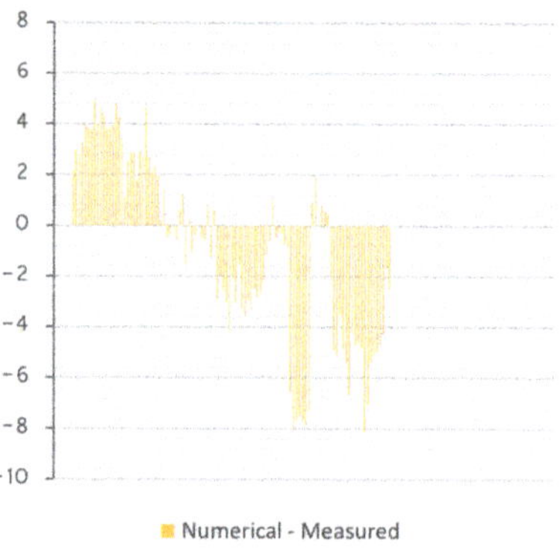

Figure 8. Difference between numerical model predictions and measured temperatures for a 90 mm asphalt pavement in Pretoria during the summer period.

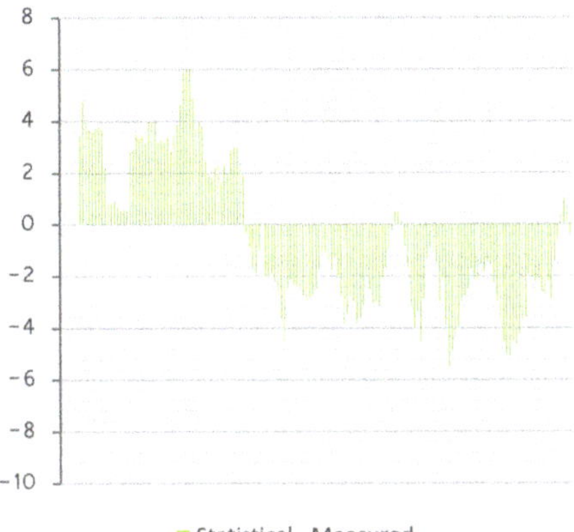

Figure 9. Difference between statistical model predictions and measured temperatures for 90 mm asphalt pavement in Pretoria during the summer period.

2.1. Climatic Factors

The main forms of heat transfer at a pavement's surface are conduction, convection and radiation. These can be modelled mathematically by the heat transfer equation and rely largely on the presiding weather conditions and the road's material properties.

Empirical models do not provide precise results but assist in developing simplified relationships between weather data and pavement temperatures. Therefore, a representative dataset is required to collect pavement temperature measurements based on the geographical location of the area to be mapped.

Weinert used the N-value, which accounts for the evaporation of the warmest month and the precipitation of an area to develop the map commonly referred to in pavement engineering to categorize South African climates into three primary zones, namely (a) wet, (b) moderate and (c) dry, where the majority of the country is considered as "dry" according to the classification [20]. However, further research has shown that this needs to be revisited and has suggested an increase in the number of climatic regions, considering the effects of climate change, to better represent the country [21].

To gain a better global perspective, the Köppen–Geiger climate classification system is one of the methods used to categorize climate regions across the world. This particular system is made of five main climate groups, namely tropical (A), dry (B), temperate (C), continental (D) and polar (E), while sub-categories describe levels of precipitation and heat [22].

New global climate maps based on data obtained between 1980 and 2016 have been established [23] where the continent is divided into 22 climatic zones (See Figure 10), with most of the continent falling under the "hot arid desert" (BWh) as well as the "tropical savannah category" (Aw). Therefore, most parts of the continent can be described either as "tropical" or "dry".

Sub-Saharan Africa shows the highest climatic variation per unit area in comparison to North Africa, which fits almost exclusively into the "hot arid desert" category. Most parts of East and West Africa show a roughly even split between "hot arid desert" and "tropical savannah". The majority of Central Africa can be described as a "tropical savannah" climate according to the climatic map.

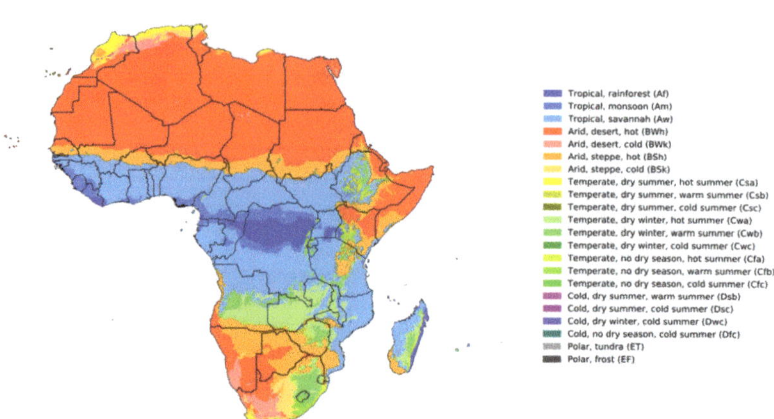

Figure 10. Köppen–Geiger climate classification map for Africa: 1980–2016 [23].

With respect to pavement temperatures, important climatic factors for modelling pavement temperatures have been determined to be solar radiation, air temperature, cloud cover, humidity, sunshine hours and wind speed [13,24,25]. In order to cater for the location in relation to the position of the sun, the zenith angle [18,26–29], phase of incident solar radiation [30], solar declination [31], inclination angle [12] or solar azimuth [32] have all been used depending on the selected pavement temperature model. Understandably, these variables can have a significant impact on the pavement temperature calculation and ultimately the selected binder for a given region in Africa.

2.2. Material Properties

Barber recommended that the thermal properties of pavement materials be taken into account for determining accurate temperatures [14]. According to a sensitivity analysis performed by Viljoen [18], the calculated pavement temperature was most sensitive to the surface material's conductivity more than any of the other material properties [18], as presented in Table 1. The reported values for asphalt thermal conductivity range from 0.7 to 2.0. Ref. [33], however, stated that albedo (solar reflectance as a ratio of incoming radiation) is the parameter that has the highest effect on pavement temperature in comparison to emissivity, thermal conductivity and specific heat.

Table 1. Sensitivity analysis of the variables used in the heat transfer equation [18].

Variable	Range	T_{smax}–T_{airmax} (°C)	Default Value
Surface absorptivity	0.85–0.93	4	0.9
Air-transmission coefficient	0.60–0.81	13	0.8
Zenith angle	1–11	−1	-
Air-emissivity	0.60–0.80	4	0.7
Asphalt-emissivity	0.70–0.90	−5	0.9
Heat transfer coefficient (W/m^2°C)	17.0–22.7	−5	19.9
Thermal conductivity (W/m^2°C)	0.7–2.0	−7	1.4

Qin and Hiller [11] found that the regression coefficients used in their model relied on the surface albedo and the pavement's thermal inertia. The ageing and weathering of an asphalt pavement have also been shown to have an increasing effect on the surface albedo, thus contributing to the change in the temperature profile within the pavement structure over time [29,33,34].

3. Effect of Urban Heat Islands on Pavement Temperatures

Most temperature models are simplified as a one-dimensional heat transfer problem, which is sufficient for pavements in unobstructed areas. However, other research has suggested that a coupled heat exchange model be developed for the heat transfer between the road and the surrounding environment [13]. This would better simulate the effect of urban heat islands on pavement temperatures in built-up areas.

Urban heat islands (UHI) could be considered as a form of meso or micro-climate occurring in urban landscapes due to the higher land coverage of paved surfaces. The UHI effect causes increased air temperatures in metropolitan areas in comparison to neighbouring rural areas especially during night-time, while the effect on daytime air temperatures is minimal. This effect tends to be more pronounced in calm, clear weather and during the winter months, and urban-rural temperature differences of up to 5.9 °C have been observed at midnight in Adelaide, Australia [35]. Air temperatures up to 12 °C warmer in cities have also been observed compared to less built-up surrounding areas on a clear, calm night [36].

The maximum values used to produce temperature maps typically occur around midday, and UHI effects may not have a significant impact on the daily extremes used for material mix-designs. However, this may cause a reduction in the 7-day-averaged maximum pavement temperature typically used for binder selection as a result of the altered diurnal cycles caused by elevated night-time temperatures. Due to the material thermal properties and the response to the daily cyclic variations of temperature, this would influence the overall structural performance and would require further research.

Most conventional construction materials tend to store heat during the day and release it at night, thus significantly contributing to urban heat islands. A surface material's absorptivity governs the amount of solar radiation absorbed by the pavement structure and the typical values for asphalt vary from 0.85 to 0.93, which results in a 4 °C difference on the calculated pavement temperatures [18].

Various studies have been conducted on the UHI phenomenon and its effect on pavement temperatures. There are different strategies to reduce the phenomenon, and the development of cool pavements has been identified as a potential strategy to alleviate the effect. The different methods to achieve a "cool pavement" can fall into one of the following categories [11]:

- Reflective pavements—aim to increase the surface albedo through different techniques in order to reduce pavement temperatures as well as the sensible heat released from the pavement.
- Evaporative cooling pavements—pavement systems that are designed to store water within the pavement layers. Latent heat from the pavement surface can be greatly reduced provided the pavement is wet and there is a steep enough humidity gradient to allow sufficient evaporation. These pavements need to be carefully designed to ensure their thermal performance.
- Thermally modified pavements—material and structural modifications to alter the thermal conductivity or even to self-regulate pavement temperatures. Although this type of pavement system can be effective in reducing daytime temperatures, it tends to have higher night-time temperatures and should therefore be carefully designed.

An example of a reflective pavement was presented by Qin and Hiller [11], who investigated how different pavement properties influenced pavement temperatures and the effect on urban heat islands. They used an energy balance method to model the heat exchange at the pavement's surface. Similar to Williamson's [24] model, their model accounts for incoming solar radiation and energy loss through convection, conduction and radiation. However, Qin and Hiller's [11] model also accounts for energy loss by means of evaporation. This allowed them to investigate the energy partition in relation to the development of cool pavement systems. They found that most of the absorbed energy is released as latent heat and long-wave radiation, particularly during the summer months, and so concluded that increasing the surface albedo and "enhancing the evaporative flux" was effective in reducing the latent heat and thus creating cool pavements.

An example of a thermally modified pavement was presented by Chen et al., who proposed an analytical approach to evaluate temperature profiles of thermally modified asphalt pavements as well as their effect on urban heat islands [37]. The proposed model is based on Green's function to predict temperatures within a multi-layered pavement system and was validated with outdoor field experiments. They were able to reduce pavement temperatures at various depths by increasing the pavement surface albedo, which had the effect of increasing the heat output intensity during the day and reducing it during the night. They found that combining a highly conductive surface material and base course showed the best thermal performance to alleviate the effect of UHI by reducing the pavement's overall maximum heat output intensity.

Adebayo demonstrated how different land uses generate different microclimates in Nigeria [38]. This was attributed to the complex interaction of incident solar radiation with surface materials and corresponding thermal properties. It was also argued that the interpolation techniques used for most "heat island" isoline maps are not ideal for modelling urban heat islands because the land use between known points is often overlooked with traditional approaches. This highlights the importance of the local terrain and climate knowledge for producing representative temperature maps.

There are hotspots within urban areas that are more susceptible to heat islands than other regions, as demonstrated through various studies in Africa [38–40]. This may need to be taken into consideration during the material selection of the asphalt mix design process, and a 6 °C difference between urban and city air temperatures could warrant a higher PG bitumen for certain metropolitan areas. In order to account for this, the data source needs to be at a high enough resolution and of sufficient quantity to capture the effect of urban heat islands.

Multiple climatic factors and material properties need to be considered for accurate pavement temperature prediction, as demonstrated above. Obtaining an extremely precise output is not always possible and, depending on the application, not necessary. However, certain considerations that significantly affect the pavement temperatures (such as albedo and the transmission co-efficient of unit air mass, which is dependent on the cloud cover and zenith angle) need to be considered for the development of pavement temperature maps. Another aspect that may warrant further investigation is the influence of air pollutants on incoming solar radiation, especially in densely populated urban areas. For instance, Peters et al. found a reduction of about 11.5% for incoming solar radiation for PV panels in Delhi as a result of air pollution [41]. A similar investigation was conducted in 16 other countries and found insolation reductions between 2.0% and 9.1% depending on the city.

UHI may be detectable from weather data if there are enough active weather stations within a built-up metropolitan area that are used in producing temperature maps. However, this is often not the case, especially when the operating weather stations are limited and few compared to high resolution weather data from satellite or re-analysis models.

Temperature maps showing the spatial distribution of pavement temperatures across the city of Tshwane, South Africa, were produced during this investigation using UHI weather data. The dataset was provided at a 1-km resolution within the city's boundary. The 7-day hottest pavement temperatures were calculated according to the Viljoen [18] algorithms for the period between 2007 and 2017, and the 97.5th percentile of the data set is shown in Figure 11.

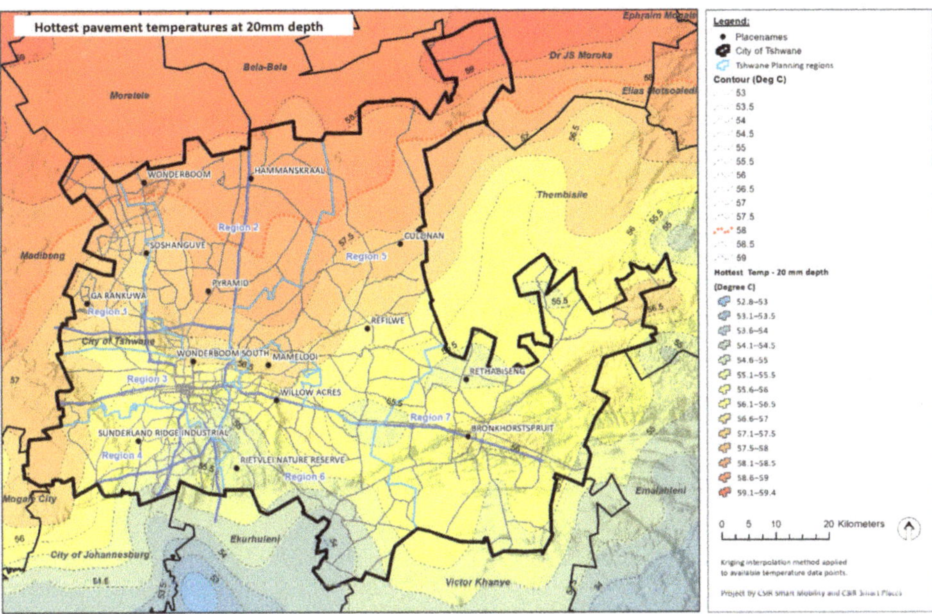

Figure 11. Hottest pavement temperatures at a depth of 20 mm for the city of Tshwane.

The temperature contours are consistent with previous findings for maximum temperatures in the period 2000–2020 [42], where the spatial temperature distribution in Figure 11 can now be observed at a higher resolution at a municipal planning scale. It is also observed that the temperature contour for a PG 58 is slowly moving south in comparison to the maximum temperature map used in the national specification for PG bitumen in South Africa [43]. This is indicative of the city's warming climate over time, considering that the UHI data used for the study are more recent than those used to develop the specification.

During the analysis, the effect of UHI was more detectable in the minimum pavement temperatures compared to the maximum pavement temperatures. The minimum pavement temperature map is shown in Figure 12.

From the analysis, relatively high minimum pavement temperatures for the city were noted in comparison to the pavement temperature maps for the period 2000–2020, where a previous study [42] illustrates a minimum pavement temperature classification of PG 2 for the Tshwane area, whereas the temperatures calculated using the UHI data show that the city is classified as a PG 8 region, with the exception of the hotspots that warrant a PG 12 minimum pavement temperature binder.

Three hotspots were detected for the city, namely within Planning Regions 1, 2 and 3, which are among the most densely populated regions. The most densely populated region is Region 1, which is situated in the North-West corner of the city's jurisdiction and accommodates 811,570 (28%) residents out of a population of 2,921,488, according to the 2011 census. The region has 28 wards and includes Soshanguve, Mabopane, Winterveld, Ga-Rankuwa and Pretoria North, as well as a provincial road (R80) which is a Class 1 primary metropolitan distributor for that location.

The higher pavement temperatures were observed to coincide with the built-up areas, according to the locality map which illustrates the various general land uses for the city, as seen in Figure 13. Lower pavement temperatures were observed for the Southeast areas, which also have a lower population density and more rural land coverage compared to the other planning regions.

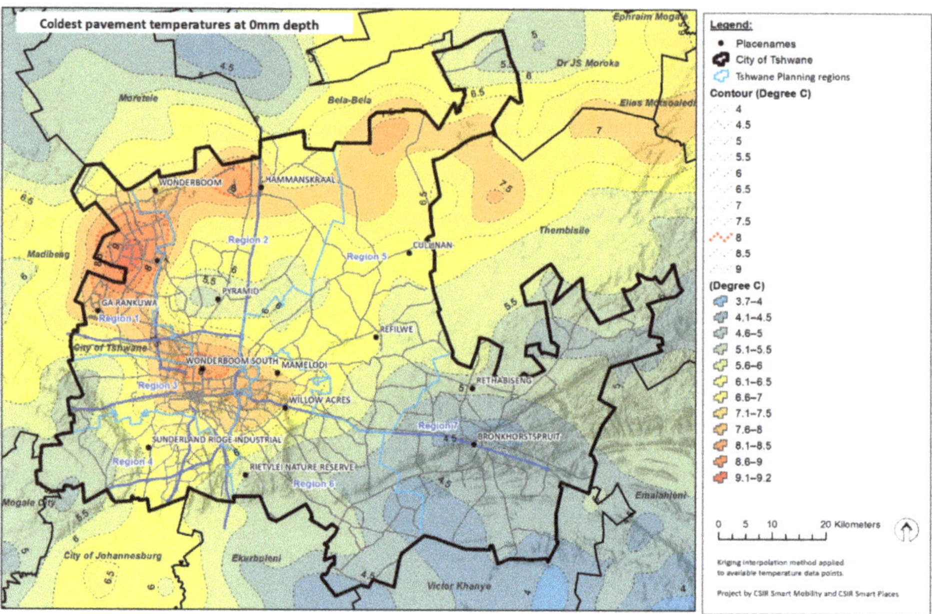

Figure 12. Coldest surface pavement temperatures for the city of Tshwane.

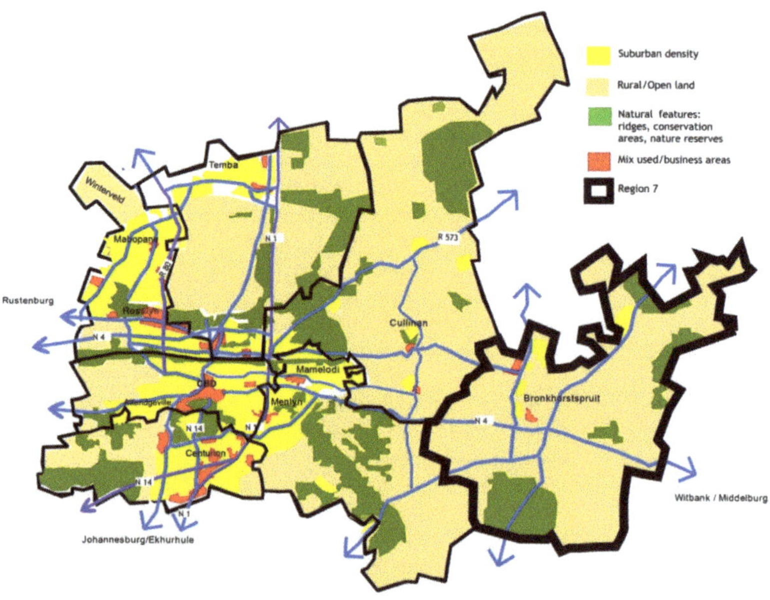

Figure 13. City of Tshwane's land use map per planning region (city of Tshwane, 2015).

The effect of UHI would need to be similarly investigated for other metropolitan cities in the continent to determine the direct effect on bituminous binder selections. While it is acknowledged that the continent primarily consists of natural land cover, i.e., trees, shrubs, cropland, grassland and areas with sparse to no vegetation [44], there are concentrated areas across the continent, usually in and around metropolitan cities that can have a potentially significant UHI effect. Therefore, this may need to be assessed in order to understand the impact of heat islands, particularly for the high-volume road design and construction in these areas. The projections by [45] also indicated rapid rates of urbanization and population growth, above 50%, for some of the fastest growing cities in Africa—namely, Dar es Salaam (Tanzania), Nairobi (Kenya), Kinshasa (Democratic Republic of Congo), Luanda (Angola), Addis Ababa (Ethiopia), Abidjan (Côte d'Ivoire) and Dakar (Senegal).

The high urbanization rates can subsequently lead to increased air and road temperatures, and thus facilitate the need to adjust how construction materials are selected. This may be true to maintain the integrity of roads but can also be managed to assist in reducing air temperatures through the implementation of "cool pavements" as a potential strategy.

4. Climate Data

The data source for air temperatures along with their collection method needs to be carefully considered, as each source is associated with a specific set of assumptions that need to be understood before using.

For correlations and comparisons, the method and the location of data collections need to be standardized. The ideal situation would be to compare two 20-year data sets from the same geographical location, i.e., comparing air temperature data and measured pavement temperature data. However, this becomes complicated considering the differences of terrain between the two points and the presence of microclimates, thus making the comparison unrealistic.

The different types of air temperature data sources for the temperature mapping discussed in this chapter include:

1. Weather station data
2. Satellite data
3. Reanalysis data
4. Global Climate Models

4.1. Weather Station Data

Data from weather stations can be collected from manual or automated weather stations. Spatially distributed and incomplete datasets from weather stations have been encountered, affecting the accuracy of the final output [6,46]. Weather stations vary in accuracy because they are prone to human errors, such as data recording and archiving errors. The location and placement of weather stations will also affect the type of data recorded and their relevance in developing pavement temperature maps. In African countries with a handful of weather stations, such data are only useful for calibration purposes. In addition, only a select number of weather stations measure air temperature, as rainfall data are the most recorded.

4.2. Satellite Data

Satellite data are often used where on-site measurements are not possible, unreliable or simply unavailable [47], or where satellite-based precipitation data and reanalysis data were evaluated in comparison with the observed rain-gauge readings. [47] found that the satellite data used in their investigation exhibited the highest correlation and the lowest deviation from the observed rain-gauge measurements. The importance of the temporal coverage was also highlighted during their investigation. Satellite-based information is usually subject to calibration errors, but it is usually a relatively accurate source of data [48].

4.3. Reanalysis Models

Reanalysis models are derived from observed data (either from satellites or weather stations) and are useful for areas where observed data are unavailable. The MERRA (Modern-Era Retrospective Analysis for Research and Applications) database is an example of data obtained from a reanalysis model and is available on the LTPP InfoPave^{TM} [49]. It combines computed model fields with ground-based, ocean-based, atmospheric-based and satellite-based observations [50]. MERRA data were compared to weather station data and reported to be "as good and, in many cases, substantially better than equivalent ground-based weather station data" [50]. A comparison was also conducted during the development of Tanzanian pavement temperature maps, and a variation ranging from <1 °C to 10.3 °C was observed [6]. The MERRA database showed relatively higher temperatures and was therefore considered as a conservative option for setting the rutting PG temperature criteria for Tanzania. Given that pavements fail at extreme conditions, the measured historic air temperature data for South Africa and Tanzania were used to quantify the number of days with extreme pavement temperatures over the years. The maps developed for Tanzania used MERRA data and are shown in Figures 14 and 15.

Reanalysis data are advised for African countries where missing weather station data is a common occurrence. Due to the nature of reanalysis data, bias correction for daily extreme points is recommended in order to obtain representable temperatures for binder selection.

Unless special precautions have been taken into consideration during the data collection process, the sources do not account either for complex landscapes or their potential effects on air temperatures.

The method used to obtain air temperature data will also affect the quality of the raw data. For instance, weather stations utilize thermometers placed inside Stevenson boxes to measure air temperature, whereas satellites measure radiances in different wavelength bands and mathematically invert them to obtain temperature inferences [51,52]. Even sources with common attributes will show differences in the data, as certain formulations and calibrations will differ and alter the data produced [53]. It is also important to consider the sensitivity of the equipment used by a source, as it could also affect the data produced.

Data from the various sources can vary significantly due to inherent methodical, operator and instrumental errors leading to more pronounced effects on pavement temperature calculations depending on the model used.

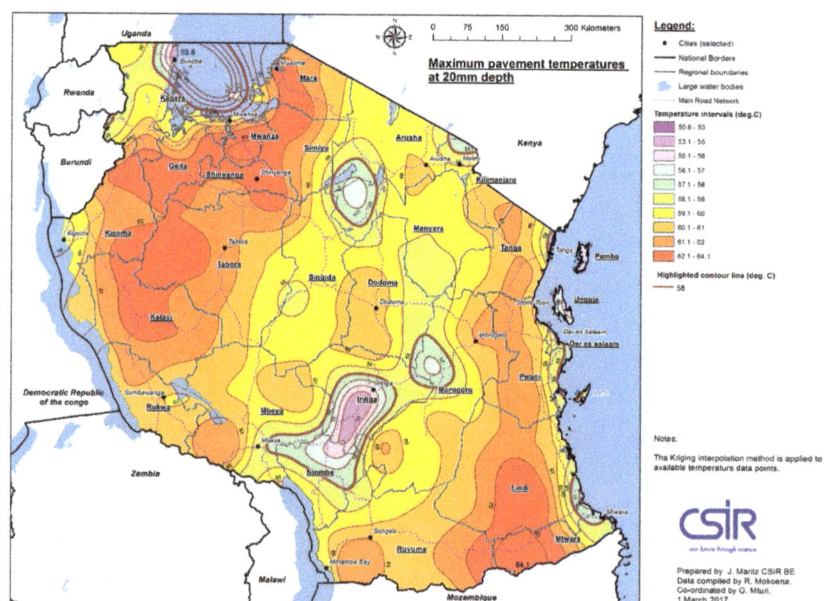

Figure 14. Tanzanian maximum pavement temperatures at a depth of 20 mm [6].

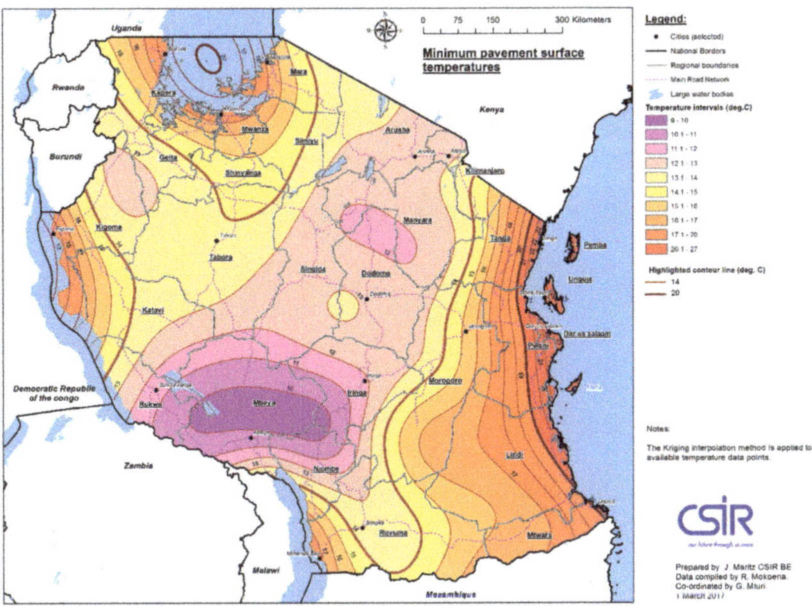

Figure 15. Tanzanian minimum surface pavement temperatures [6].

4.4. Global Climate Models

Climate models are fundamentally based on the laws of physics and use mathematical equations to simulate global climates. Downscaling techniques are employed to obtain a higher geographical resolution and to better represent the regional climates usually required for road design procedures. Even with downscaling techniques, micro or meso-climates are often not well represented by climate models [54] and would require a customized approach for specific road sections.

To obtain high-resolution climate data, climate model simulations covering the historic and future climate (1960–2100) were performed at a 50 km horizontal resolution globally, using the conformal-cubic atmospheric model (CCAM). The CCAM model is a variable-resolution global climate model (GCM) developed by the Commonwealth Scientific and Industrial Research Organization (CSIRO) [55–57]. The CCAM model is run coupled with a dynamical land-surface model known as the CSIRO Atmosphere Biosphere Land Exchange (CABLE). For the global runs, the two GCM models were used to force CCAM-CABLE with bias-adjusted sea-surface temperatures (SSTs) and sea-ice concentration. The GCMs in this study are the Community Climate and Earth-System Simulator (ACCESS1-0) and the National Centre for Meteorological Research Coupled Global Climate Model, version 5 (CNRM-CM5). Both GCMs' outputs, used as CCAM-CABLE forcing data, formed part of the Coupled Model Intercomparison Project Phase Five (CMIP5) and the Assessment Report Five (AR5) of the Intergovernmental Panel on Climate Change (IPCC).

The obtained CCCAM-CABLE native model outputs were further downscaled to a high resolution of 8 km (0.08° degrees in latitude and longitude) using a spectral-nudging strategy. Subsequently, the 8 km horizontal was further downscaled to a 1 km resolution using CCAM-CABLE coupled with an urban canopy scheme that is based on the Town Energy Budget approach [58]. The native CCAM-CABLE data used in this case derived from the proceeding downscaling experiment with CNRM-CM5 model forcing data. All the high-resolution downscaling model experiments were conducted with CCAM-CABLE prescribed with atmospheric CO_2, sulphate and ozone under the Representative Concentration pathway 8.5 (RCP8.5), which is commonly referred to as the low-mitigation scenario for the 21st century. The ability of CCAM-CABLE to simulate climate over Southern Africa has been showcased in several scientific outputs [59–64].

The 8-km horizontal downscaling based on ACCESS1-0 forcings is used to demonstrate South Africa's wide long-term average pavement spatial pattern temperatures, while the spatial features of the urban pavement temperature at 1 km are depicted using CCAM native outputs obtained from the CNRM-CM5 model downscaling.

Depending on the data collection method, data cleaning is required particularly along the coastal regions, as shown in Figure 16, and within the proximity of other large water bodies prior to processing, so that the road temperatures are accurately represented. This is essential in order to avoid skewed results leading to lower temperature grades of bitumen being allocated to certain regions.

Figure 16. Pre-processing of climate model data for points over water.

4.5. Designing for Current vs. Future Scenario

Past research has shown that, due to climate change, there is a necessity to plan and design pavements for expected climatic conditions rather than historic climates [65–69]. This can be challenging because there are limited standardized procedures to do so. It is also not a simple exercise because climate models are not predictions or forecasts, but rather projections according to the likelihood of future CO_2 emissions. Therefore, using exact outputs from the models for engineering design purposes may not be appropriate. Simulation outputs are best used to allow informed planning that can provide indicators for the development of guidelines.

The importance of considering climate change projections in this process and incorporating the expected future increases in temperatures has been previously demonstrated, due to the potentially adverse effects that climate change may have on bituminous binder selection [42]. The authors incorporated a high-resolution model simulation of present-day climate as well as future climate projections for South Africa. With coupling and corrections for specific terrains, the conformal-cubic atmospheric model (CCAM) [55–57] has a proven ability to realistically simulate the present-day Southern African climate. Although a low-mitigation scenario climate model was used based on average air temperatures, the observed impact appeared significant and prompted the South African pavement industry resolution to research mitigation measures.

They demonstrated an increase of up to 7 °C in maximum pavement temperatures between four 20-year periods spanning between 1980 and 2060. Figure 17 shows a trend depicting an increase in extreme temperatures.

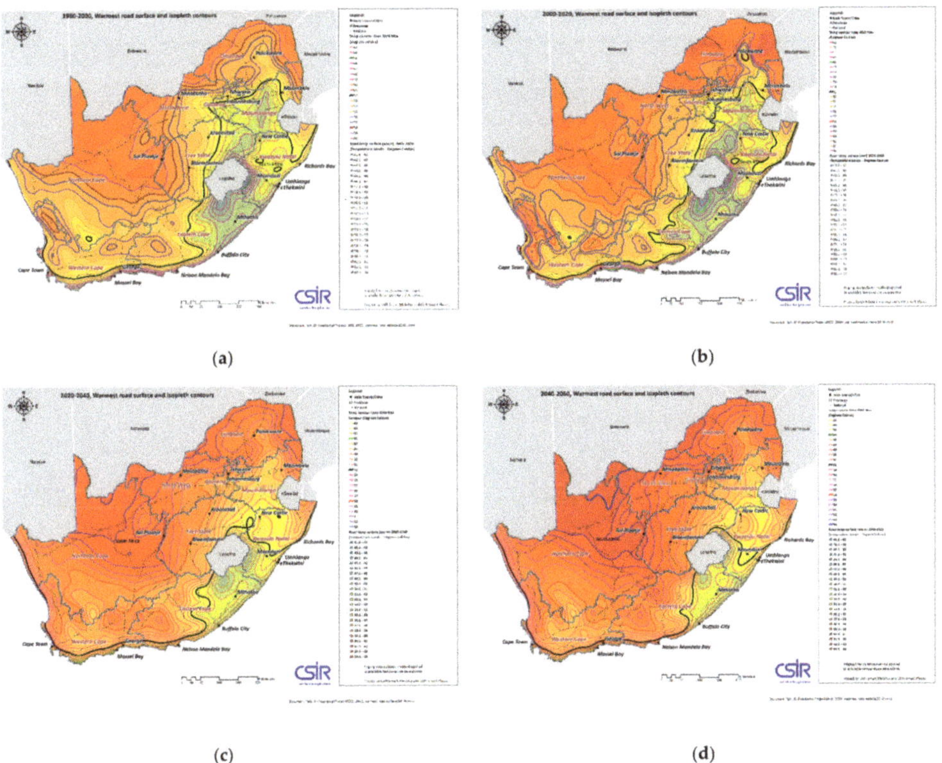

Figure 17. Effect of climate change on maximum pavement temperatures over 80 years between (**a**) 1980 and 200; (**b**)2000 and 2020; (**c**) 2020 and 2040; (**d**) 2040 and 2060 [42].

Given the impacts of climate change on the air and, consequently, on the pavement temperatures, there is a need to develop a customized approach for Africa with projections of extreme pavement temperatures into future scenarios, for incorporation into designs and bitumen selection specifications.

5. Mapping Considerations

To assist during the process of specifying the asphalt bitumen, country pavement temperature maps are generated from calculated pavement temperature points according to the latitude and longitude co-ordinates.

In order to extend these data points across a country, an interpolation process is applied. The selected general interpolation technique is the Kriging method or a Gaussian process of regression [43,70]. Kriging assumes that the distance or direction between the sample points reflects a spatial correlation that can be used to explain the variation in the surface. The Kriging tool fits a mathematical function to a specified number of points, or all points within a specified radius, to determine the output value for each location. Like most interpolation techniques, Kriging is built on the basis that things that are close to one another are more alike than those farther away (quantified here as spatial autocorrelation).

The input dataset obtained from weather stations is not always comprehensive but is still considered to be the most representative. Given the limited number of data points from weather stations and the Kriging interpolation used to generate the maps, based on the interpolation technique, the most accurate pavement temperatures are obtained within the proximity of the existing weather stations. This contrasts with temperature maps developed from climate modelling data, where the calculated points are based on grid points obtained from downscaling techniques and not from weather station locations.

The factors to consider, which affect the reliability of pavement temperature maps include the number and distribution of data points as well as the accuracy of the Kriging interpolation method at points between known values.

Therefore, there is a need for a statistical methodology to define the distribution with respect to the road network.

Figure 18 shows the weather stations superimposed onto the current South African temperature map based on the data from 122 weather stations [43], where the national road network is illustrated in blue. Given the difference in nature of both data sources (weather station vs. climate model), a direct comparison of the spatial distribution between both maps cannot be made.

Figure 19 shows the comparison of the distribution of the weather station points vs. the 8 × 8 km grid acquired from the climate model across South Africa. The current temperature maps [43] are based on measured data but limited to only a few sites in comparison to the climate model. This results in less accuracy, particularly in areas where weather stations are far from one another.

A comparison was therefore made between the results obtained from an 8 × 8 km grid and selected data points from the same model. The reduced points were located at the same position as the weather stations used for producing the current maps.

Figures 20 and 21 illustrate the difference in temperature maps depending on the number of points as well as the distribution of those points. It is demonstrated that a change in the number and distribution of points affects the distance between the contour plots and the resolution of the maps, which becomes more important for regional road planning, design and construction. In areas with few and distant spatialized weather station data, the bitumen performance-grading requirement of local roads can be inaccurate. This effect is demonstrated by the steep contour plots observed in the Western Cape region, where there are several weather stations located near each other in comparison to the rest of the country. An increase in the PG binder requirement is observed in Polokwane as well as along the Northern/Western Cape border, with a higher resolution data source. The introduction of a PG 58 zone in the Western/Eastern Cape is also observed for the same reason.

The Kernell Density "interpretation" process, which was used for developing the maps, relies on more points to increase the accuracy of the maps, thus making the modelled data more advantageous due to its higher resolution and regular distribution across the country.

For the foreseeable future, the use of climate models and mapping for material selection presents a promising outlook for engineering design, where expected weather events resulting from a changing climate can be accommodated for in the provision of resilient road infrastructure. The effects of climate change usually affect a region spanning multiple countries, which may also influence the process of using downscaled climate models.

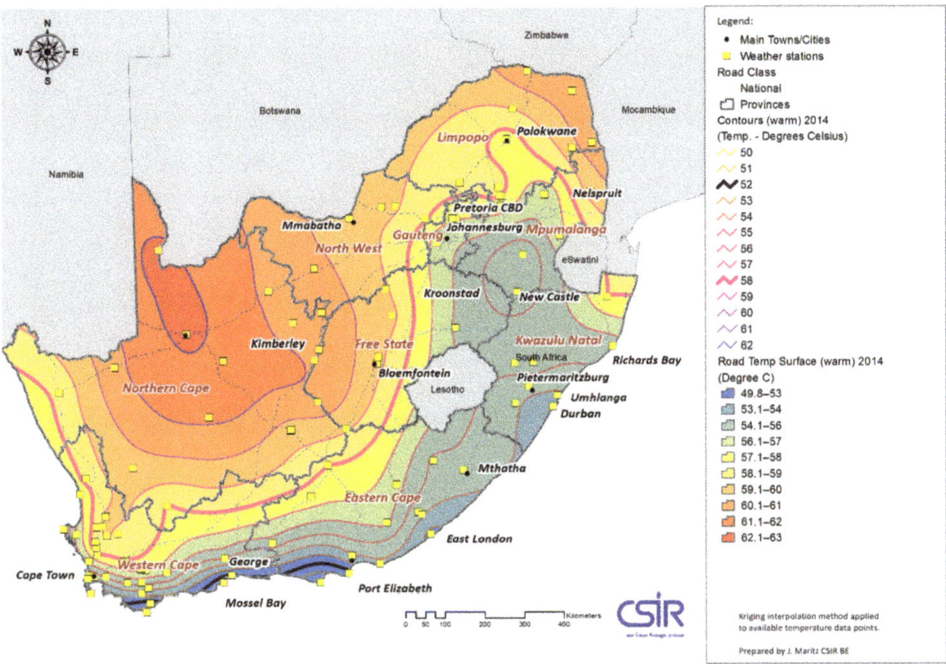

Figure 18. South African 7-day-average maximum pavement temperature map with weather station locations.

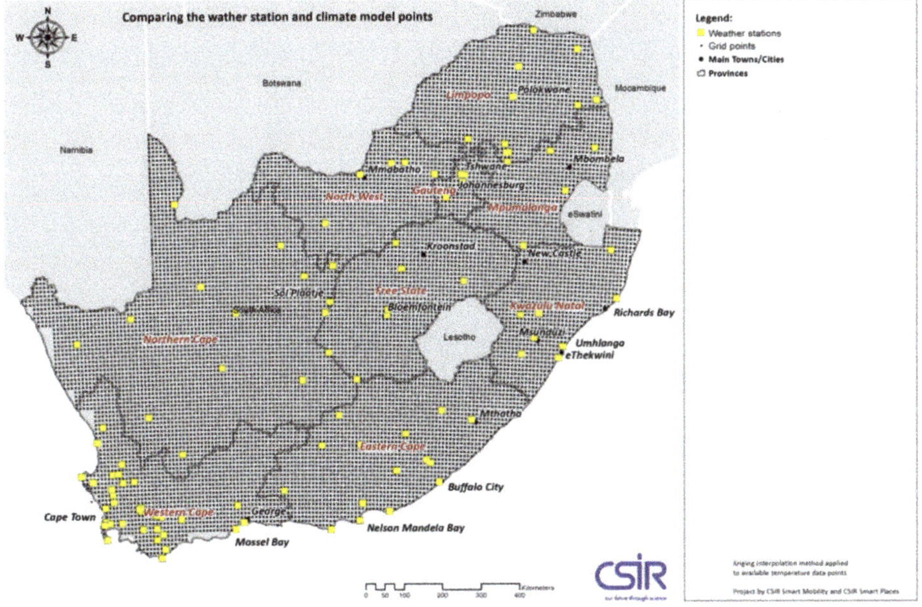

Figure 19. Comparison of South African weather station points and climate model points.

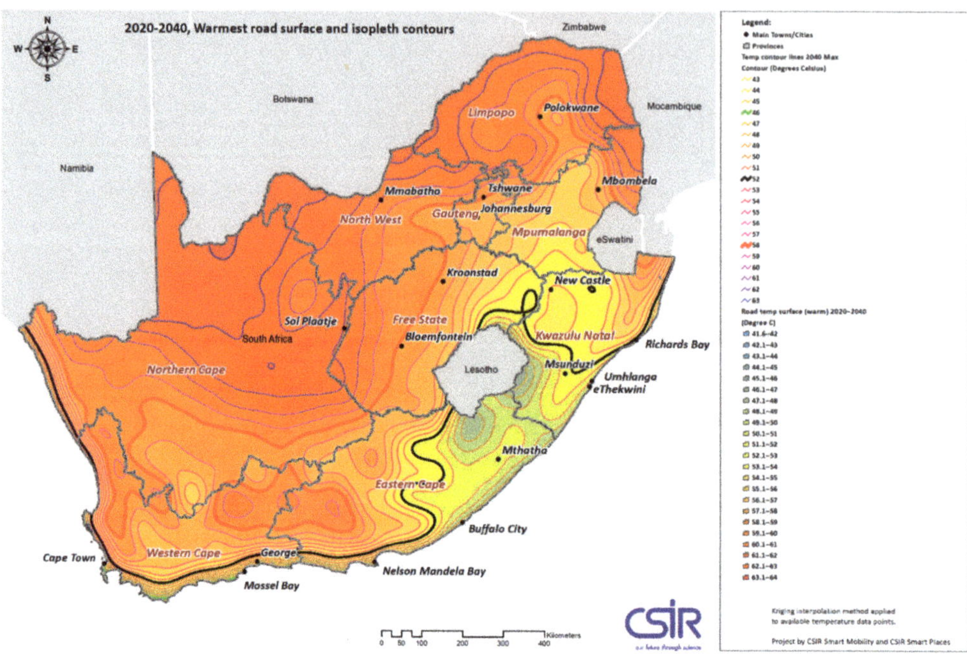

Figure 20. 2020–2040 map and national road network from the 8 × 8 km grid.

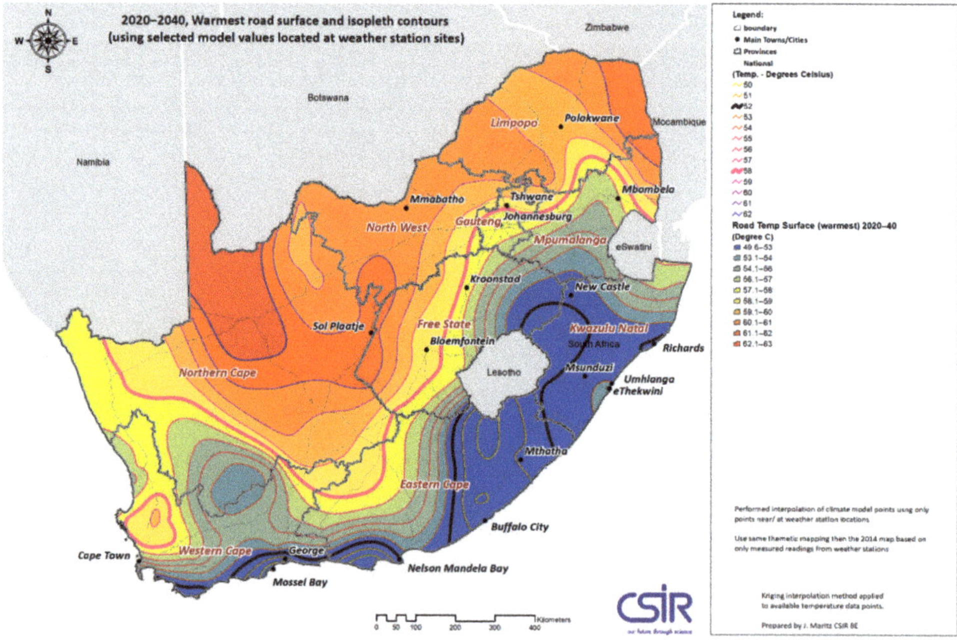

Figure 21. 2020–2040 map and national road network generated using limited climate model points.

6. Conclusions

The African case studies presented in this paper show the critical factors for the development of the temperature maps that are necessary for the performance-graded bitumen selection used in asphalt pavement design and construction. The findings highlight the influence of the following identified parameters:

1. The selection of a pavement temperature model
2. The effect of urban heat islands on bituminous binder selection
3. Source of climate data
4. Mapping considerations and scale of use

The calculation of pavement temperatures is the foundation of temperature zoning, and, depending on the model, it must take into consideration the pavement structure and the material's thermal properties, as well as the weather conditions, which are dependent on the prevailing regional climate. Most pavement temperature models are fundamentally based on the theory of heat transfer and can be categorized based on the method of calculation. For the purpose of producing national temperature maps, empirical models can be used and are usually preferred over analytical and numerical models, since less computation is required and they usually provide a simplified relationship between pavement temperatures and key weather parameters. With the use of empirical models, the calibration against other areas within the country is also recommended in order to obtain pavement temperatures that are representative of the various climatic zones prevalent in the country.

For this paper, an urban heat island model for a local South African municipality was used to develop corresponding pavement temperature maps, and minimal differences were observed in the maximum pavement temperatures. However, the formation of heat island hotspots in the city had a significant effect on the minimum pavement temperatures, where an increase by two performance grades was observed in comparison to the calculated minimum pavement temperatures at the national scale. The findings indicated that a conservative PG allocation for minimum temperatures can be expected in areas where urban heat islands occur.

The ideal air temperature source for pavement temperature calculations is provided by the most recent observed measurements. This paper showed that due to the sparse distribution of weather stations and the sometimes inconsistent datasets produced by the weather stations, parameter outputs from reanalysis or satellite models are a more convenient solution, given that they are based on observed measurements. Engineering designs are often based on historical and extreme weather observations, but this is restricting when infrastructure is designed and built to withstand expected conditions. With a changing global climate, it is necessary to make informed planning and design decisions based on the expected changes in climatic conditions. This paper discussed the usefulness of downscaled Global Climate Models for adaptive engineering design, as they can simulate expected climates for specified emission scenarios.

Finally, the paper also presented and discussed geographical mapping considerations for the purpose of selecting performance-graded bitumen. Pavement temperature visualization in the form of geographical maps is a particularly useful tool for engineers in selecting the most appropriate binder for a particular location. This paper showed that, depending on the specific scale and use required, the choice of data source will affect the resolution and accuracy of the map.

Author Contributions: Conceptualization, G.M.; methodology, G.M. and R.M.; software, J.M.; validation, P.K.; formal analysis, R.M., J.M., M.M. and P.K.; investigation, R.M. and G.M.; resources, R.M., M.M. and P.K.; data curation, R.M.; writing—original draft preparation, R.M. and G.M.; writing—review and editing, R.M., G.M., J.M. and M.M.; visualization, J.M.; supervision, G.M.; project administration, R.M.; funding acquisition, R.M. All authors have read and agreed to the published version of the manuscript.

Funding: This research was funded by the Council for Scientific and Industrial Research (CSIR) through the Parliamentary Grant funding programme. A component of the dataset used for this study was funded by the City of Tshwane Climate Action Plan (Project EECM138).

Institutional Review Board Statement: Not applicable.

Informed Consent Statement: Not applicable.

Data Availability Statement: Restrictions apply to the availability of these data. Data was obtained from the Council for Scientific and Industrial Research (CSIR) and the City of Tshwane (CoT) and are available from the corresponding author with the permission of the CSIR and the CoT where applicable.

Acknowledgments: In addition to the CSIR, the authors would like to acknowledge and thank the following: (i) Yvette van Rensburg for the software support on data handling and processing, (ii) The City of Tshwane for permission to use urban heat island data from the Climate Action Plan project (EECM138) and (iii) Colin Fisher for providing technical support with pavement temperature measurements.

Conflicts of Interest: The authors declare no conflict of interest.

References

1. Kaplan, S.D.; Teufel, F. The role of road networks in addressing fragility and building resilience. *Afr. Econ. Brief.* **2016**, *7*, 1–8.
2. Karlaftis, M.; Kepaptsoglou, K. Performance measurement in the road sector: A cross-country review of experience. In Proceedings of the International Transport Forum, Athens, Greece, 25–26 October 2012.
3. Bredenhann, S.J.; Myburgh, P.A.; Jenkins, K.J.; O'Connell, J.S.; Rowe, G.M.; D'Angelo, J. Implementation of a performance-grade bitumen specification in South Africa. *J. S. Afr. Inst. Civ. Eng.* **2019**, *61*, 20–31. [CrossRef]
4. O'Connell, J. *Interim Report: Evaluation of South African Bituminous Binders in Terms of a Proposed Performance Related Specification*; CSIR Built Environment: Pretoria, South Africa, 2012.
5. Cominsky, R.J.; Huber, G.A.; Kennedy, T.W.; Anderson, M. Strategic highway research program. In *The Superpave Mix Design Manual for New Construction and Overlays, SHRP-A-407*; National Academy of Sciences: Washington, DC, USA, 1994.
6. Mokoena, R.; Mturi, G. *Temperature Mapping Procedure for Selection of Tanzanian Bituminous Binders*; CSIR Built Environment: Pretoria, South Africa, 2017.
7. Gustavsson, T. Thermal mapping—a technique for road climatological studies. *Meteorol. Appl.* **1999**, *6*, 385–394. [CrossRef]
8. Mahura, A.; Petersen, C.; Sass, B.H.; Holm, P.; Pedersen, T.S. *Thermal Mapping Data Measurements: Road Weather SeaSons 2008–2011*; Danish Meteorological Institute: Copenhagen, Denmark, 2011.
9. Marchetti, M.; Chapman, L.; Khalifa, A.; Buès, M. New role of thermal mapping in winter maintenance with principal components analysis. *Adv. Meteorol.* **2014**, 1–11. [CrossRef]
10. Williamson, R.H. *Environmental Effects in Road Pavements—Pavement Temperatures: Their Measurement and Prediction*; National Institute for Road Research: Pretoria, South Africa, 1972.
11. Qin, Y.; Hiller, J.E. Understanding pavement-surface energy balance and its implications on cool pavement development. *Energy Build.* **2014**, *85*, 389–399. [CrossRef]
12. Huang, K.; Zollinger, D.G.; Shi, X.; Sun, P. A developed method of analyzing temperature and moisture profiles in rigid pavement slabs. *Constr. Build. Mater.* **2017**, *151*, 782–788. [CrossRef]
13. Chen, J.; Wang, H.; Xie, P. Pavement temperature prediction: Theoretical models and critical affecting factors. *Appl. Therm. Eng.* **2019**, *158*, 1359–4311. [CrossRef]
14. Barber, E.S. Calculation of Maximum Pavement Temperatures from Weather Reports. *Highw. Res. Board Bull.* **1957**, *168*, 1–8.
15. Williamson, R.H.; Kirby, W.T. *Measurement of Pavement and Air Temperatures: A Basic Discussion*; National Institute for Road Research: Pretoria, South Africa, 1970.
16. Williamson, R.H.; Marais, C.P. Pavement temperatures in Southern Africa. *Die Siviele Ing. Suid-Afr.* **1975**, *17*, 203–214.
17. Everitt, P.R.; Dunbar, R.; Swart, G.; Rossman, D. Current pavement temperature monitoring in South Africa and its applicability to superpave and FWD requirements. In CD-ROM Proceedings of the 7th Conference on Asphalt Pavements in Southern Africa (CAPSA), Victoria Falls, Zimbabwe, 29 August–2 September 1999.
18. Viljoen, A.W. *Estimating Asphalt Temperatures from Air Temperatures and Basic Sky Parameters*; CSIR Transportek: Pretoria, South Africa, 2001.
19. Denneman, E. The Application of Locally Developed Pavement Temperature Prediction Algorithms in Performance Grade (PG) Binder Selection. In Proceedings of the 26th Annual Southern African Transport Conference (SATC), Pretoria, South Africa, 9–12 July 2007.
20. Weinert, H.H. *The Natural Road Construction Materials of Southern Africa*; National Institute for Transport and Road Research: Pretoria, South Africa, 1980.

21. Mndawe, M.B.; Ndambuki, J.; Kupolati, W. Revision of the macro climatic regions of Southern Africa. *Int. J. Sustain. Dev.* **2013**, *6*, 37–44.
22. Chen, D.; Chen, H.W. Using the Köppen classification to quantify climate variation and change: An example for 1901–2010. *Environ. Dev.* **2013**, *6*, 69–79. [CrossRef]
23. Beck, H.E.; Zimmermann, N.E.; McVicar, T.R.; Vergopolan, N.; Berg, A.; Wood, E.F. Present and future Köppen-Geiger climate classification maps at 1 km resolution. *Sci. Data* **2018**, *5*, 180214–180225. [CrossRef] [PubMed]
24. Williamson, R.H. *The Estimation of Pavement Temperatures from Finite Difference Considerations*; National Institute for Road Research: Pretoria, South Africa, 1971.
25. Kršmanc, R.; Slak, A.Š.; Demšar, J. Statistical approach for forecasting road surface temperature. *Meteorol. Appl.* **2013**, *20*, 439–446. [CrossRef]
26. Mrawira, D.M.; Luca, J. Thermal properties and transient temperature response of full-depth asphalt pavements. *Transp. Res. Rec. J. Transp. Res. Board* **2002**, *1809*, 160–171. [CrossRef]
27. Minhoto, M.J.C.; Pais, J.C.; Pereira, P.A.A.; Picado-Santos, L.G. Predicting asphalt pavement temperature with a three-dimensional finite element method. *Transp. Res. Rec.* **1919**, *2005*, 96–110. [CrossRef]
28. Ho, C.-H; Romero, P. Low design temperatures of asphalt pavements in dry-freeze regions Predicting by means of solar radiation, transient heat transfer, and finite element method. *Transp. Res. Rec.* **2009**, *2127*, 60–71.
29. Qin, Y. A review on the development of cool pavements to mitigate urban heat island effect. *Renew. Sustain. Energy Rev.* **2015**, *52*, 445–459. [CrossRef]
30. Qin, Y. Pavement surface maximum temperature increases linearly with solar absorption and reciprocal thermal inertial. *Int. J. Heat Mass Transf.* **2016**, *97*, 391–399. [CrossRef]
31. Diefenderfer, B.K.; Al-Qadi, I.L.; Diefenderfer, S.D. Model to predict pavement temperature profile: Development and validation. *J. Transp. Eng.* **2006**, *132*, 162–167. [CrossRef]
32. Qin, Y. Urban canyon albedo and its implication on the use of reflective cool pavements. *Energy Build.* **2015**, *96*, 86–94. [CrossRef]
33. Li, H.; Harvey, J.; Kendall, A. Field measurement of albedo for different land cover materials and effects on thermal performance. *Build. Environ.* **2013**, *59*, 536–546. [CrossRef]
34. Sen, S.; Roesler, J. Aging albedo model for asphalt pavement surfaces. *J. Clean. Prod.* **2016**, *117*, 169–175. [CrossRef]
35. Soltani, A.; Sharifi, E. Daily variation of urban heat island effect and its correlations to urban greenery: A case study of Adelaide. *Front. Arch. Res.* **2017**, *6*, 529–538. [CrossRef]
36. Lee, K.W.; Kohm, S. Cool pavements. In *Climate Change, Energy, Sustainability and Pavements*; Gopalakrishnan, K., Jvd, M., Steyn, W., Harvey, J., Eds.; Springer: Heidelberg, Germany, 2014; pp. 439–454.
37. Chen, J.; Wang, H.; Zhu, H. Analytical approach for evaluating temperature field of thermal modified asphalt pavement and urban heat island effect. *Appl. Therm. Eng.* **2017**, *113*, 739–748. [CrossRef]
38. Adebayo, Y.R. Land-use approach to the spatial analysis of the urban heat island in Ibadan, Nigeria. *Weather* **1987**, *42*, 273–280. [CrossRef]
39. Tyson, P.D.; du Toit, W.J.F.; Fuggle, R.F. Temperature structure above cities: Review and preliminary findings from the Johannesburg Urban Heat Island Project. *Atmos. Environ.* **1972**, *6*, 533–542. [CrossRef]
40. Monama, T.E. *Evaluating the Urban Heat Island over the City of Tshwane Metropolitan Municipality Using Remote Sensing Techniques*; University of Johannesburg: Johannesburg, South Africa, 2016.
41. Peters, I.M.; Karthik, S.; Liu, H.; Buonassisi, T.; Nobre, A. Urban haze and photovoltaics. *Energy Environ. Sci.* **2018**, *11*, 3043–3054. [CrossRef]
42. Mokoena, R.; Mturi, G.A.J.; Maritz, J.; Malherbe, J.; O'Connell, J.S. Adapting Asphalt Pavements to Climate Change Challenges. In Proceedings of the 12th Conference on Asphalt Pavements for Southern Africa (CAPSA), Sun City, South Africa, 13–16 October 2019.
43. SATS 3208. *Technical Specification—Performance Grade (PG) Specifications for Bitumen in South Africa*; South African Bureau of Standards: Pretoria, South Africa, 2021.
44. European Space Agency, African Land Cover. 2021. Available online: http://2016africalandcover20m.esrin.esa.int/viewer.php (accessed on 20 September 2021).
45. African Development Bank Group. *Tracking Africa's Progress in Figures*; Phoenix Design Aid: Randers, Denmark, 2014.
46. O'Connell, J.; Mturi, G.A.J.; Zoorob, S. A Review of the Development of the Non-Recoverable Compliance, J_{nr}, for Use in South Africa. In Proceedings of the 11th Conference on Asphalt Pavements for southern Africa (CAPSA), Sun City, South Africa, 16–19 August 2015.
47. Pfeifroth, U.; Mueller, R.; Ahrens, B. Evaluation of satellite-based and reanalysis precipitation data in the tropical pacific. *J. Appl. Meteorol. Clim.* **2013**, *52*, 634–644. [CrossRef]
48. Durand, W. Long Term Forecasting of Climate & Rainfall—The Importance of Current Data. In Proceedings of the AgBIZ Grain Mini Symposium, Centurion, South Africa, 10 August 2016.
49. FHWA, Federal Highway Administration, U.S. Department of Transportation. LTPP InfoPave. Available online: https://infopave.fhwa.dot.gov (accessed on 30 April 2017).
50. Schwartz, C.W.; Elkins, G.E.; Li, R.; Visintine, B.A.; Forman, B.; Rada, G.R.; Groeger, J.L. *Evaluation of LTPP Climatic Data for Use in Mechanistic-Empirical Pavement Design Guide Calibration and Other Pavement Analysis*; US Department of Transportation, Federal

Highway Administration, Research, Development, and Technology, Turner-Fairbank Highway Research Center: McLean, VA, USA, 2015.

51. Jenkins, G. Observing the Weather. 2010. Available online: https://www.rmets.org/sites/default/files/inline-files/simweameasurements.pdf (accessed on 27 October 2021).
52. Uddstrom, M.J. Retrieval of atmospheric profiles from satellite radiance data by typical shape function maximum a posteriori simultaneous retrieval estimators. *J. Appl. Meteorol.* **1988**, *27*, 515–549. [CrossRef]
53. Colston, J.M.; Ahmed, T.; Mahopo, C.; Kang, G.; Kosek, M.; de Sousa Junior, F.; Shrestha, P.S.; Svensen, E.; Turab, A.; Zaitchik, B. Evaluating meteorological data from weather stations, and from satellites and global models for a multi-site epidemiological study. *Environ. Res.* **2018**, *165*, 91–109. [CrossRef] [PubMed]
54. Dawson, A. Anticipating and responding to pavement performance as climate changes. In *Climate Change, Energy, Sustain-Ability and Pavements*, 1st ed.; Gopalakrishnan, K., Jvd, M., Steyn, W., Harvey, J., Eds.; Springer: Heidelberg, Germany, 2014; pp. 127–158.
55. McGregor, J.L. *C-CAM: Geometric Aspects and Dynamical Formulation*; CSIRO Atmospheric Research No. 70: Melbourne, Australia, 2005.
56. Mcgregor, J.L.; Dix, M.R. The CSIRO conformal-cubic atmospheric GCM. In *IUTAM Symposium on Advances in Mathematical Modelling of Atmosphere and Ocean Dynamics*; Springer: Dordrecht, The Netherlands, 2001; pp. 197–202.
57. McGregor, J.L.; Dix, M.R. An updated description of the conformal-cubic atmospheric model. In *High Resolution Numerical Modelling of the Atmosphere and Ocean*; Springer: New York, NY, USA, 2008; pp. 51–75.
58. Thatcher, M.; Hurley, P. Simulating Australian urban climate in a mesoscale atmospheric numerical model. *Bound.-Layer Meteorol.* **2012**, *142*, 149–175. [CrossRef]
59. Engelbrecht, F.A.; McGregor, J.L.; Engelbrecht, C.J. Dynamics of the conformal-cubic atmospheric model projected climate-change signal over southern Africa. *Int. J. Climatol.* **2009**, *29*, 1013–1033. [CrossRef]
60. Engelbrecht, F.A.; Landman, W.A.; Engelbrecht, C.J.; Landman, S.; Bopape, M.M.; Roux, B.; McGregor, J.L.; Thatcher, M. Multi-scale climate modelling over Southern Africa using a variable-resolution global model. *Water SA* **2011**, *37*, 647–658. [CrossRef]
61. Engelbrecht, C.J.; Engelbrecht, F.A.; Dyson, L.L. High-resolution model-projected changes in mid-tropospheric closed-lows and extreme rainfall events over southern Africa. *Int. J. Clim.* **2013**, *33*, 173–187. [CrossRef]
62. Malherbe, J.; Engelbrecht, F.A.; Landman, W.A. Projected changes in tropical cyclone climatology and landfall in the Southwest Indian Ocean region under enhanced anthropogenic forcing. *Clim. Dyn.* **2013**, *40*, 2867–2886. [CrossRef]
63. Winsemius, H.C.; Dutra, E.; Engelbrecht, F.A.; Van Garderen, E.A.; Wetterhall, F.; Pappenberger, F.; Werner, M.G.F. The potential value of seasonal forecasts in a changing climate in southern Africa. *Hydrol. Earth Syst. Sci.* **2014**, *18*, 1525–1538. [CrossRef]
64. Engelbrecht, F.; Adegoke, J.; Bopape, M.-J.; Naidoo, M.; Garland, R.; Thatcher, M.; McGregor, J.; Katzfey, J.; Werner, M.; Ichoku, C.; et al. Projections of rapidly rising surface temperatures over Africa under low mitigation. *Environ. Res. Lett.* **2015**, *10*, 085004. [CrossRef]
65. Meagher, W.; Daniel, J.S.; Jacobs, J.; Linder, E. Method for evaluating implications of climate change for design and performance of flexible pavements. *J. Transp. Res. Board* **2012**, *2305*, 111–120. [CrossRef]
66. Jacobs, J.M.; Kirshen, P.H.; Daniel, J.S. Considering climate change in road and building design. In Proceedings of the First Infrastructure and Climate Network Steering Committee Workshop, Durham, NH, USA, 8–9 January 2013.
67. Daniel, J.S.; Jacobs, J.M.; Douglas, E.; Mallick, R.B.; Hayhoe, K. Impact of climate change on pavement performance: Preliminary lessons learned through the infrastructure and climate network (ICNet). In Proceedings of the International Symposium on Climatic Effects on Pavement and Geotechnical Infrastructure, Fairbanks, AK, USA, 4–7 August 2013.
68. Meyer, M.; Flood, M.; Keller, J.; Lennon, J.; McVoy, G.; Dorney,, C. NCHRP Report 750: Strategic issues facing transportation. In *Volume 2: Climate Change, Extreme Weather Events, and the Highway System: Practitioner's Guide and Research Report*; The National Academies Press: Washington, DC, USA, 2014. [CrossRef]
69. Steyn, W.J. Climate change scenarios and their potential effects on transportation infrastructure systems. In *Climate Change, Energy, Sustainability and Pavements*; Springer: Heidelberg, Germany, 2014; pp. 159–172.
70. Anochie-Boateng, J.; O'Connell, J.; Komba, J. Interim guideline for the design of hot-mix asphalt. In *The United Republic of Tanzania: Ministry of Works, Transport and Communication*; TANROADS: Dar Es Salaam, Tanzania, 2018.

MDPI
St. Alban-Anlage 66
4052 Basel
Switzerland
Tel. +41 61 683 77 34
Fax +41 61 302 89 18
www.mdpi.com

Sustainability Editorial Office
E-mail: sustainability@mdpi.com
www.mdpi.com/journal/sustainability

www.ingramcontent.com/pod-product-compliance
Lightning Source LLC
LaVergne TN
LVHW070134170726
843515LV00007B/1796

* 9 7 8 3 0 3 6 5 4 7 4 5 9 *